AF686336

Recent Advances in Geographic Information System for Earth Sciences

Recent Advances in Geographic Information System for Earth Sciences

Special Issue Editor

Yosoon Choi

MDPI • Basel • Beijing • Wuhan • Barcelona • Belgrade • Manchester • Tokyo • Cluj • Tianjin

Special Issue Editor
Yosoon Choi
Pukyong National University
Korea

Editorial Office
MDPI
St. Alban-Anlage 66
4052 Basel, Switzerland

This is a reprint of articles from the Special Issue published online in the open access journal *Applied Sciences* (ISSN 2076-3417) (available at: https://www.mdpi.com/journal/applsci/special_issues/GIS_earth_sciences).

For citation purposes, cite each article independently as indicated on the article page online and as indicated below:

LastName, A.A.; LastName, B.B.; LastName, C.C. Article Title. *Journal Name* **Year**, *Article Number*, Page Range.

ISBN 978-3-03936-489-3 (Hbk)
ISBN 978-3-03936-490-9 (PDF)

© 2020 by the authors. Articles in this book are Open Access and distributed under the Creative Commons Attribution (CC BY) license, which allows users to download, copy and build upon published articles, as long as the author and publisher are properly credited, which ensures maximum dissemination and a wider impact of our publications.
The book as a whole is distributed by MDPI under the terms and conditions of the Creative Commons license CC BY-NC-ND.

Contents

About the Special Issue Editor . . . . **vii**

Preface to "Recent Advances in Geographic Information System for Earth Sciences" . . . . . . **ix**

Yosoon Choi
Recent Advances in Geographic Information System for Earth Sciences
Reprinted from: *Appl. Sci.* **2020**, *10*, 3847, doi:10.3390/app10113847 . . . . **1**

Yuke Zhou, Shaohua Wang and Yong Guan
An Efficient Parallel Algorithm for Polygons Overlay Analysis
Reprinted from: *Appl. Sci.* **2019**, *9*, 4857, doi:10.3390/app9224857 . . . . **5**

Giao N. Pham, Son T. Ngo, Anh N. Bui, Dinh V. Tran, Suk-Hwan Lee and Ki-Ryong Kwon
Vector Map Random Encryption Algorithm Based on Multi-Scale Simplification and Gaussian Distribution
Reprinted from: *Appl. Sci.* **2019**, *9*, 4889, doi:10.3390/app9224889 . . . . **25**

Zdena Dobesova
Evaluation of Effective Cognition for the QGIS Processing Modeler
Reprinted from: *Appl. Sci.* **2020**, *10*, 1446, doi:10.3390/app10041446 . . . . **41**

Sehrish Malik and DoHyeun Kim
Geo-Sensor Framework and Composition Toolbox for Efficient Deployment of Multiple Spatial Context Service Platforms in Sensor Networks
Reprinted from: *Appl. Sci.* **2019**, *9*, 4993, doi:10.3390/app9234993 . . . . **69**

Yosoon Choi, Jieun Baek and Sebeom Park
Review of GIS-Based Applications for Mining: Planning, Operation, and Environmental Management
Reprinted from: *Appl. Sci.* **2020**, *10*, 2266, doi:10.3390/app10072266 . . . . **91**

Hui Liu, Shanjun Mao, Mei Li and Shuangyong Wang
A Tightly Coupled GIS and Spatiotemporal Modeling for Methane Emission Simulation in the Underground Coal Mine System
Reprinted from: *Appl. Sci.* **2019**, *9*, 1931, doi:10.3390/app9091931 . . . . **117**

Daeryong Park, Huan-Jung Fan, Jun-Jie Zhu, Sang-Eun Oh, Myoung-Jin Um and Kichul Jung
Evaluation of Reliable Digital Elevation Model Resolution for TOPMODEL in Two Mountainous Watersheds, South Korea
Reprinted from: *Appl. Sci.* **2019**, *9*, 3690, doi:10.3390/app9183690 . . . . **135**

Nan Wang, Yunyan Du, Fuyuan Liang, Jiawei Yi and Huimeng Wang
Spatiotemporal Changes of Urban Rainstorm-Related Micro-Blogging Activities in Response to Rainstorms: A Case Study in Beijing, China
Reprinted from: *Appl. Sci.* **2019**, *9*, 4629, doi:10.3390/app9214629 . . . . **153**

Abhirup Dikshit, Raju Sarkar, Biswajeet Pradhan, Samuele Segoni and Abdullah M. Alamri
Rainfall Induced Landslide Studies in Indian Himalayan Region: A Critical Review
Reprinted from: *Appl. Sci.* **2020**, *10*, 2466, doi:10.3390/app10072466 . . . . **169**

Xia Zhao and Wei Chen
GIS-Based Evaluation of Landslide Susceptibility Models Using Certainty Factors and Functional Trees-Based Ensemble Techniques
Reprinted from: *Appl. Sci.* **2020**, *10*, 16, doi:10.3390/app10010016 . . . . **193**

Hui Xiang, Qing-Yuan Yang, Kang-chuan Su and Zhong-Xun Zhang
Spatiotemporal Dynamics and Obstacles of the Multi-Functionality of Land Use in Xiangxi, China
Reprinted from: *Appl. Sci.* **2019**, *9*, 3649, doi:10.3390/app9183649 . . . . **213**

Wenhao Yu, Menglin Guan and Zhanlong Chen
Analyzing Spatial Community Pattern of Network Traffic Flow and Its Variations across Time Based on Taxi GPS Trajectories
Reprinted from: *Appl. Sci.* **2019**, *9*, 2054, doi:10.3390/app9102054 . . . . **231**

About the Special Issue Editor

Yosoon Choi received a B.S. at the School of Civil, Urban, and Geosystem Engineering, Seoul National University, Korea, in 2004. He received a Ph.D. degree from the Department of Energy Systems Engineering, Seoul National University in 2009. He was a Post-Doc fellow at the Department of Energy and Mineral Engineering, Pennsylvania State University, USA, in 2010. He has been a Professor at the Department of Energy Resources Engineering, Pukyong National University, Korea, since 2011. He has been working in the area of smart mining; renewables in mining; AI, IoT, cloud, big data, mobile (AICBM) convergence; mining engineering; geographic information systems (GISs); 3D geo-modeling; operations research; and solar energy engineering.

Preface to "Recent Advances in Geographic Information System for Earth Sciences"

Geographic information systems (GISs) are computer-based technologies and methodologies for collecting, managing, analyzing, modeling, and presenting geospatial data for a wide range of applications. In recent decades, GISs have played a vital role in Earth sciences by providing a powerful means of observing the world and various tools for solving complex problems. The scientific community has used GISs to reveal fascinating details about the Earth and other planets.

This book on recent advances in GIS for Earth sciences includes 12 publications from esteemed research groups around the world. The research and review papers in this book belong to the following broad categories: Earth science informatics (geoinformatics), mining, hydrology, natural hazards, and society.

Yosoon Choi
Special Issue Editor

applied sciences

MDPI

Editorial

Recent Advances in Geographic Information System for Earth Sciences

Yosoon Choi

Department of Energy Resources Engineering, Pukyong National University, Busan 48513, Korea; energy@pknu.ac.kr; Tel.: +82-33-570-6313

Received: 14 May 2020; Accepted: 29 May 2020; Published: 1 June 2020

1. Introduction

Geographic Information System (GIS) is a computer-based technology and methodology for collecting, managing, analyzing, modeling, and presenting geospatial data for a wide range of applications. GIS plays a vital role in Earth sciences by providing a powerful means of observing the world and various tools for solving complex problems. The scientific community has used GIS to reveal fascinating details about the Earth and other planets.

This special issue on recent advances in GIS for Earth sciences includes 12 publications from esteemed research groups worldwide. The research and review papers in this special issue belong to the following broad categories: Earth science informatics (geoinformatics), mining, hydrology, natural hazards, and society.

2. GIS for Geoinformatics

GIS is an important tool used to solve complex spatial problems in geoinformatics. Several articles dealing with basic algorithms for spatial data analysis are included in this special issue. Zhou et al. [1] propose an efficient parallel algorithm for polygon overlay analysis. Overlay analysis is a fundamental operator in spatial data analytics and is widely used in Earth science applications. The proposed algorithm includes procedures for active-slave spatial index decomposition for intersection, multi-strategy Hilbert ordering decomposition, and parallel spatial union. The application of their new parallel algorithm to a land-use map of China consisting of multiple polygons with 15,615 elements and 886,547 points shows that the algorithm can perform polygon overlay analysis with high efficiency. Therefore, the study contributes to geoinformatics by allowing the processing of large scale spatial data for spatial data analytics.

Vector maps in GIS have been widely used in various fields, including Earth science. Currently, huge volumes of vector map data can be easily stolen and distributed without permission from the original data providers. Pham et al. [2] propose a random encryption algorithm based on multi-scale simplification and the Gaussian distribution to encrypt vector map data before it is stored and transmitted. Their experiment using vector maps of Scotland at different scales shows that the proposed algorithm provides higher security and computational efficiency of storage and transmission of vector map data than previous methods. Therefore, the algorithm can be applied to improve the security of online and offline Earth science map services.

QGIS [3], an open-source GIS software, has been utilized in the Earth science community. Dobesova [4] assesses the visual notation of QGIS's Processing Modeler, a graphical editor for workflow design, using the Physics of Notations theory in combination with eye-tracking measurements. The results from this study provide several practical recommendations to improve the effective cognition of the QGIS Processing Modeler, including changing the fill color of symbols, increasing the size and variety of inner icons, removing functional icons, using a straight connector line instead of a curved line, and providing a supplemental preview window for the entire model.

Geo-sensor networks produce large amounts of Earth science data that can be processed using GIS for different purposes and for intelligent decision making. Malik and Kim [5] propose a geo-sensor framework that can be used by multiple clients to deploy their own geo-sensor networks, bind their sensor objects to desired locations, generate geo-sensor services for the uploaded networks, and manage the services with a geo-sensor composite toolbox. The framework is implemented based on the RESTful and SOAP web services [6]. Performance analysis shows that the lightweight RESTful web service is the best choice for ease of implementation and access.

3. GIS for Mining

Systematic and strategic mine planning, operation, and environmental management are necessary to improve mineral productivity, operational efficiency, and stability in the mining environment. To accomplish these objectives, GIS has been effectively used to design and optimize mine development. Choi et al. [7] review GIS-based methods and applications utilized in mine development, especially for mine planning, operation, and environmental management. They observe that GIS-based methods, including database management, spatial analysis, mapping, and visualization, are effectively used for all stages of mine development at global, regional, and local scales. In the mine planning phase, GIS-based methods are adopted for ore reserve estimation, open-pit boundary optimization, mine infrastructure design, and potential conflict analysis. Various mine operation systems based on GIS have been implemented in mining sites for ore haulage operations, wireless communication, ore management, safety monitoring, underground ventilation, and drainage systems. Moreover, various GIS applications have been developed to support decision-making in mine reclamation planning and re-utilization designs.

As an example of a GIS application for mining, Liu et al. [8] present a spatiotemporal model tightly coupled with GIS for simulating methane emissions in underground coal mines. Such a tight coupling approach is achieved by developing a lattice Boltzmann method (LBM)-based turbulent model with an underlying shared FluentEntity model within the GIS. A case study demonstrates that the proposed GIS-based model is capable and effective in providing functionalities for lattice domain decomposition, simulation, visualization, and analyses, as well as improving computational efficiency compared with traditional computational fluid dynamics (CFD) methods. The tight coupling approach for integrating GIS and simulation models is applicable to underground coal mine disasters.

4. GIS for Hydrology

In hydrological studies, GIS has facilitated the development of a dynamic model for analyzing runoff phenomena as well as a distributed parameter model that considers spatial variability in parameters related to the runoff process. The topography-based hydrological MODEL (TOPMODEL) is a distributed parameter model that uses a digital elevation model (DEM) in GIS. However, TOPMODEL is affected by the resolution of the DEM used. A reliable DEM grid-size resolution that exhibits low sensitivity to changes in input parameters during runoff simulations is investigated by Park et al. [9]. A case study in the Dongkok and Ieemokjung watersheds in South Korea shows that the efficiency of TOPMODEL rarely changes up to a DEM grid-size resolution of approximately 40 m, but changes more noticeably with coarser resolution. The findings of this study are important for understanding and quantifying the modeling behaviors of TOPMODEL under the influence of varying DEM resolution.

Social media data collected through Twitter, Facebook, Flicker, and Weibo can be used to improve understanding of urban hydrology. Wang et al. [10] examine rainstorm-related micro-blogging activities in response to rainstorms in an urban environment at fine spatial and temporal scales. The study collected hourly precipitation data and a total of 3.32 million Weibo blogs geotagged with Beijing, China from June to September 2017. The consistency between rainfall amount and human activities can be explained by the distribution of water ponding sites and major transportation hubs. The results show that human responses to the rainstorm event are consistent, though with certain time lags, in virtual and physical spaces at both grid and city scales.

5. GIS for Natural Hazards

Advances in GIS have popularized its application to spatial analysis of natural hazards. In particular, GIS has been widely used for landslide susceptibility mapping. Landslide susceptibility maps generated by GIS can be effectively used for future land planning and monitoring. Dikshit et al. [11] review studies of rainfall-induced landslides in the Indian Himalayan region to provide a reference point for the first time for researchers working in this region, and a summary of the improvements most urgently needed to better address landslide hazard research and management. Their study reveals that the inclusion of climate change factors and the acquisition of basic input data of the highest quality for computational models is critical for landslide susceptibility mapping.

Zhao and Chen [12] present an example of GIS-based landslide susceptibility mapping using ensemble techniques of functional tree-based bagging, rotation forest, and dagging (functional trees (FT), bagging-functional trees (BFT), rotation forest-functional trees (RFFT), and dagging-functional trees (DFT)). A landslide inventory map with 263 landslide events is established for Zichang County, China, and 14 landslide conditioning factors selected to analyze the correlation between the conditioning factors and the occurrence of landslides. The results show that the prediction rate of the BFT model is the highest when compared with the accuracy of the four ensemble models.

6. GIS for Society

GIS plays an important role in society, especially for land-use planning. The land is a complex system providing food, fresh water, and other material resources for humans. It is essential for habitation, transport, leisure, and other activities. For land-use planning, various factors such as topography, soil, hydrology, biology, and climate will be considered simultaneously. Xiang et al. [13] use GIS to assess the spatiotemporal dynamic multi-functionality of land use and to analyze obstacle indicators in Xiangxi, China using two methods (analytic hierarchy and hierarchical weighting). The study finds that spatial heterogeneity of land use in Xiangxi is increasingly clear. The production function of land use in Xiangxi is slowly increasing, with more rapid growth in the southern and central regions than in the northern regions. Three types of obstacles preventing efficient land use in Xiangxi are identified by GIS-based spatial analysis.

Different land uses are connected by transport networks to improve accessibility for human activities. Yu et al. [14] analyze the traffic flow network using GIS to understand the properties of spatial connectivity, spatial aggregation, and spatial dynamics. The study conducted a series of experiments to explore the transport system in Beijing city using taxi trajectory points recorded by the global positioning system (GPS). The results indicate that the interactions of land use show different characteristics over different time periods. Aggregation patterns of functional areas are dynamic over time and are strongly associated with the travel behaviors of residents in the city.

Funding: This work was supported by Basic Science Research Program through the National Research Foundation of Korea (NRF) funded by the Ministry of Education (2018R1D1A1A09083947).

Acknowledgments: This special issue would not be possible without the contributions of professional authors and reviewers, and the excellent editorial team of Applied Sciences.

Conflicts of Interest: The authors declare no conflict of interest.

References

1. Zhou, Y.; Wang, S.; Guan, Y. An Efficient Parallel Algorithm for Polygons Overlay Analysis. *Appl. Sci.* **2019**, *9*, 4857. [CrossRef]
2. Pham, G.N.; Ngo, S.T.; Bui, A.N.; Tran, D.V.; Lee, S.-H.; Kwon, K.-R. Vector Map Random Encryption Algorithm Based on Multi-Scale Simplification and Gaussian Distribution. *Appl. Sci.* **2019**, *9*, 4889. [CrossRef]
3. QGIS. Available online: https://qgis.org/en/site/index.html (accessed on 12 May 2020).
4. Dobesova, Z. Evaluation of Effective Cognition for the QGIS Processing Modeler. *Appl. Sci.* **2020**, *10*, 1446. [CrossRef]

5. Malik, S.; Kim, D. Geo-Sensor Framework and Composition Toolbox for Efficient Deployment of Multiple Spatial Context Service Platforms in Sensor Networks. *Appl. Sci.* **2019**, *9*, 4993. [CrossRef]
6. World Wide Web Consortium. Web Services Architecture. Available online: https://www.w3.org/TR/2004/NOTE-ws-arch-20040211/#relwwwrest (accessed on 12 May 2020).
7. Choi, Y.; Baek, J.; Park, S. Review of GIS-Based Applications for Mining: Planning, Operation, and Environmental Management. *Appl. Sci.* **2020**, *10*, 2266. [CrossRef]
8. Liu, H.; Mao, S.; Li, M.; Wang, S. A Tightly Coupled GIS and Spatiotemporal Modeling for Methane Emission Simulation in the Underground Coal Mine System. *Appl. Sci.* **2019**, *9*, 1931. [CrossRef]
9. Park, D.; Fan, H.-J.; Zhu, J.-J.; Oh, S.-E.; Um, M.-J.; Jung, K. Evaluation of Reliable Digital Elevation Model Resolution for TOPMODEL in Two Mountainous Watersheds, South Korea. *Appl. Sci.* **2019**, *9*, 3690. [CrossRef]
10. Wang, N.; Du, Y.; Liang, F.; Yi, J.; Wang, H. Spatiotemporal Changes of Urban Rainstorm-Related Micro-Blogging Activities in Response to Rainstorms: A Case Study in Beijing, China. *Appl. Sci.* **2019**, *9*, 4629. [CrossRef]
11. Dikshit, A.; Sarkar, R.; Pradhan, B.; Segoni, S.; Alamri, A.M. Rainfall Induced Landslide Studies in Indian Himalayan Region: A Critical Review. *Appl. Sci.* **2020**, *10*, 2466. [CrossRef]
12. Zhao, X.; Chen, W. GIS-Based Evaluation of Landslide Susceptibility Models Using Certainty Factors and Functional Trees-Based Ensemble Techniques. *Appl. Sci.* **2020**, *10*, 16. [CrossRef]
13. Xiang, H.; Yang, Q.-Y.; Su, K.-C.; Zhang, Z.-X. Spatiotemporal Dynamics and Obstacles of the Multi-Functionality of Land Use in Xiangxi, China. *Appl. Sci.* **2019**, *9*, 3649. [CrossRef]
14. Yu, W.; Guan, M.; Chen, Z. Analyzing Spatial Community Pattern of Network Traffic Flow and Its Variations across Time Based on Taxi GPS Trajectories. *Appl. Sci.* **2019**, *9*, 2054. [CrossRef]

© 2020 by the author. Licensee MDPI, Basel, Switzerland. This article is an open access article distributed under the terms and conditions of the Creative Commons Attribution (CC BY) license (http://creativecommons.org/licenses/by/4.0/).

applied sciences

MDPI

Article

An Efficient Parallel Algorithm for Polygons Overlay Analysis

Yuke Zhou [1], Shaohua Wang [2,*] and Yong Guan [3]

1 Key Laboratory of Ecosystem Network Observation and Modeling, Institute of Geographic Sciences and Natural Resources Research, Chinese Academy of Sciences, Beijing 100101, China; zhouyk@igsnrr.ac.cn
2 Department of Geography, University of California, Santa Barbara, CA 93117, USA
3 Department of Information System and Technology, Claremont Graduate University, Claremont, CA 91711, USA; yong.guan@cgu.edu
* Correspondence: shaohua@geog.ucsb.edu; Tel.: +1-805-284-3507

Received: 29 September 2019; Accepted: 11 November 2019; Published: 13 November 2019

Abstract: Map overlay analysis is essential for geospatial analytics. Large scale spatial data pressing poses challenges for geospatial map overlay analytics. In this study, we propose an efficient parallel algorithm for polygons overlay analysis, including active-slave spatial index decomposition for intersection, multi-strategy Hilbert ordering decomposition, and parallel spatial union algorithm. Multi-strategy based spatial data decomposition mechanism is implemented, including parallel spatial data index, the Hilbert space-filling curve sort, and decomposition. The results of the experiments showed that the parallel algorithm for polygons overlay analysis achieves high efficiency.

Keywords: parallel algorithm; map overlay analysis; Hilbert ordering decomposition; spatial analysis

1. Introduction

Map overlay analysis is a fundamental operator in geospatial analytics and widely used in geospatial applications [1,2]. Large scale spatial data pressing poses challenges for geospatial map overlay analytics [3].

The parallel GIS algorithm is an efficient way to conduct map overlay analysis [4–7]. Spatial data decomposition is the basis of parallel computing architecture based on the spatial data partitioning mechanism [8]. Spatial data domain decomposition in parallel GIS refers to the decomposition of object sets in the study area according to a certain granularity and is assigned to different computing units for processing to achieve high concurrency. Spatial data domain decomposition from the perspective of geographic data storage mainly refers to the database domain to allocate spatially adjacent geographical elements to the same physical medium storage according to a certain decomposition principle. The feature elements form different groups in space in the form of clusters, and the spatially separated clusters are divided into different storage areas to realize parallelized spatial data extraction mode. The parallelized map overlay analysis algorithm technology route is based on data division and behavior.

Parallel spatial data decomposition needs to take into consideration the data storage and geo-computation in each child node from the perspective of the spatial distribution of feature objects, while spatial data has multidimensionality [9]. In the process of parallel overlay analysis, the core of parallelization is a fast intersection judgment of geometric objects and the interactive communication between geospatial data [10,11]. Therefore, the critical principle of layer element decomposition is to maintain the spatial proximity of data. The main feature of geographic data is that it has a strong spatial correlation, and its data parallel strategy should be compatible with spatial data types. The spatial feature is the difference between ordinary numerical parallel computing and the key technology of parallel GIS system [12,13]. The purpose of spatial data decomposition is to implement the local process of spatial analysis operations (to reduce the synchronization operation between computing nodes). The modeling of spatial elements can be accelerated by computational localization. With the improvement of computer hardware performance and the increasing cost of storage, the usual strategy is to exchange storage space for computing time [12,14].

In this paper, we propose an efficient parallel algorithm for polygons overlay analysis. We decompose vector space data based on space filling-curve (Hilbert curve) to better maintain the spatial proximity of data decomposition, which is conducive to the parallelization of spatial proximity and sensitive overlay operations, and with the basis of the original Hilbert ordering decomposition, the spatial data is decomposed using different sorting strategies. This paper is organized as follows. Section 2 introduces related work. Section 3 shows the methods of a parallel algorithm for polygons overlay analysis. The experimental results are given in Section 4. Following this, the last section contains the conclusion and further work.

2. Related Work

The most time-consuming operation of polygon overlay analysis continues to be the intersection of line segments. Ching studied the load balancing problem for the parallel map overlay calculation based on the line segment set [15] but did not deeply discuss the communication and merge-efficiency of each node after parallel calculation, and cannot guarantee the constant acceleration ratio of the whole process.

Parallel spatial data region decomposition needs to take into consideration the storage and calculation in each child node from the perspective of the spatial distribution of feature objects while spatial data has multidimensionality. In traditional data partitioning methods, such as token rotation, hash table segmentation, and simple region partitioning, the spatial relationship between objects is split during decomposition, which does not reflect the proximity between spatial data.

The purpose of using Hilbert space sorting is to maximize the mapping of high-dimensional data to low-dimensional data [16,17] and to close the geographically adjacent points in computer storage to accelerate data extraction and improve the efficiency of data operations in the first-level storage. The access to spatial data in memory is performed randomly. For spatial data with unbalanced distribution, if the point data comparison is too dense in a certain area, data redundancy in the index sub-node is caused. In order to maintain the uniqueness of spatial data mapping, a more detailed division of the index grid is required. However, if the division is too detailed, it will increase the difficulty and computational complexity of spatial sorting coding and also increase the size of the spatial query.

The spatial data decomposition can be divided by the spatial indexing mechanism. Based on the spatial index, the search space of the candidate dataset in the overlay analysis can be effectively reduced. At the same time, the false intersection can be further filtered in the proximity analysis of the candidate geometry data into a map overlay object with real intersections. The data decomposition in the parallel superposition analysis method is based on the vector topology data model. The key difficulty is how to assign the elements with large spatial proximity to the same node. The vector data capacity is usually between megabytes to gigabytes, and the current computer hard disk is measured by the terabyte level storage capacity. The equalization of storage capacity is not a critical issue. The key problem is also how to effectively equalize the computational tasks of vector data and reduce unnecessary intersection detection and parameter communication between distributed nodes. Because the input spatial data will have different feature density, the division of spatial data into distributed nodes in the GIS parallel algorithm does not have a conventional experience.

Most spatial decomposition methods are based on planar space, where one point on one side of the plane defines one area while the other side determines another area. However, as points on the plane can be arbitrarily divided into one area, using the plane to divide the space recursively will eventually generate a Binary Space Partition Tree (BSP Tree) [18]. Using spatial data decomposition structures to store objects can easily speed up specific geometric query operations, such as conflict detection, to determine if two objects are close together or if a ray intersects the object. The quadtree index belongs to the vertical decomposition mode of the plane [19]. The generation process is to recursively divide the geospatial by four points until the termination condition is set by itself (for example, each area contains no more than two points, if it exceeds four points). Finally, a hierarchical quadtree is formed. Quadtree decomposition is a typical planar recursive decomposition [20]. This method is suitable for the case where the data isomorphism and distribution are relatively balanced [21].

The decomposition of the spatial index is determined by geometric objects (object priority). Decomposing spatial regions into sub-regions according to spatial objects are also called bucket. Therefore, this method is usually called bucket partitioning. The object-oriented data decomposition method needs to follow certain principles. The most classic strategy is that the B-tree rule uses a separate point or line to decompose the spatial extent recursively [22]. Another classic decomposition principle is to keep the outer bounding rectangle of the object to be the smallest, and R-tree is its important implementation [23]. Use the R-tree spatial data domain to decompose the main packing or bucket mode of the R-tree. R-tree optimization is mainly for the packing and sorting of sub-node index packets. Kamel and Faloutsos proposed the MBR sorting method based on the Hilbert curve [24]. Roussopoulos proposed a sorting method based on Nearest-X in a certain direction. In the classic R-tree implementation, Guttman (1984) proposed two heuristic tree node decomposition methods, including Quadratic-Split and Linear-Split [23]. The performance of the R-tree depends on the quality of the data outsourcing the rectangular clustering algorithm in the node. The Hilbert R-tree uses the space filling curve, especially the Hilbert curve, to organize the data outsourcing the rectangle linearly. The Hilbert R-tree comes in two forms: one is a static database form, and the other is a dynamic database form. In the paper research scheme, the Hilbert fill curve is used to achieve better ordering of high-dimensional data in nodes. This order ensures that similar objects outsourcing rectangles are grouped into groups, keeping the area and perimeter of the resulting outer rectangle as small as possible, so the Hilbert curve is a good sorting method in the sense of this layer. The compact Hilbert R-tree is suitable for static databases [25], and there are few update operations in the static database or no update operations at all. The dynamic Hilbert R-tree is suitable for dynamic databases [26–28], which require real-time insert, delete, and update operations. The dynamic Hilbert R-tree uses an elastic segmentation delay mechanism to increase space utilization. Each node in the tree has a well-defined set of sibling nodes. By establishing the order of the nodes in the R-tree and adjusting the Hilbert R-tree data partitioning strategy, the ideal space utilization degree can be achieved. The Hilbert R-tree is ordered based on the Hilbert value of the center point of the object rectangle, and the point Hilbert value is the length from the start of the Hilbert curve to the point. In contrast, other R-tree variants do not have control over space utilization. Leutenger proposed a new R-tree variant, the Sort-Tile-Recursive tree (STR-tree). The algorithm uses the recursive idea. For the set of spatial rectangles with r in the k-dimensional plane, let the maximum capacity of the leaves of the R-tree be n, and the rectangles are sorted according to the x value of the center point. The concept of the tile is to use $\sqrt{(r/n)}$ vertical cuttings. The line divides the sorted rectangle so that each strip can be loaded close to $\sqrt{(r/n)}$ nodes. Each slice continues to be sorted according to the y value of the center point of the rectangle, and a leaf node is pressed every n rectangle; the top-down reclusiveness processes the slice to generate the entire R-tree. One of the measures of efficiency and accuracy of the R-tree index is the area and perimeter of the sub-node MBR in the tree. The smaller the area and perimeter, the higher the spatial aggregation. Therefore, the analysis of the R-tree proposed by Guttman (1984) has some shortcomings: long loading time, insufficient subspace optimization, and long data extraction time for the spatial query.

No matter the equilibrium grid decomposition, quadtree decomposition, and traditional R-tree decomposition, the problem of large spatial distribution and density imbalance cannot be avoided. The regularized partitions of these decomposition methods are divided into different degrees.

In the algorithm of parallel overlay analysis, a lot of frequent data extraction from the cluster environment and the intersection of geometric objects are involved [6]. In the single-disk and single-processor environment, the traditional spatial data extraction method uses the index structure of the spatial database. However, the single-point spatial index storage and access mechanism in multi-disk and multi-processor environments cannot meet the requirements of high-performance computing for data extraction speed [29]. Therefore, it is necessary to implement a fast filtering and extraction mechanism for spatial data matching with a distributed shared-nothing mode in the computing environment of parallel overlay analysis. Spatial index is an important criterion for measuring the quality of the spatial database. In the spatial database, there are usually millions of data tables. If the traditional spatial indexing method of the database is adopted, the data query efficiency will be seriously reduced. At the same time, spatial data has spatial object uncertainty, and the intersection, inclusion, and separation are complex. It is difficult to classify and sort spatial data by processing ordinary data. The spatial index of the spatial database field can be divided into two types: embedded and external [30]. The embedded spatial index structure is incorporated into the database as part of the database itself, while the plug-in database is usually in the form of middleware, which performs a similar proxy and forwarding mechanism at the data request and data layers. For example, the default indexing methods for Oracle spatial and PostGIS are R-tree and B-tree [31,32]; ESRI ArcSDE is an external spatial database management mechanism, which has no specific spatial data carrier but is based on the traditional RDBMS system. The extension, such as ArcSDE, can implement spatial data management and indexing mechanisms based on SqlServer and Oracle. The spatial database established by ArcSDE is called Geodatabase [33]. The default indexing strategy for geodatabase is a spatial grid index for the feature class. Secondly, in the aspect of distributed spatial index data decomposition, Kamel (1993) and other research applied R-tree to the deployment of single-processor and multi-disk hardware structures and implemented an R-tree-based parallel query mechanism. Zhang et al. used a multivariate R-tree (Multiplexed-R-tree) structure to optimize R-tree in combination with proximity index constraints [34]. Experiments show that parallel domain query performance is better when dealing with spatially balanced data. In order to improve the efficiency of massive spatial data management and parallelization processing in the distributed parallel computing environment, Tanin et al. implemented the distributed Quadtree-based spatial query function in the peer–to peer network environment [35].

The parallel spatial index has gradually formed an essential branch of the spatial index family with the development of high-performance parallel GIS applications, which can solve the problem of the simple data decomposition method in this study. The most typical parallel spatial index is the MC-R-tree (Master-client R-tree) method proposed by Schnitzer [36]. The method is characterized in that all non-leaf nodes in the spatial index tree are stored in the main cluster. In the node, each subtree in the index tree is stored in the sub-computing node of the cluster. The disadvantage is that the space utilization of the conventional R-tree is low, and the number of MBR overlaps higher when the subtree index are assigned to the child nodes.

3. Methods

This section includes active-slave spatial index decomposition for intersection, multi-strategy Hilbert ordering decomposition, and the parallel spatial union algorithm.

3.1. Active-Slave Spatial Index Decomposition for Intersection

From the dynamic nature of the overlay analysis, the superimposed layers are divided into the active layer (overlay layer) and passive layer (base layer). The point of parallel acceleration is the fast query of the geometric elements in the active layer. The intersection part of the layer and the spatial index is the key technology to achieve acceleration. According to the characteristics of the parallel overlay analysis in this paper, the storage of spatial data adopts a completely redundant mechanism, and each child node maintains a complete set of spatial data tables to be superimposed.

A data decomposition method is proposed for the parallel intersection operation. The process is to spatially decompose the active layer in advance and send the partition location information to the corresponding child node according to the FID (The name primarily used in the spatial data layer), the child node then locally extracts the active layer data in the range, then, the geographic elements in the passive layer all participate in the establishment of the whole spatial index tree; the elements in the active layer query the passive layer index and perform superposition operations on the intersecting candidate sets. Since the process of spatial query and the intersection operation in the overlay analysis are mostly the same operation, the data decomposition mode is adapted to be combined with the parallel intersection operation. In Figure 1, O_obj represents the entity object in the active overlay layer O_lyr. After the decomposition strategy, it is distributed to each child node, and the base layer B_lyr is parallelized.

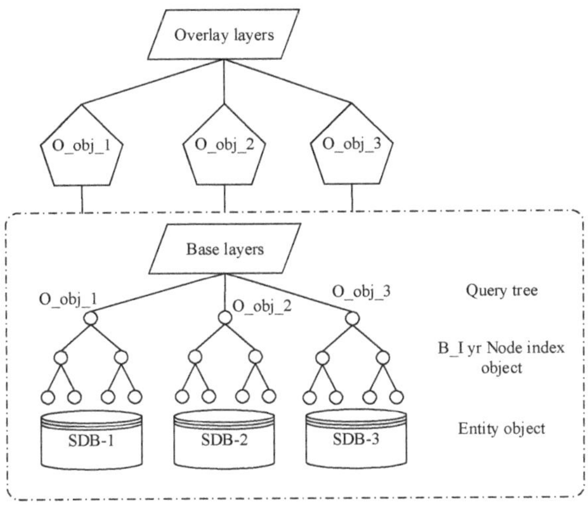

Figure 1. Parallel data decomposition using R-tree.

The actual data decomposition is only the division and treatment of the overlay layer O_lyr, and the base layer is still a query as a whole. This solution enables acceleration in a reasonable computational complexity in small to medium data scale applications. The data decomposition characteristics of the above map overlay algorithm shows that if each child node resides in a base layer, the spatial index of the entire tree has two defects: first, when processing a layer with a large number of geographic features, each node needs to consume a lot of time and memory resources; second, each query is for the traversal of the entire tree, and there are a large number of invalid queries in the intersecting query of a single geometric object.

To improve the efficiency of massive spatial data management and processing in a parallel environment, an improved master-slave spatial indexing method MC-STR-tree was designed according to the current research results. The rapid management of data management and geometric intersections provides conditions.

The basic index structure of the index adopts STR-tree, and the modification of the R-tree object node is improved, and the conflict of data distribution among the nodes is reduced. It has good spatial clustering characteristics and extremely low spatial index packet overlap rate, so it has excellent parallelization characteristics. STR-tree is a highly balanced tree with simple and maximized space utilization.

The processing steps of the improved master-slave parallel space index MC-STR-tree decomposition (Figure 2) are as follows:

(1) Data preprocessing

The master node divides the MBR of the spatial data index node into n parts according to specific rules. The commonly used rules are Hilbert ordering and STR partitioning method, and n is an integer not larger than the cluster computing node. The master node needs to record the node tag to which each subtree is sent and the root node MBR of the subtree. In this paper, the leaf nodes are divided according to the STR rule, and each computing node corresponds to a stripe in the STR. To control the number of subtrees less than the number of computable nodes, node Vol is used to set the maximum number of indices sub-nodes allowed for each computed node in the STR-tree.

(2) Sending of the subtree index

The primary node sends the index entries to the child nodes according to a certain replication and allocation policy. The data structure uses an in-memory spatial index. Unlike the traditional hard disk-based spatial index, the in-memory spatial index has the characteristics of dynamic construction and fast query.

(3) The master node sets up the Master-R-tree

The master node builds a new spatial index into all non-leaf nodes in the original entire R-tree index tree in computer memory. The index tree records the index item information allocated by each computing node, and its function is equivalent to an overall index controller. The intersecting query of the active overlay layer is first filtered by the overall index tree.

(4) Computing node to build a Client-tree

According to the agreed STR-tree node organization rule, the index items sent by the master node are built into the client terminal tree. The entity object stored in the computer node is associated with the index item of the subtree, and the entity space data is extracted according to the index item MBR.

The parallel R-tree index decomposition based on the STR rule is analyzed. The essence is to replace the random partition in the form of elemental fid with the domain decomposition partition considering the neighboring spatial rule. The master node controls the overall spatial index R-tree of a large node, where the id information of the spatial data domain decomposition block in the cluster is recorded. The active layer first performs a query operation with the overall index tree. If one or more intersecting subdomains are queried, the master node notifies the child nodes having the subdomains to extract the active layer locally. The element is further intersected with the subtree, and the process will be analyzed in the parallelization of the intersection.

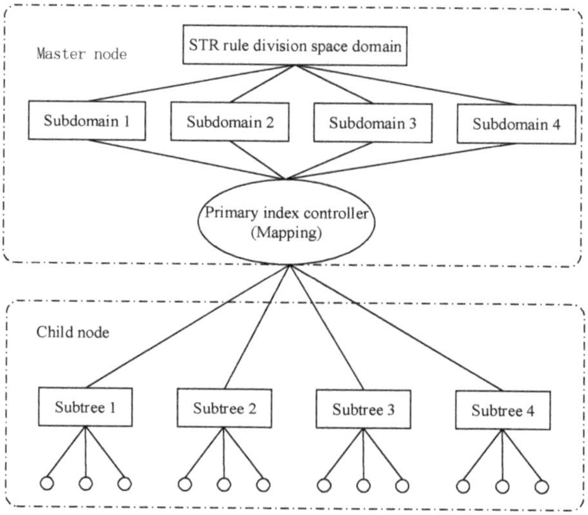

Figure 2. Sort-Tile-Recursive (STR)-tree data decomposition.

3.2. Multi-Strategy Hilbert Ordering Decomposition

It can be seen from the process of generating Hilbert code values that the determination of the code value is determined by the center point of the space grid, which is called the order method of the midpoint strategy. This section studies the difference between the middle strategy and the median strategy. Hilbert curve sorting is used to optimize the problem so that the sorting method can better take into account the spatial proximity of map features. The spatial sort is the reversible 1:1 correspondence between spatial feature points and consecutive integers in one-dimensional space as key values.

The middle strategy is easier to understand. In the process of dividing the spatial hierarchy, each cell is divided according to a fixed size, that is, at the middle point of the cell. The midpoint strategy is suitable for point set objects with a more balanced distribution in the low dimension in practical applications.

Unlike the middle strategy, the median strategy divides the cells in the space according to the median point in the x or y direction of the point to be sorted (Figure 3). The geometric midpoint is an effective solution to the positioning of resource rules. In the high dimensional space, the median point can better represent the centrality of the clustering point. The median strategy is adapted to high-dimensional point sets or point set objects with uneven distribution.

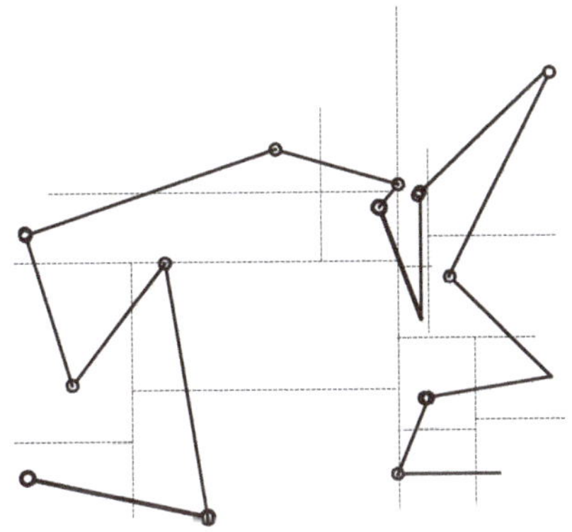

Figure 3. The demonstration of the Hilbert median sort demo.

From the definition of the median and the application context, the characteristics of maintaining clustering in Hilbert ordering can be better understood. The median is a concept in sequential statistics. In a set of n elements, the nth order statistic is the i-th small element in the set. For example, in a set of elements, the minimum is the first order statistic (i = 1), and the maximum is the nth order statistic (n = 1). An informal definition of a median is the "midpoint element" of the collection it is in. When n is an odd number, the set has a unique median at $i = ((n + 1))/2$ out; when n is even, it has two medians, which appear at $i = n/2$ and $i = n/(2 + 1)$. If not considering the parity of n, the median always appears at $i = \lceil((n + 1))/2\rceil$ (upper median) and $i = \lfloor((n + 1))/2\rfloor$ (lower median). In our study, the Hilbert median strategy sorting adopts the lower median by default. The CGAL library is used to assist Hilbert in sorting in the decomposition process of spatial data. Because the spatial data cannot be extracted from the data source directly according to the Hilbert code value, it must be the main processing data in parallel computing. After the node establishes the mapping relationship with the spatial database, the data is decomposed. Figure 4 shows the world map divided by two different Hilbert sorting methods. Each color denotes the different classes for the result of Hilbert sorting methods.

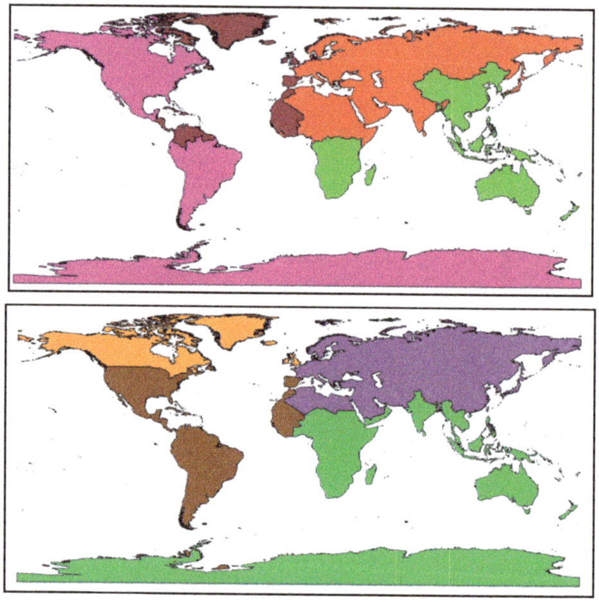

Figure 4. Hilbert middle and median sort for world map (4 nodes).

The process based on Hilbert's sorting decomposition is shown in Figure 5. First, the feature points are extracted from the feature elements, and then the appropriate Hilbert space sorting strategy is selected according to the distribution characteristics of the spatial data. After Hilbert sorting, the main computing nodes follow the order of the one-dimensional Hilbert curve according to the multi-channel form, and the data-polling type is generated to each sub-node according to the interval division. Because the storage of vector data in this paper takes the form of a spatial database, what is referred to here is the location tag of the object in the spatial database table.

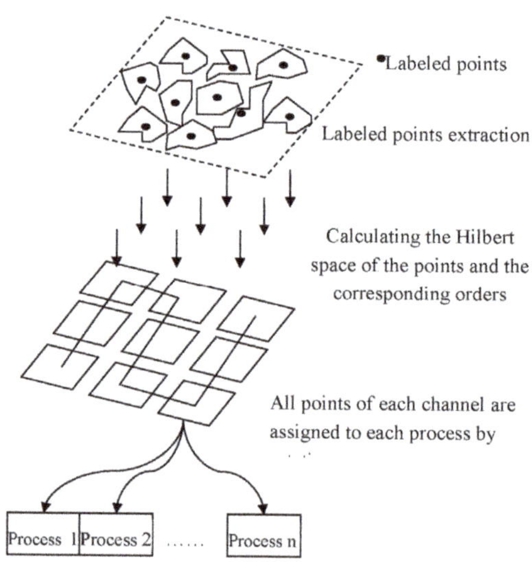

Figure 5. Spatial data decomposition using Hilbert sort.

The most crucial purpose of spatial data domain decomposition in parallel GIS computing systems is that each child node can perform spatial data extraction or query work at the same time, without pulling information from the global control node; however, the traditional data partitioning method is based on the static clustering method. Reasonable spatial data partitioning should consider the spatial location distribution and length uncertainty of geographic data. It can be considered from two aspects whether the decomposition strategy of Hilbert ordering is the best, including whether each piece of sub-data can maintain spatial proximity to the maximum extent and whether the storage capacity of the data blocks in each node is equalized.

3.3. Parallel Spatial Union Algorithm

The parallel spatial union algorithm process is as follows:

(1) Polygon data feature point extraction

The Hilbert curve sorted objects are point objects with integer values, and the ordering of polygons was converted to the ordering of their representative feature points. The feature points of a polygon usually have an outer center point of the rectangle or four corner points, a geometric center point, and a mass center point. Since the calculation of the center point and the center of gravity of the polygon is time-consuming, it is not conducive to the parallel acceleration of the map overlay analysis. Therefore, the feature point of the outsourced rectangular is the center point. All the layers S are regarded as a set of all the layers L, and the feature point extraction is represented as, among them, a feature point set. The feature point set is constructed by the mapping function map.

(2) Feature point Hilbert space ordering

The function corresponding to the Hilbert curve sort is described as follows. Among them are integer coordinate values, and d is the Hilbert code value. For floating-point data in practical applications, it is necessary to scale to the nearest integer value. Constructing a Hilbert curve in a two-dimensional space maps the interval to a unit square interval, and each cell is recursively divided into four parts until each cell contains only one feature point. The computational complexity of constructing the Hilbert sorting curve algorithm is, where n is the number of hierarchical divisions of the Hilbert space. In the calculation process of Hilbert coded values in this paper, the value of n is not forced to be constrained so as to ensure that the point object clustering is denser and still maintains a good mapping unity.

In the parallel overlay analysis, Hilbert curve ordering is performed according to the distribution characteristics of the feature elements using two strategies: middle strategy and median strategy.

(3) Distribution of nearby polygons

In the parallel overlay analysis system, the Master node is responsible for loading all the map layers involved in the calculation and establishing the Hilbert sort number for its content. The elements in each layer are evenly distributed to each node according to the number of compute nodes. The number of layer elements and number of processors is n. It is necessary to map the code on the Hilbert curve string to the FID of the map element. The generation of the one–to–one correspondence depends on the previous Hilbert ordering process in which there are two mapping relationships, and the two mapping relationships are connected by feature point coordinates. The specific process can be described by the following formula, which is the Hilbert code value, the FID marked element, and the data extraction condition as the overlay query.

The implementation process of the Hilbert joint algorithm is as follows:

(i) A data partitioning strategy. The experiment uses three data distribution strategies in the order of fid, Hilbert_median, and Hilbert_middle. First, the master node determines the number of nodes participating in the calculation, and then equally distributes the ordered sets separately.

(ii) Independent calculations, generating intermediate results. Each computing node is responsible for the joint operation of the local polygons and writes the generated intermediate results to a temporary table in the MySQL spatial database.

(iii) Results are collected in conjunction. The master node loads the middle layer into the local memory in the order in which the child nodes complete the task, and then jointly outputs the final result layer to these layers.

The previous part is shown the operation of a single layer, the data decomposition hierarchy object is single, and it is easier to divide and conquer. For the overlaying and merging of two or more layers, it is necessary to consider the spatial data distribution of multiple layer species, such as the case of the line surface overlaying, to prevent the layer overlaying of the spatial distribution imbalance.

From the perspective of spatial data decomposition parallelism, the parallel stacking of multiple layers can use two methods.

(i) According to the above design, the Hilbert sorting decomposition parallel method is based on the spatial ordering of multi-level two-dimensional plan layers through high-dimensional Hilbert curves. For the processing of two-dimensional plane map data, the specific process can be divided into two processes of reducing the dimension and raising the dimension, but the two are not contradictory. Dimensionality reduction is for a polygon layer, extracting a single point from a polygon in the layer, such as the center point of the outer frame, a corner point, or the Centre's CentreID point to uniquely identify the polygon as a mapping to the one-dimensional Hilbert function.

The polygon objects in the two layers were loaded into the same container, and the geometric objects in the container were manipulated according to the Hilbert parallel sorting algorithm to merge the individual layers. The advantage of this method is that it is not necessary to consider the combined correspondence of polygons in the container, and it can be divided according to Hilbert ordering. Polygons that do not want to intersect in the original single layer may be indirectly connected due to the addition of new layer features, as shown in Figure 6. Indirect connectivity increases the complexity of parallel association. If the polygons in the two layers are only a one–to–one relationship, they can be directly merged in batch form. The polygons may be spread in pieces when considering indirect connectivity. In this way, the parallel joint process needs to continue the binary tree merging of the data decomposition results.

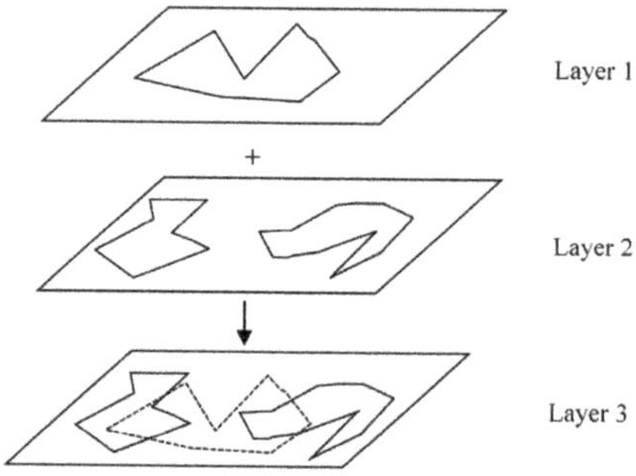

Figure 6. Connect relationship between two layers.

(ii) A combination of multi-layer filtering mechanisms using parallel master-slave indexes.

The master node establishes a coarse-grained spatial index for the two layers, which is equivalent to establishing a distribution for each layer to establish m and n spatial index subtrees. First, test the intersection of the root nodes of these subtrees and filter the separated subtrees. The information of the intersecting subtrees is sent to the child nodes for the merge operation of the independent subtrees. The merge mode can be understood as spatial analysis of different spatial data of different spatial granularity levels, which can be performed based on spatial data with lower spatial granularity level, and spatial analysis of different spatial data of the same spatial granularity level must be superimposed to generate Spatial data with a lower spatial granularity level, which is then based on spatial data with a lower spatial granularity level.

According to the implementation principle of the parallel master-slave index union, with data decomposition based on a spatial index, it is difficult to completely eliminate the redundancy of index objects, so different nodes will have the same geometric object, and the subtree merge is a recursive consumption. At the time of operation, the last master still needs to recycle and combine the intermediate results. The sprawling characteristics of joint operations determine that they cannot be parallelized in the form of batch processing. The parallel combination of master-slave index tree mode is difficult to implement in practical applications. The parallel joint research in this paper uses Hilbert as the experimental subject for the two layers as a whole, and the parallel master-slave index joint will be further studied in the next step.

In our study, two joint layers are treated as an overall virtual layer and parallelized according to the strategy of polygon merging in the same layer (Figure 7). The specific process is as follows:

(i) First, create a data structure featureInfo for each feature in each layer. The main member variables are fid, layer_num, hilbert_code, and mbr_pt. The fid is used to record the cursor of the feature in the layer. As a query condition for quickly extracting data in the future, layer_num records which layer is one of the filtering conditions of the data query. The hilbert_code defaults to a null value when the FeatureInfo is initialized, which is used to record the encoding of the Hilbert space sorting. Mbr_pt records the center point of the desired MBR as a feature point for Hilbert ordering.

```
typedef struct
{
    long fid, hilber_code;
    GTPoint* mbr_pt;
} featureInfo;
```

(ii) Put the featureInfo in the two layers into the vector array Vector<featureInfo*> and use the Hilbert spatial sorting function hilbert_sort() to sort and encode the elements in the array.

(iii) According to the number k of available computing nodes, the total n elements are sent to one node every n/k according to the Hilbert coding order. The polling method only sends the Hilbert space sorting number interval instead of sending the specific geometry object.

(iv) The child node receives the Hilbert space sorting coding interval, extracts the geometric objects of the feature from the local database according to fid and layer_num in featureInfo, and starts to merge the intersecting polygons in the interval.

(v) Continue the merge in the form of a single layer binary tree merge until all intersecting polygons are merged.

According to this layer, the overall layer is added to perform Hilbert sorting decomposition. The advantage is that the data is divided into non-redundant polygons of each node. The unique characteristics of Hilbert coded values can ensure the low coupling of joint calculation between processes. The binary tree parallel combination can reduce the pressure of the main node recovery. Therefore, the scheme is feasible, and theoretically, an acceleration effect can be obtained. The following is an experimental verification of the multi-layer Hilbert joint method using actual data.

The computational complexity of this algorithm is O(N+K), which is derived from the count of polygons for the overlayed map layers. Given N polygons in each layer, if using the brute-force overlay test, the complexity of this algorithm should be $O(N^2)$. However, we performed an intersecting filter process that can exclude the unnecessary computing for polygons that do not overlay to each other in the space. Thus, there would be K polygons in the overlayed layer that participate in overlay computation. Ultimately, the complexity of this algorithm should be O(N+K).

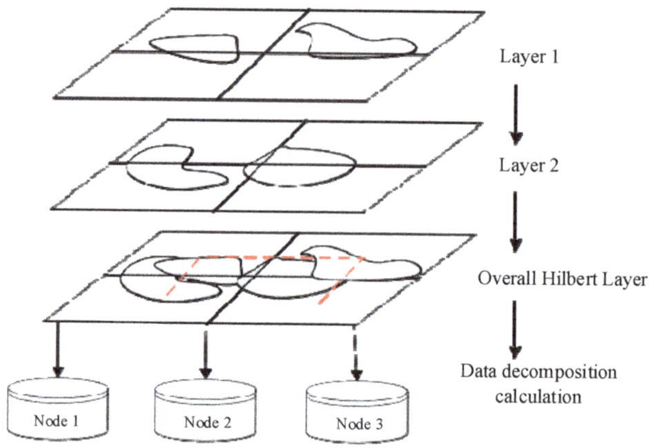

Figure 7. Hilbert decomposition for two layers.

4. Experiments

The parallel algorithm for polygons overlay analysis was implemented, and the performance of the parallel spatial union was verified. Multi-strategy based spatial data decomposition mechanism was implemented, including parallel spatial data index, the Hilbert space-filling curve sort, and spatial data decomposition. The programming language was C++. The operating system of each computation node was 64-bit Ubuntu Linux (16.04), and run on Intel CPU based hardware architecture (Model: Intel Core i7-3960X, 3.8 GHz, 12 cores).

4.1. The Experiment for Parallel Spatial Union

The experiment used a spatial union in map overlay analysis as an example to conduct parallelization experiments. The experimental data shows the Chinese 1:4 million land use map. The data type was multi-polygon, the number of elements was 15,615, and the number of points was 886,547 (Figure 8). The visualization strategy in Figure 8 is FID based strategy. The experimental dataset is a type of data commonly used in GIS applications, representing the general characteristics of spatial data decomposition. The spatial distribution of these maps data is irregular and unbalanced, and the amount of data is large. It is representative of experiments and has guiding significance for practical applications. Figures 9–11 are the results for the parallel spatial union using fid decomposition, median decomposition, and middle decomposition. The number of the results for the decomposition method is 2, 4, 8, and 16, respectively. Different color denotes different class in these maps.

Figure 8. The landuse map of China.

Figure 9. Parallel spatial union using the fid decomposition method.

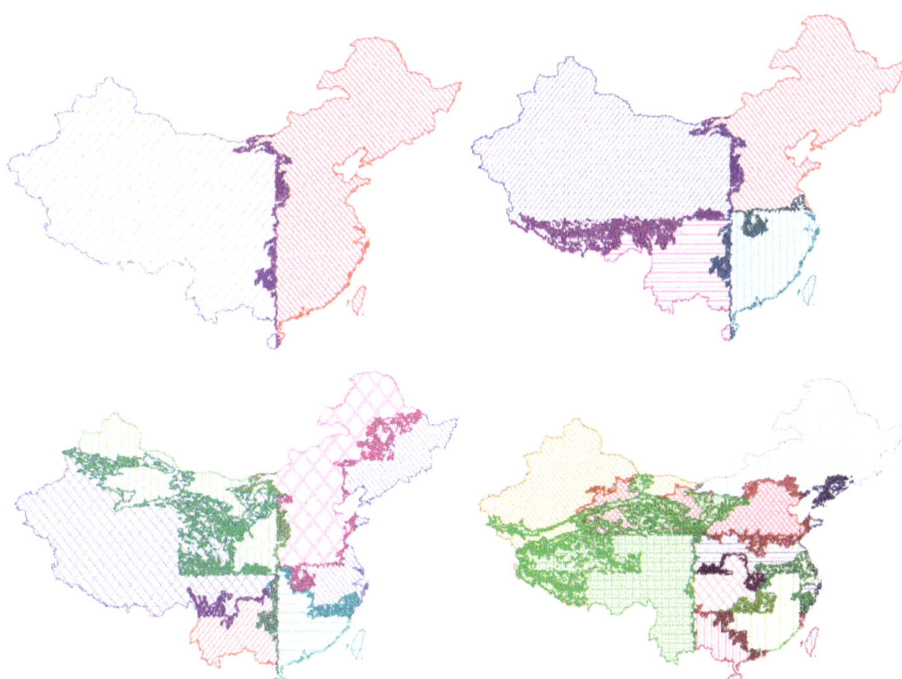

Figure 10. Parallel spatial union using the median decomposition method.

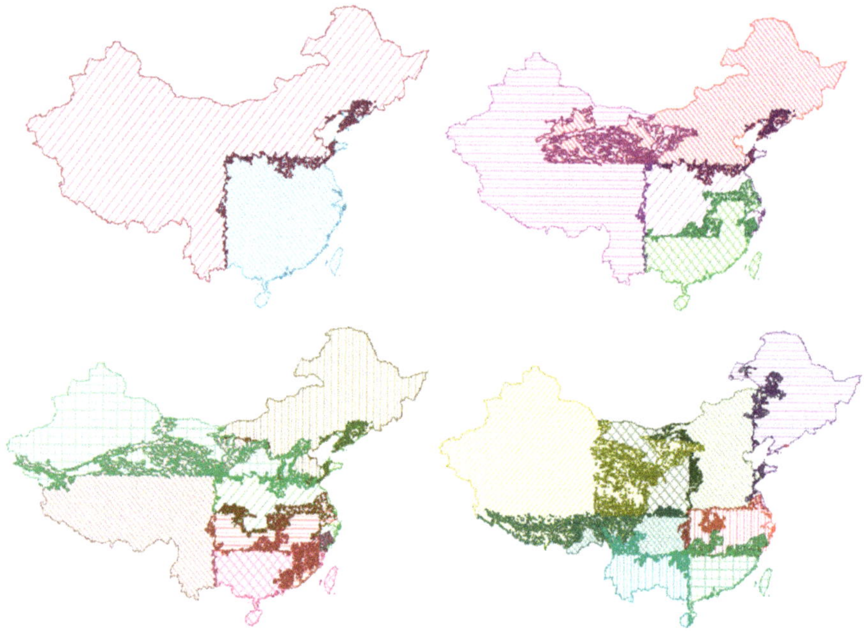

Figure 11. Parallel spatial union using the middle decomposition method.

Figure 12 shows the resource consumption of a computer when the parallel overlay is used to implement Hilbert ordering using the MPI method. The system starts a total of 10 computing processes. It can be seen from the figure that the CPU usage of each process is basically the same in the parallel computing process, and the memory usage averages about 1.3%.

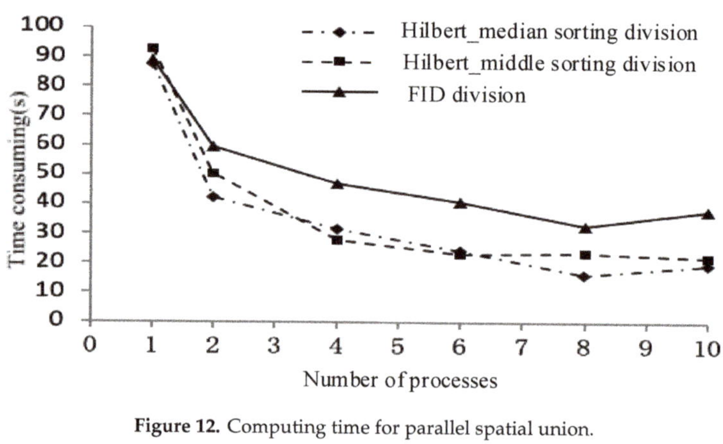

Figure 12. Computing time for parallel spatial union.

According to the experimental results, the parallel merging strategy is analyzed. From the comparison of FID partitioning and Hilbert partitioning, the results of FID partitioning are scattered. The merging results in the same sub-node are not necessarily adjacent. Hilbert divides the merging result better, for example, if the FID is merged, even if there are only two computing nodes, the combined result of the first node also includes part of the combined result of the second node. At the same time, it can be observed that the combined results of FID partitioning are basically horizontal strips, and the Hilbert partitioning results are basically blocky. It can also be observed from the merger of the Taiwan Province that the fid division strategy will be divided into multiple blocks (the whole map), the islands will be divided into multiple blocks for processing, and the Hilbert spatial sorting combination will ensure a small area and approximate shape. The circular polygonal area maintains functional integrity.

The parallel computing time and acceleration ratio effects of the three merge modes are shown in Figures 13 and 14. The parallel computing time is the sum of the data decomposition time, the independent calculation time, and the last merge time of the master node. The calculation time has decreased with the increase of the number of processes; however, the acceleration effect is obvious when the acceleration effect is less than four processes, which is higher than the linear acceleration ratio, but the acceleration effect is significantly reduced when there are more than four processes because of two aspects: one is that if there are multiple processes in one child node, the data is concurrent with I/O, and the other is that multiple child nodes send writing database requests to the master node. The number of polygons that the master node needs to merge increases, and the database is locked as the writing process takes time. When the number of threads is greater than eight, the median strategy shows a trend of decreasing the acceleration ratio, indicating that the consumption of the collection and consolidation of the primary node is greater than the acceleration of the parallel computing. The solution is to increase the number of physical machines and try to keep each child node running a small number of processes, reduce read-write conflicts, allow as many binary tree merges as possible between child nodes, and reduce the number of merged polygons at the primary node.

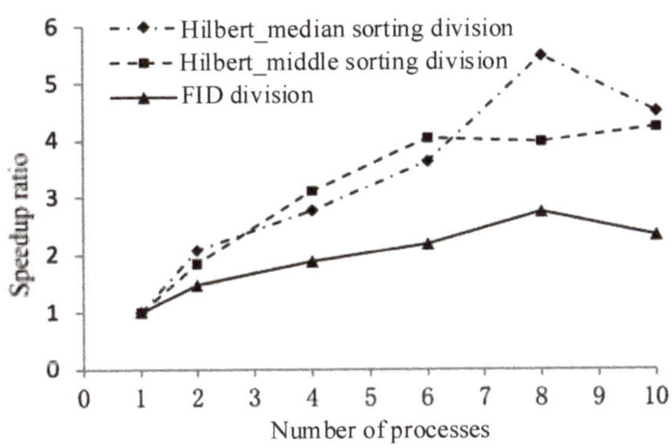

Figure 13. Speedup ratio for parallel polygon union.

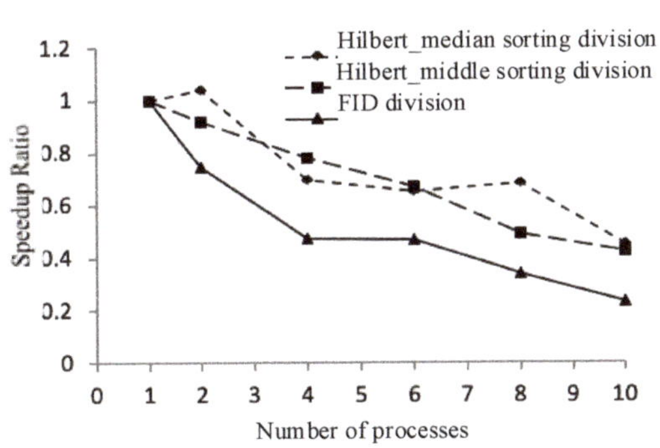

Figure 14. Speedup ratio for parallel polygon union.

From the accelerated efficiency analysis of parallelization, the acceleration efficiency is the lowest according to the strategy of the fid division. The acceleration efficiency change from 1 process to 10 processes is from 1 to 0.234, which is a large change. The reason is that the polygon data is allocated according to fid. The number of nodes actually intersecting is small, which causes many polygons to merge again when the master node collects, causing the overload of the master node to affect the overall acceleration efficiency. The acceleration efficiency of the Hilbert_median partitioning strategy varies between 2~4 processes and 8~10 processes, and the acceleration efficiency is basically unchanged in 4~8 processes. The Hilbert_middle strategy divides the efficiency with the number of processes. The increase has a linear downward trend. From the perspective of the accelerated efficiency degradation of these parallel joints, although the data decomposition strategy for estimating the spatial proximity characteristics can achieve a specific acceleration effect, there is still room for further optimization in the parallelization algorithm of the data in the reduction phase of the data.

4.2. The Experiment for Parallel Hierarchical Spatial Union

In the experimental data, the basic layer is the planning map of a certain urban area, and the overlayed layer is the post-translation planning diagram (Figure 15). The result of the spatial union operation is shown in Figure 16.

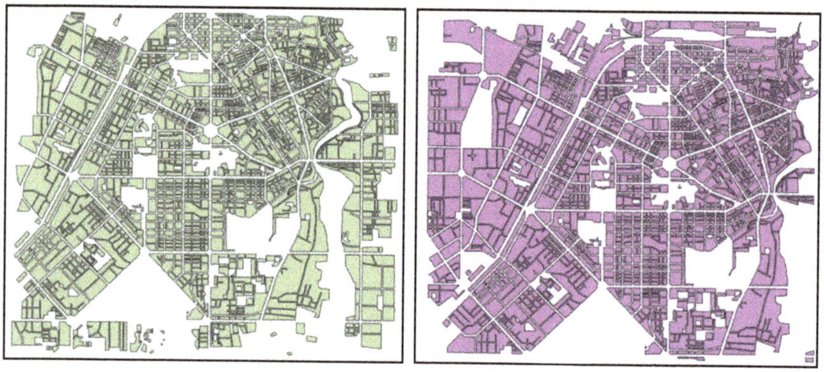

Figure 15. Test data for the Hilbert decomposition of two layers.

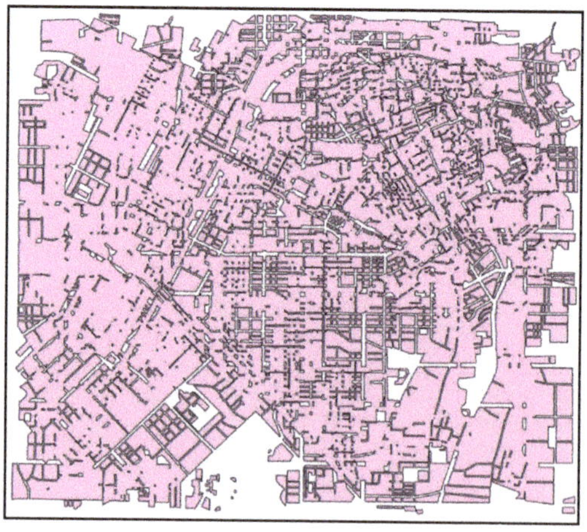

Figure 16. Two data layers' union result.

5. Conclusions and Future Work

In this study, we propose an efficient parallel algorithm for polygons overlay analysis, including active-slave spatial index decomposition for intersection, multi-strategy Hilbert ordering decomposition, and a parallel spatial union algorithm. A multi-strategy based spatial data decomposition mechanism is implemented, including parallel spatial data index, the Hilbert space-filling curve sort, and data decomposition. The results of the benchmark experiments showed that the parallel algorithm for polygons overlay analysis achieves high efficiency.

However, there are some limitations to this study. Firstly, in this study, we have not discussed cloud computing and edge computing for map overlay analysis. The new computing framework can improve the efficiency of spatial analysis for large-scale geospatial data [37–39]. Secondly, graphics processing unit (GPU)-based geo-computation is not considered. GPU is a potential factor that could improve the speedup ratio [40]. In the future, we will improve the study by addressing those limitations. Firstly, the map overlay algorithm based on cloud computing will be proposed. Secondly, we would take account of GPU-based parallel algorithms for map overlay in the future.

Author Contributions: Conceptualization, S.W.; data curation, Y.Z., S.W., and Y.G.; funding acquisition, Y.Z.; investigation, Y.Z. and S.W.; methodology, Y.Z. and S.W.; project administration, Y.Z. and S.W.; software, Y.Z., S.W. and Y.G.; supervision, Y.Z. and S.W.; validation, Y.G.; visualization, Y.Z. and Y.G.; writing—original draft, Y.Z., S.W., and Y.G.

Funding: This research was funded by the National Natural Science Foundation of China (grant number 41601478) and the National Key Research and Development Program (grant numbers 2018YFB0505301, 2016YFC0500103).

Acknowledgments: We would like to acknowledge the Institute of Geographic Sciences and Natural Resources Research of the Chinese Academy of Science for providing the research grant to conduct this work.

Conflicts of Interest: The authors declare no conflict of interest.

References

1. Wang, S.; Zhong, E.; Lu, H.; Guo, H.; Long, L. An effective algorithm for lines and polygons overlay analysis using uniform spatial grid indexing. In Proceedings of the 2015 2nd IEEE International Conference on Spatial Data Mining and Geographical Knowledge Services (ICSDM), Fuzhou, China, 8–10 July 2015; pp. 175–179.
2. Zhao, K.; Jin, B.; Fan, H.; Song, W.; Zhou, S.; Jiang, Y. High-performance overlay analysis of massive geographic polygons that considers shape complexity in a cloud environment. *ISPRS Int. J. Geo Inf.* **2019**, *8*, 290. [CrossRef]
3. You, S.; Zhang, J.; Gruenwald, L. Large-scale spatial join query processing in cloud. In Proceedings of the 2015 31st IEEE International Conference on Data Engineering Workshops, Seoul, Korea, 13–17 April 2015; pp. 34–41.
4. Zhou, Q.; Zhong, E.; Huang, Y. A parallel line segment intersection strategy based on uniform grids. *Geo Spat. Inf. Sci.* **2009**, *12*, 257–264. [CrossRef]
5. Puri, S.; Agarwal, D.; He, X.; Prasad, S.K. Mapreduce algorithms for gis polygonal overlay processing. In Proceedings of the 2013 IEEE International Symposium on Parallel & Distributed Processing, Workshops and Phd Forum, Cambridge, MA, USA, 20–24 May 2013; pp. 1009–1016.
6. Agarwal, D.; Puri, S.; He, X.; Prasad, S.K. A system for gis polygonal overlay computation on linux cluster-an experience and performance report. In Proceedings of the 2012 IEEE 26th International Parallel and Distributed Processing Symposium Workshops & PhD Forum, Shanghai, China, 21–25 May 2012; pp. 1433–1439.
7. Zhou, C.; Chen, Z.; Li, M. A parallel method to accelerate spatial operations involving polygon intersections. *Int. J. Geogr. Inf. Sci.* **2018**, *32*, 2402–2426. [CrossRef]
8. Wang, S.; Armstrong, M.P. A quadtree approach to domain decomposition for spatial interpolation in grid computing environments. *Parallel Comput.* **2003**, *29*, 1481–1504. [CrossRef]
9. Kantere, V.; Skiadopoulos, S.; Sellis, T. Storing and indexing spatial data in p2p systems. *IEEE Trans. Knowl. Data Eng.* **2008**, *21*, 287–300. [CrossRef]
10. Wang, F. A parallel intersection algorithm for vector polygon overlay. *IEEE Comput. Graphics Appl.* **1993**, *13*, 74–81. [CrossRef]

11. Agarwal, D.; Puri, S.; He, X.; Prasad, S.K.; Shi, X. *Crayons—A Cloud Based Parallel Framework for Gis Overlay Operations*; Distributed & Mobile Systems Lab: New York, NY, USA, 2011.
12. Healey, R.; Dowers, S.; Gittings, B.; Mineter, M.J. *Parallel Processing Algorithms for Gis*; CRC Press: Boca Raton, FL, USA, 1997.
13. Mineter, M.J. A software framework to create vector-topology in parallel gis operations. *Int. J. Geogr. Inf. Sci.* **2003**, *17*, 203–222. [CrossRef]
14. Xiao, N. *GIS Algorithms*; Sage: Thousand Oaks, CA, USA, 2015.
15. Ching, Y.-T. Load balancing for the parallel map overlay-operation in the geographic information system. *J. Inf. Sci. Eng.* **1999**, *15*, 441–449.
16. Xun, L.; Wenfeng, Z. Parallel spatial index algorithm based on hilbert partition. In Proceedings of the 2013 International Conference on Computational and Information Sciences, Shiyang, China, 21–23 June 2013; pp. 876–879.
17. Papadopoulos, A.; Manolopoulos, Y. Parallel bulk-loading of spatial data. *Parallel Comput.* **2003**, *29*, 1419–1444. [CrossRef]
18. Naylor, B. Binary space partitioning trees as an alternative representation of polytopes. *Comput. Aided Des.* **1990**, *22*, 250–252. [CrossRef]
19. Samet, H. *The Design and Analysis of Spatial Data Structures*; Addison-Wesley Reading: Reading, MA, USA, 1990; Volume 85.
20. Peuquet, D.J. A conceptual framework and comparison of spatial data models. *Cartogr. Int. J. Geogr. Inf. Geovis.* **1984**, *21*, 66–113. [CrossRef]
21. Minasny, B.; McBratney, A.B.; Walvoort, D.J. The variance quadtree algorithm: Use for spatial sampling design. *Comput. Geosci.* **2007**, *33*, 383–392. [CrossRef]
22. Chen, S.; Ooi, B.C.; Tan, K.-L.; Nascimento, M.A. St 2 b-tree: A self-tunable spatio-temporal b+-tree index for moving objects. In Proceedings of the 2008 ACM SIGMOD International Conference on Management of Data, Ancouver, CB, Canada, 9–12 June 2008; pp. 29–42.
23. Guttman, A. *R-Trees: A Dynamic Index Structure for Spatial Searching*; ACM: New York, NY, USA, 1984; Volume 14.
24. Kamel, I.; Faloutsos, C. Hilbert R-Tree: An Improved R-Tree Using Fractals; College Park, MD, USA. 1993. Available online: https://www.researchgate.net/publication/2303411_Hilbert_R-Tree_An_Improved_R-Tree_Using_Fractals (accessed on 13 November 2019).
25. Hamilton, C.H.; Rau-Chaplin, A. Compact hilbert indices for multi-dimensional data. In *Proceedings of the First International Conference on Complex, Intelligent and Software Intensive Systems (CISIS'07), Vienna, Austria, 10–12 April 2007*; IEEE: New York, NY, USA, 2007; pp. 139–146.
26. Arge, L.; Hinrichs, K.H.; Vahrenhold, J.; Vitter, J.S. Efficient bulk operations on dynamic r-trees. *Algorithmica* **2002**, *33*, 104–128. [CrossRef]
27. Leutenegger, S.T.; Lopez, M.A.; Edgington, J. Str: A simple and efficient algorithm for r-tree packing. In Proceedings of the 13th International Conference on Data Engineering, Birmingham, UK, 7–11 April 1997; pp. 497–506.
28. Zhu, Q.; Gong, J.; Zhang, Y. An efficient 3d r-tree spatial index method for virtual geographic environments. *ISPRS* **2007**, *62*, 217–224. [CrossRef]
29. Ferhatosmanoglu, H.; Tuncel, E.; Agrawal, D.; El Abbadi, A. Vector approximation based indexing for non-uniform high dimensional data sets. In Proceedings of the 9th International Conference on Information and Knowledge Management, McLean, VA, USA, 6–11 November 2000; pp. 202–209.
30. Zhang, C.; Pan, M. A study on topological reconstruction of gis vector data based on grid index. *Geogr. Geo Inf. Sci.* **2006**, *22*, 20–24.
31. Obe, R.O.; Hsu, L.S. *Postgis in Action*; Manning Greenwich: New York, NY, USA, 2011; Volume 2.
32. Hu, Y.; Ravada, S.; Anderson, R.; Bamba, B. Topological relationship query processing for complex regions in oracle spatial. In Proceedings of the 20th International Conference on Advances in Geographic Information Systems, Redondo Beach, CA, USA, 6–9 November 2012; pp. 3–12.
33. Gaskill, J.; Brooks, D. Understanding arcsde. In Proceedings of the Annual ESRI International User Conference, Paris, France, 3–7 July 2000.

34. Zhang, M.; Lu, F.; Cheng, C. A forced transplant algorithm for dynamic r-tree implementation. In *Proceedings of the International Conference on Database and Expert Systems Applications, Kraków, Poland, 4–8 September 2006*; Springer: Berlin/Heidelberg, Germany, 2006; pp. 459–466.
35. Tanin, E.; Harwood, A.; Samet, H. Using a distributed quadtree index in peer-to-peer networks. *VLDB J.* **2007**, *16*, 165–178. [CrossRef]
36. Schnitzer, B.; Leutenegger, S.T. Master-client r-trees: A new parallel r-tree architecture. In Proceedings of the Eleventh International Conference on Scientific and Statistical Database Management, Cleveland, OH, USA, 30 July 1999; pp. 68–77.
37. Wang, S.; Zhong, Y.; Wang, E. An integrated gis platform architecture for spatiotemporal big data. *Future Gener. Comput. Syst.* **2019**, *94*, 160–172. [CrossRef]
38. Gui, Z.; Yang, C.; Xia, J.; Huang, Q.; Liu, K.; Li, Z.; Yu, M.; Sun, M.; Zhou, N.; Jin, B. A service brokering and recommendation mechanism for better selecting cloud services. *PLoS ONE* **2014**, *9*, e105297. [CrossRef] [PubMed]
39. Yang, C.; Yu, M.; Hu, F.; Jiang, Y.; Li, Y. Utilizing cloud computing to address big geospatial data challenges. *Comput. Environ. Urban Syst.* **2017**, *61*, 120–128. [CrossRef]
40. Heitzler, M.; Lam, J.C.; Hackl, J.; Adey, B.T.; Hurni, L. Gpu-accelerated rendering methods to visually analyze large-scale disaster simulation data. *J. Geovis. Spat. Anal.* **2017**, *1*, 3. [CrossRef]

© 2019 by the authors. Licensee MDPI, Basel, Switzerland. This article is an open access article distributed under the terms and conditions of the Creative Commons Attribution (CC BY) license (http://creativecommons.org/licenses/by/4.0/).

applied sciences

MDPI

Article

Vector Map Random Encryption Algorithm Based on Multi-Scale Simplification and Gaussian Distribution

Giao N. Pham [1,2,*], Son T. Ngo [1], Anh N. Bui [1], Dinh V. Tran [3], Suk-Hwan Lee [4] and Ki-Ryong Kwon [5]

1 Department of Computing Fundamentals, FPT University, Hanoi 10000, Vietnam; sonnt69@fe.edu.vn (S.T.N.); anhbn5@fe.edu.vn (A.N.B.)
2 Advanced Analytics Center, FPT Software Co., Ltd., Hanoi 10000, Vietnam
3 Department of Computer, University of Freiburg, Freiburg 79098, Germany; Dinh@informatik.uni-freiburg.de
4 Department of Information Security, Tongmyong University, Busan 608-711, Korea; skylee@tu.ac.kr
5 Department of IT Convergence and Application Engineering, Pukyong National University, Busan 608-737, Korea; krkwon@pknu.ac.kr
* Correspondence: giaopn@fe.edu.vn or giaopn@fsoft.com.vn

Received: 2 October 2019; Accepted: 13 November 2019; Published: 14 November 2019

Abstract: In recent years, GIS (Geographical Information System) vector maps are widely used in everyday life, science, and the military. However, the production process of vector maps is expensive, and a large volume of vector map data is easily stolen and illegally distributed. Therefore, original providers desire an encryption solution to encrypt GIS vector map data before being stored and transmitted in order to prevent pirate attacks and to ensure secure transmission. In this paper, we propose an encryption algorithm for GIS vector map data for preventing illegal copying, and ensuring secured storage and transmission. Polyline/polygon data of GIS vector maps are extracted to compute a backbone object. The backbone object is then selectively simplified by the multi-scale simplification algorithm in order to determine the feature vertices of the backbone object. The feature vertices of the backbone object are encrypted by the advanced encryption standard and the secret key. Finally, all vertices of the backbone object are randomized by the random Gaussian distribution algorithm to obtain the encrypted GIS vector map. Experimental results show that the entire map is altered completely after the encryption process. The proposed method is responsive to the various GIS vector map data formats, and also provides better security than previous methods. The computation time of the proposed method is also significantly shorter than that of previous methods.

Keywords: GIS vector map data; GIS vector map security; selective encryption; simplification method and cryptography

---

1. Introduction

Currently, GIS vector maps are used in applications in many fields, such as science, navigation, and online or offline digital map services. GIS vector maps are created and developed by the merging of cartography, statistical analysis, and database technology based on vector models [1]. Due to the fact that GIS vector maps may have significant value, GIS vector map data may be stolen or easily purchased and redistributed or resold several times without obtaining any permission from the original data providers. In addition, some applications of vector maps by the military or some personal applications of vector maps must be secured from unauthorized users. Consequently, GIS vector map data protection is necessary to prevent illegal duplication and distribution.

GIS vector map security includes copyright protection [2–4], access control for users, and vector map data encryption in order to prevent attacks or illegal distribution during storage and transmission, and damage in the integration process of geographical information [5–14]. However, access control on the Web is unable to prevent attacks or illegal duplication and distribution, and watermarking techniques are only useful in identifying ownership and for copyright protection, while unauthorized users should not be able to see, attack, or extract GIS vector map content. Therefore, GIS vector map data encryption is necessary and suitable for protection. Moreover, conversion between data formats is vulnerable to attack and security techniques based on database management systems are not responsive to the various data formats of vector maps [10,11]. Thus, the encryption techniques for GIS vector map should be responsive to the various data formats of GIS vector maps and the security requirements.

In order to meet the issues above, we propose an encryption algorithm for GIS vector map data in this paper. Our algorithm solves the current issues of GIS vector map security by encrypting the geometric objects of GIS vector maps. The main intent of the proposed algorithm is to extract geometric objects (polylines and polygons) from GIS vector maps in order to encrypt them. Geometric objects are used to compute the backbone object of polyline/polygon. The backbone object is then selectively simplified by the multi-scale simplification algorithm in order to determine the feature vertices of the backbone object. The feature vertices of the backbone object is encrypted by the AES (Advanced Encryption Standard) algorithm and the secret key. Finally, all vertices of the backbone object are randomized by the random Gaussian distribution in order to generate the encrypted vector map. To clarify the detailed contents of the proposed algorithm, this paper is organized as follows: In Section 2, we discuss vector map data security techniques and explain the relationship between vector map data and the proposed algorithm. In Section 3, we explain the proposed algorithm in detail. We then perform experiments, discuss the experimental results, and evaluate the performance of the proposed algorithm in Section 4. The conclusion is provided in Section 5.

2. Related Works

2.1. Vector Map Data Security

The main idea of the watermarking schemes for GIS vector maps is to embed the watermark into the vector map by modifying the coordinates of vertices of geometric objects in the spatial domain [2,3], or embedding the watermark into the spectrum coefficients of a sequence of vertices or topologies in the frequency domain [4]. Thus, vector map watermarking is not suitable for preventing attacks and illegal copying.

The security techniques for GIS vector map data have been proposed in recent years [5–9]. Authors have mostly explained various methods to define an access control system for spatial data on the Web, or mentioned the security requirements for geospatial database management systems and privacy policies. However, access control on the Web and the management of databases do not prevent attacks, illegal duplication, and distribution. Wu et al. [10] proposed a compound chaos-based encryption algorithm for vector data by considering the storage characteristics and the parameters of a chaos-based system; however, this method is not responsive to the various data formats of vector maps. Li et al. [11] encrypted the spatial index of a set of vector data in an external Oracle database management system when it is transmitted to the client; however, this algorithm does not ensure the security of vector maps because the key length is very short. Yasser et al. [12] also described an encryption algorithm that combined AES and RSA (Rivest–Shamir–Adleman) cryptography with a simple watermarking technique in order to protect the copyright protection of vector maps in online and offline services. This method encrypts all parts of a shape-file using an AES block cipher operator of 256 bits. This approach is typical because it encrypts the data stream of a shape-file using the AES cipher, and the computation time is very long. Jang et al. [13] proposed a perceptual encryption method that is combined with the compression process for vector map data. This method only encrypted the direction and position of data units in the compression process. This method has low security.

Bang et al. [14] proposed a selective encryption method for vector map data based on a chaotic map in the frequency of discrete wavelet transform (DWT); however, this method only selects some objects and encrypts DC (Direct Current) values in the DWT domain by a common secret key. This method is very simple and weak because it does not encrypt all contents of vector maps and uses a common key. Moreover, this research did not present performance or security evaluations. In summary, the weakness of previous methods is solved by the proposed algorithm, which is presented in Section 3.

2.2. *Vector Map Data-Based Encryption*

GIS vector map data is stored in layers. Each layer contains a number of geometric objects, such as points, polylines, and polygons, and annotations, such as symbols. This notation is shown in Figure 1a. Annotation is used to display notes on vector maps, while geometric objects are used to represent geographical objects on vector maps. Points are used to represent simple objects, such as position, while polygons and polylines are used to represent complex objects. A polyline is a set of ordered vertices used to represent objects such as roads, contour lines, rivers, and railways. A polygon is a set of connected polylines used to represent objects such as buildings, areas, lakes, and boundaries. Thus, polylines and polygons are considered the important components of a vector map.

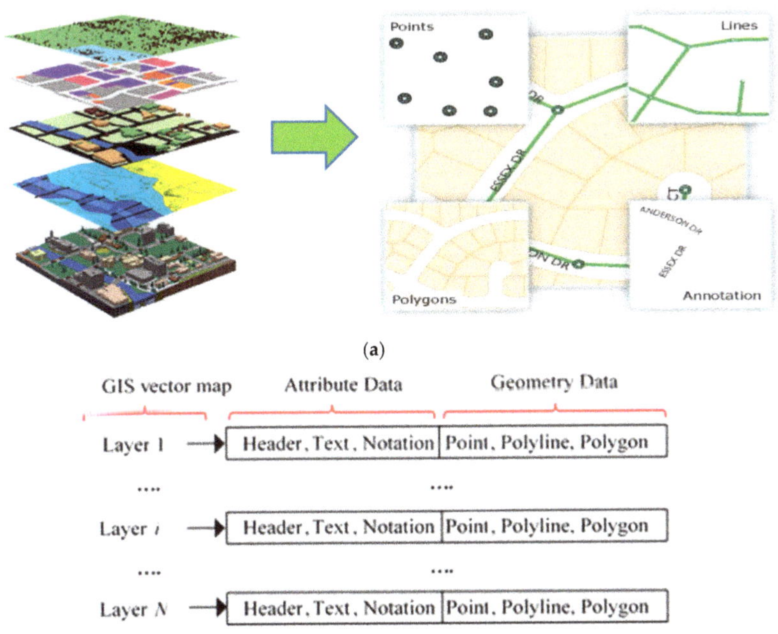

Figure 1. (**a**) Vector map data model, and (**b**) data components of a vector map.

In addition to annotation and geometric objects, vector maps also contain the storage information as header and text. Because points, polylines, and polygons determine the content of vector maps, we consider these geometry data, while the annotation, header, and text are considered attribute data. Figure 1b shows the data components of a GIS vector map. The attribute data of a GIS vector map does not contain geographical information or determines the shape of the GIS vector map; thus, it is only required to extract polylines and polygons to perform the random encryption process for GIS vector maps.

3. The Proposed Method

3.1. Overview

The proposed algorithm is shown in detail in Figure 2. To begin, each geometric object (polyline/polygon) is extracted from the GIS vector map to compute a backbone object. The backbone object is then selectively simplified by the multi-scale simplification algorithm with differential scale factors in order to obtain new geometric objects. If the backbone object cannot be simplified, it is then encrypted by a key value and randomized by a set of random Gaussian numbers to generate an encrypted object. The key value is generated by a hashing function with a user's key input. New geometric objects, which are simplified from the backbone object, are then used to compute the feature vertices of the backbone object. Here, the backbone object can be divided into two parts: feature vertices and the remaining vertices. The feature vertices of the backbone object will be continuously encrypted by a key value in order to generate the encrypted feature vertices. Finally, the remaining vertices and encrypted feature vertices of the backbone object will be randomly encrypted by a set of random Gaussian numbers in order to obtain the encrypted object. The encrypted GIS vector map is a set of the encrypted geometric objects.

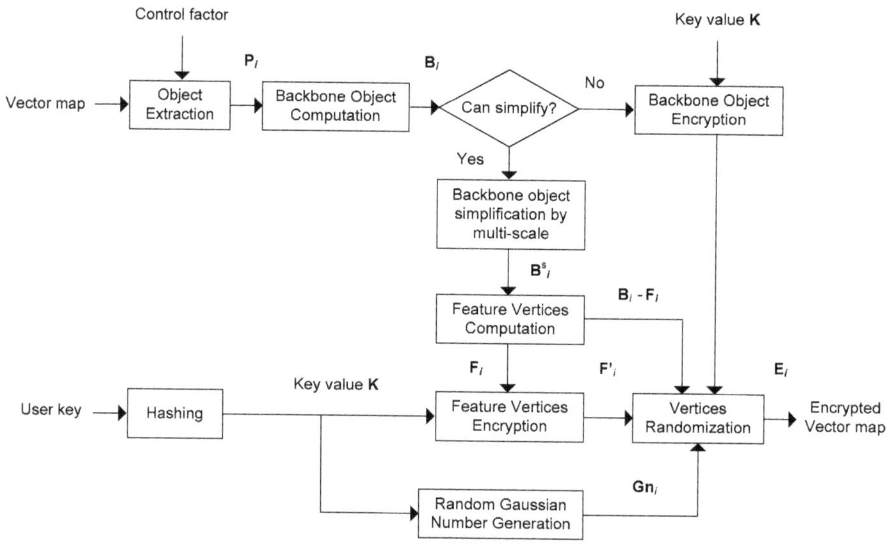

Figure 2. The proposed algorithm.

A GIS vector map contains a number of data layers. Each layer $\mathbf{L}$ contains a number of objects (polyline/polygon) $\mathbf{L} = \{\mathbf{P}_i | i \in [1, |\mathbf{L}|]\}$ and each object contains a number of vertices $\mathbf{P}_i = \{v_{i,j} | j \in [1, |\mathbf{P}_i|]\}$. We briefly define the main notations as the following: $\mathbf{P}_i$ indicates a polyline/polygon object, and $|\mathbf{L}|$ and $|\mathbf{P}_i|$ are the cardinalities of a layer $\mathbf{L}$ and an object $\mathbf{P}_i$ respectively. Thus, $v_{i,j}$ indicates the jth vertex in the ith object of a layer L and is defined by two coordinates $v_{i,j} = (x_{i,j}, y_{i,j})$. Next, $\mathbf{B}_i$ is the backbone object of $\mathbf{P}_i$ and $\mathbf{B}_i^S$ is the simplified object of $\mathbf{B}_i$ after the multi-scale simplification process with the differential scales. $\mathbf{B'}_i$ is the changed backbone object after the feature vertices encryption process or the backbone object encryption process. $\mathbf{F}_i$ represents the feature vertices of the backbone object $\mathbf{B}_i$, $\mathbf{Gn}_i$ is a set of random Gaussian numbers, $\mathbf{K}$ is the secret key, and $\mathbf{E}_i$ is the encrypted object of $\mathbf{P}_i$. Finally, $C_P(.)$, $G_P(.)$, and $R_P(.)$ are the cipher function, the random Gaussian number function, and the randomization function, respectively.

3.2. Backbone Object Simplification and Feature Vertices Computation

The backbone object is a set of vertices, in which each vertex is the average point between two continuous vertices in that object. Thus, the backbone object of $\mathbf{P}_i$ is $\mathbf{B}_i = \{b_{i,j} | j \in [1, |\mathbf{B}_i|]\}$ where $|\mathbf{P}_i| = |\mathbf{B}_i|$ and the value of $b_{i,j}$ is computed as shown in Equation (1). Figure 3a shows a backbone object of a polyline object.

$$b_{i,j} = \begin{cases} v_{i,j} \text{ with } j = 1 \text{ or } j = |\mathbf{P}_i| \\ \frac{v_{i,j} + v_{i,j-1}}{2} \quad \text{for otherwise} \end{cases} \tag{1}$$

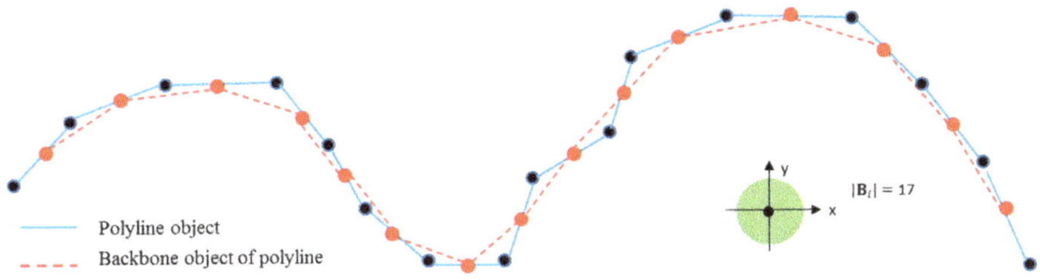

(a)

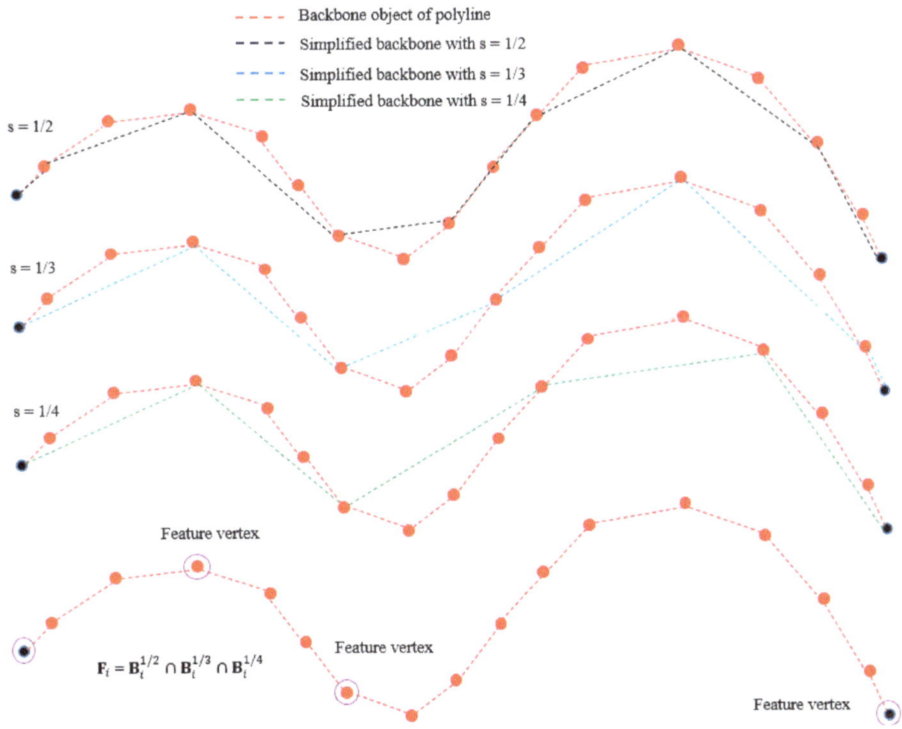

(b)

Figure 3. (**a**) Backbone object of a polyline, and (**b**) feature vertices of a backbone object.

Our purpose is to simplify the backbone object with the scale factor s in order to compute the feature vertices of the backbone object. Thus, we must check the condition before simplifying. The condition of a backbone object $\mathbf{B}_i$ can be simplified as:

$$|\mathbf{B}_i| > \frac{1}{s} \tag{2}$$

If the value of $|\mathbf{B}_i| \leq \frac{1}{s}$, that backbone object is encrypted by the cipher function $C_P(.)$ as shown in Equation (3).

$$\mathbf{B}\prime_i = C_P(\mathbf{K}, \mathbf{B}_i) = \{b'_{i,j} | j \in [1, |\mathbf{B}_i|] \tag{3}$$

The backbone object $\mathbf{B}_i$ is simplified by the multi-scale simplification algorithm in order to obtain the simplified backbone object $\mathbf{B}_i^s$ as shown in Equation (4):

$$\mathbf{B}_i^s = \{b_{i,\frac{j}{s}} \Big| j \in [1, |\mathbf{B}_i| \times s] \tag{4}$$

$$\mathbf{F}_i = \mathbf{B}_i^{1/2} \cap \mathbf{B}_i^{1/3} \cap \mathbf{B}_i^{1/4} \tag{5}$$

Here, we obtain three simplified objects $\mathbf{B}_i^{1/2}$, $\mathbf{B}_i^{1/3}$, $\mathbf{B}_i^{1/4}$ with differential scales $s = 1/2$, $s = 1/3$, and $s = 1/4$ (see Figure 3b). We then calculate the feature vertices $\mathbf{F}_i$, which is a set of the common vertices of three simplified backbone $\mathbf{B}_i^{1/2}$, $\mathbf{B}_i^{1/3}$ and $\mathbf{B}_i^{1/4}$ as shown in Equation (5). From Equations (4) and (5), we can see that $\mathbf{F}_i$ will be determined if $|\mathbf{B}_i| > 4$. Figure 3b shows three simplified backbone objects of a polyline object with differential scales 1/2, 1/3, and $\frac{1}{4}$, and the feature vertices of the backbone object.

3.3. Random Encryption

The secret key $\mathbf{K}$ is generated by the SHA-512 hashing algorithm [15] with user's key input. We use the secret key $\mathbf{K}$ to compute a set of random numbers $\mathbf{Gn}_i$ by the Gaussian distribution [16,17] as shown in Equation (6):

$$\mathbf{Gn}_i = G_P(\mathbf{K}) = \{g_{i,j} | j \in [1, |\mathbf{Gn}_i|\} \tag{6}$$

Therein, $|\mathbf{Gn}_i| = |\mathbf{P}_i|$, $g_{i,j}$ is calculated by the Gaussian function as shown in Equation (7), and x is the value of $\mathbf{K}$:

$$g_{i,j} = \frac{i \times j}{|\mathbf{P}_i|} \times \frac{1}{\sqrt{2\pi}} e^{\frac{-x^2}{2}} \tag{7}$$

As mentioned above, after the feature vertices computation process, we have a set of feature vertices $\mathbf{F}_i$ and the remaining vertices of the backbone object $(\mathbf{B}_i - \mathbf{F}_i)$. We then encrypt the feature vertices $\mathbf{F}_i$ by the cipher function $C_P(.)$ using the secret key $\mathbf{K}$, as shown in Equation (8), in order to obtain the encrypted feature vertices $\mathbf{F}\prime_i$. The cipher function $C_P(.)$ can be the AES cipher function, the DES cipher function, or others.

$$\mathbf{F}\prime_i = C_P(\mathbf{F}_i, \mathbf{K}) \tag{8}$$

The encrypted feature vertices $\mathbf{F}\prime_i$ and the remaining vertices $(\mathbf{B}_i - \mathbf{F}_i)$ are used for the random encryption process in order to obtain the encrypted object $\mathbf{E}_i$. From Section 3.2, the changed backbone object $\mathbf{B}\prime_i$ is computed by Equation (3) when $|\mathbf{B}_i| \leq \frac{1}{s}$. Here, $\mathbf{B}\prime_i$ includes the encrypted feature vertices $\mathbf{F}\prime_i$ and the remaining vertices $(\mathbf{B}_i - \mathbf{F}_i)$ as shown in Equation (9).

$$\mathbf{B}\prime_i = (\mathbf{B}_i - \mathbf{F}_i) \cup \mathbf{F}'_i \tag{9}$$

The random encryption process is performed by a randomization process using the random Gaussian numbers $\mathbf{Gn}_i$:

$$\mathbf{E}_i = R_P(\mathbf{B}'_i, \mathbf{Gn}_i) = \{e_{i,j} | j \in [1, |\mathbf{E}_i|]\} \tag{10}$$

where $|\mathbf{E}_i| = |\mathbf{P}_i|$ and $e_{i,j}$ is computed as follows:

$$e_{i,j} = b\prime_{i,j} \times g_{i,j} = b\prime_{i,j} \times \frac{i \times j}{|\mathbf{P}_i|} \times \frac{1}{\sqrt{2\pi}} e^{\frac{-x^2}{2}} \quad \forall\, j \in [1, |\mathbf{P}_i|] \tag{11}$$

Figure 4 shows the random encryption process for the backbone object $\mathbf{B}_i$ when the value of $|\mathbf{B}_i| > \frac{1}{s}$ (Figure 4a with $|\mathbf{B}_i| = 17$) and when $|\mathbf{B}_i| \leq \frac{1}{s}$ (Figure 4b with $|\mathbf{B}_i| = 4$). When the value of $|\mathbf{B}_i| > 4$, the feature vertices of $\mathbf{B}_i$ will be encrypted by the cipher function $C_P(.)$. Then, the randomization process is performed as shown in Figure 4a. In case $|\mathbf{B}_i| = 4$, the backbone object will be directly encrypted by the cipher function $C_P(.)$ and then randomized by a set of Gaussian numbers, as shown in Figure 4b.

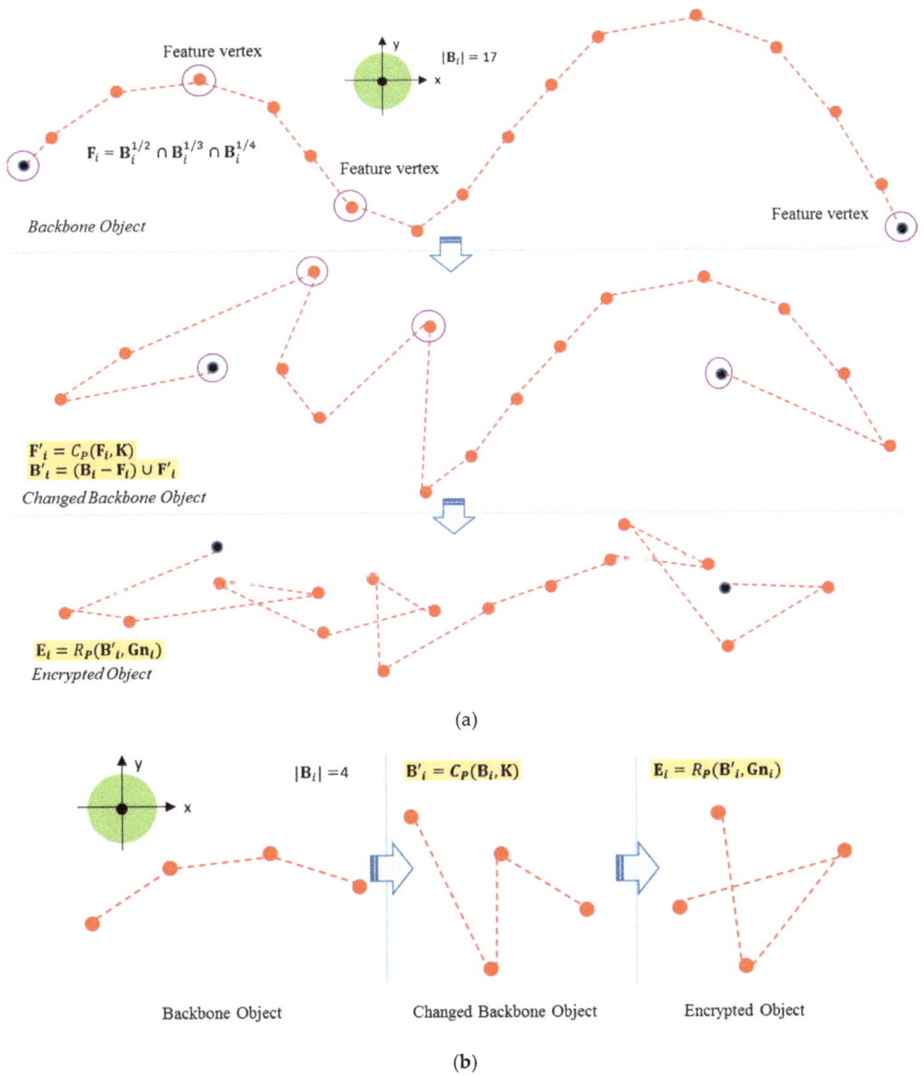

Figure 4. (**a**) Backbone object encryption, and (**b**) feature vertices encryption.

3.4. Decryption Process

The decryption process is the inverse of the encryption process. Firstly, the key value **K** us generated from the user's key by the SHA-512 hashing algorithm. The key value K is then used to compute a set of random Gaussian numbers as described by Equations (6) and (7) in Section 3.3. The encrypted objects are then extracted from the encrypted vector map and vertices are re-randomized by the random Gaussian numbers before decryption using the key value **K**. After the vertex re-randomization process, if the encrypted object cannot be simplified, it is decrypted by the key value **K** to generate the backbone object. If the encrypted object can be simplified, it is simplified to compute the encrypted feature vertices. These encrypted feature vertices are then decrypted by the key value **K** to generate the backbone object. From the backbone object, we can calculate the decrypted objects based on Equation (1). The decrypted GIS vector map is a set of the decrypted objects.

4. Experimental Results and Analysis

We used the GIS vector maps of the country of Scotland [18] with differential scales in visualization experiments, and evaluation of security and computation time. The detailed information of the GIS vector maps is shown in Table 1. The data format of the GIS vector map was the shape-file (SHP) format [19], which is a popular geographical vector data format. The proposed algorithm was applied to the polylines and polygons of the vector maps. The backbone encryption process and the feature vertices encryption process were performed by the AES algorithm. We selected the AES algorithm because its security is higher than others. Compared with conventional approaches, the proposed algorithm is more original than previous methods because it encrypts objects based on encrypting the value of feature vertices in polylines/polygons. Consequently, it does not alter or expand the size of the encrypted file, thus preventing data loss. The GIS vector maps are completely altered after the random encryption process (see Figures 5–9).

4.1. Visualization Experiments

Experimental results are shown in Figures 5–9, which show the original map and a part of the original map beside the encrypted map for comparison. Figure 5a shows the original railway map of Scotland and a part of the original railway map. The content of the railway map is presented by polylines. After encryption, polylines are altered, broken into segments, and positioned in a disorderly manner (Figure 5b). In the experiment with the land-use map of Scotland, the content of the map includes polygons (Figure 6a). After encryption, the shape of all polygons is changed to smaller polygons on the map, and the shape of the entire map is altered completely (Figure 6b). Experiments on the waterway map (Figure 7a), the nature map (Figure 8a), and the road map (Figure 9a) of Scotland also yields similar results. Waterway lines and roads on the maps are broken into shorter polylines and positioned in a disorderly manner (Figures 7b and 9b), and the shape of the original polygons are altered and moved to other positions (Figure 8b). Consequently, the content of the GIS vector maps is altered completely.

Table 1. Experimental results of Scotland maps.

Scotland Maps	Size (Kb)	# Objects	Computation Time (ms)	Entropy (dB)
Railway	1951	9724	77	128,817
Land-use	12,452	41,242	263	632,315
Waterway	27,954	53,389	668	838,434
Nature	51,417	99,835	1367	1,657,985
Road	78,136	372,138	1896	6,886,591

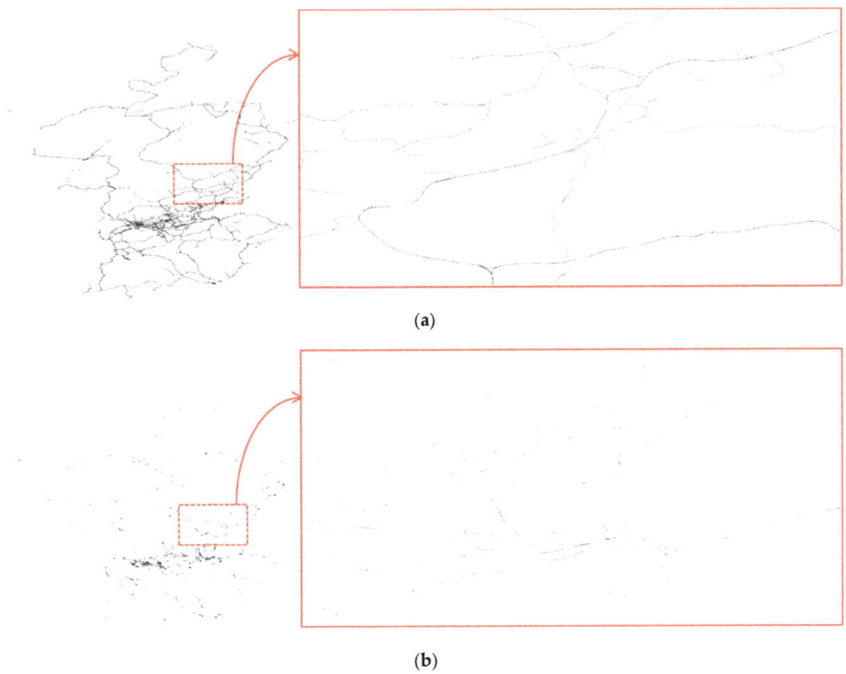

Figure 5. (a) Original railway map, and (b) encrypted railway map.

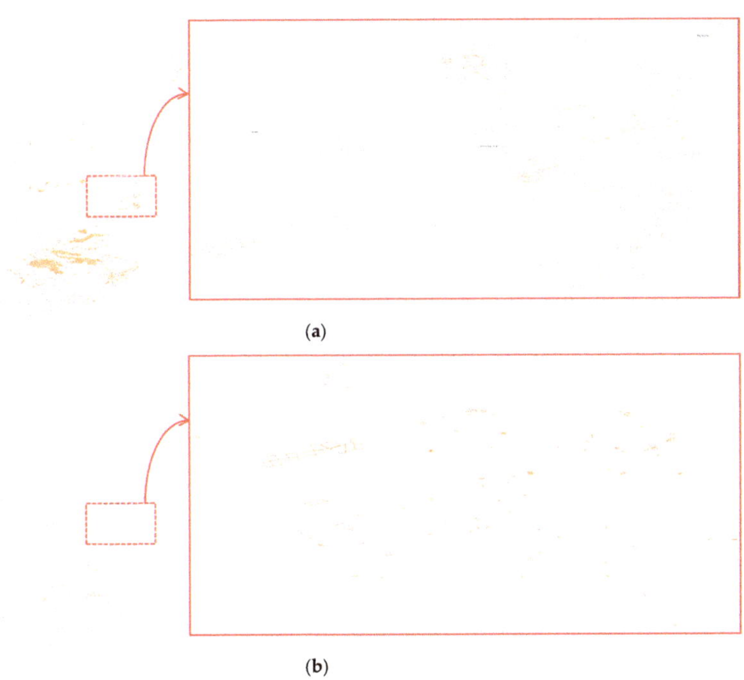

Figure 6. (a) Original land-use map, and (b) encrypted land-use map.

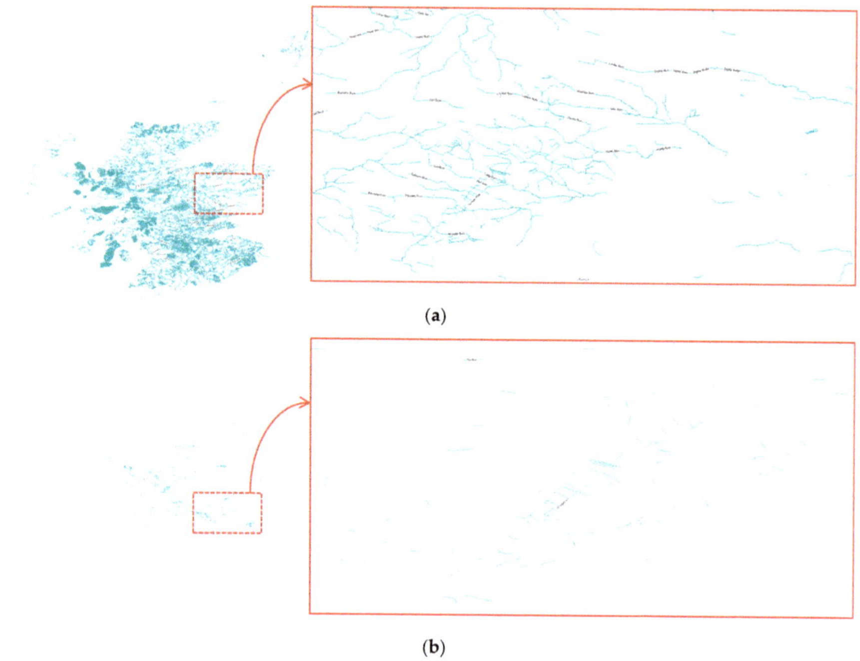

Figure 7. (a) Original waterway map, and (b) encrypted waterway map.

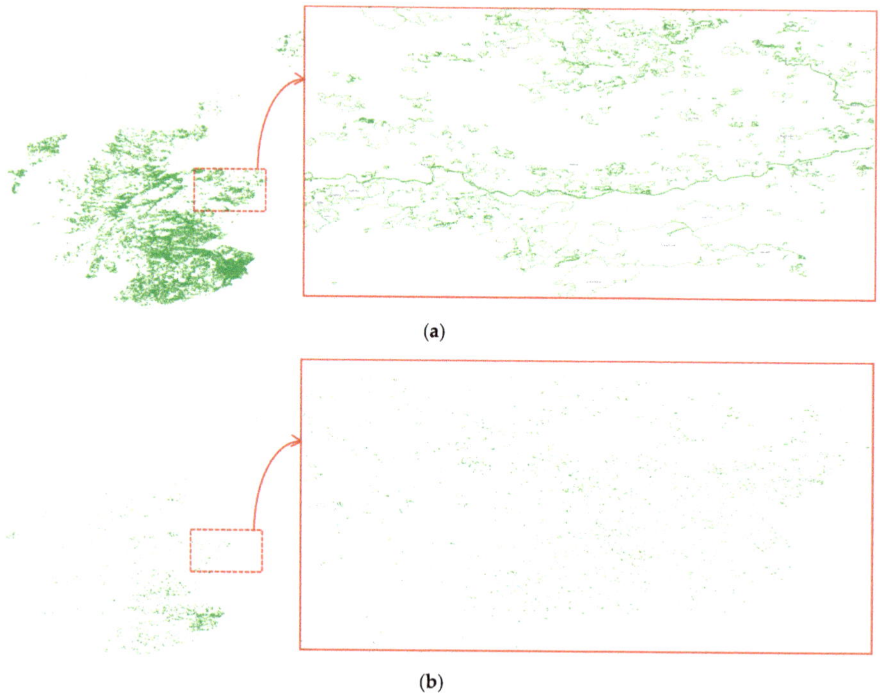

Figure 8. (a) Original nature map, and (b) encrypted nature map.

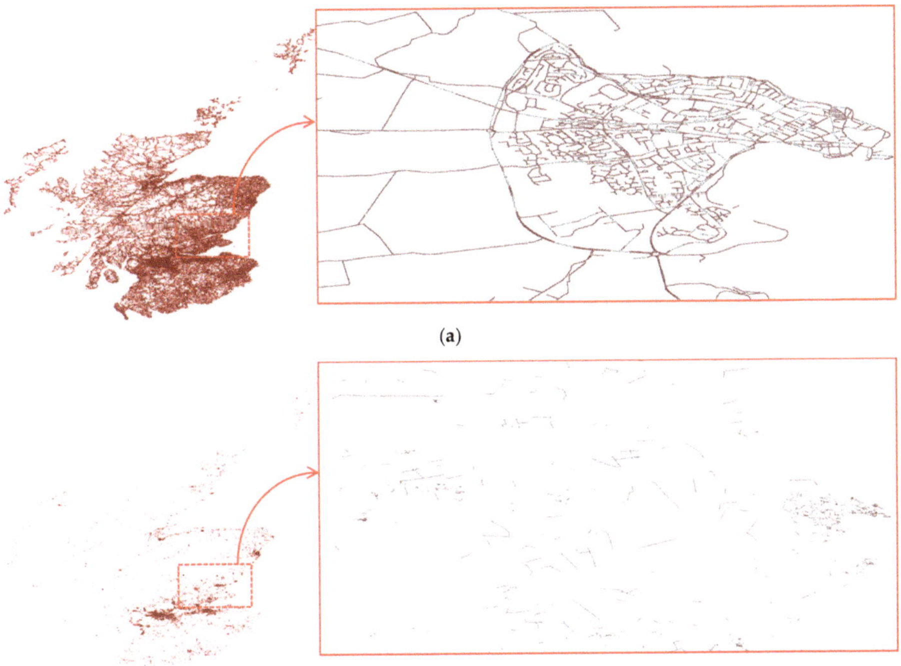

(**a**)

(**b**)

Figure 9. (**a**) Original road map, and (**b**) encrypted road map.

4.2. Security Evaluation

In order to evaluate the security of the proposed method, in this section we evaluate the randomness of the encrypted map. If the randomness of the encrypted map is high, it will also be difficult to attack. The randomness of the encrypted map is measured by its entropy. Thus, we calculate the entropy of the encrypted map to evaluate the security of the proposed method. From Section 3, we can see that the entropy of the encrypted object is dependent on the secret key $\mathbf{K}$ and the value of $|\mathbf{P}_i|$. Both $\mathbf{K}$ and $|\mathbf{P}_i|$ are discrete random variables. Thus, the entropy $H_{\mathbf{P}_i}$ of the encrypted object $\mathbf{E}_i$ is the sum of the entropies of the random variables:

$$H_{\mathbf{P}i} = H(\mathbf{K}) + H(|\mathbf{P}_i|) = |\mathbf{K}|.\log_2\left(\frac{1}{|\mathbf{K}|}\right) + |\mathbf{P}_i| log_2\left(\frac{1}{|\mathbf{P}_i|}\right) \tag{12}$$

Furthermore, the entropy $H_{\mathbf{L}}$ of the encrypted map from the original map layer $\mathbf{L}$ will be the sum of the entropies of the encrypted objects:

$$H_{\mathbf{L}} = \sum_{i=1}^{|\mathbf{L}|} H_{\mathbf{P}_i} \tag{13}$$

Clearly, $H_{\mathbf{L}}$ is dependent on the values of $|\mathbf{L}|$ and $\mathbf{K}$; the key value $\mathbf{K}$ is a random variable dependent on a user's key input. If the value of $\mathbf{K}$ is fixed, $H_{\mathbf{L}}$ is only dependent on the value of $|\mathbf{L}|$. As a result, if $|\mathbf{L}|$ is high, the entropy is high. For example, the railway map of Scotland has $|\mathbf{L}|= 9724$ and $H_{\mathbf{L}} = 128{,}817\ dB$, but the natural map of Scotland has $|\mathbf{L}| = 99{,}835$ and $H_{\mathbf{L}} = 1.66 \times 10^6\ dB$. In the vector maps of Scotland in Table 1, the entropy of the proposed method ranges from $1.3 \times 10^5\ dB$ to $6.88 \times 10^6\ dB$ with $|\mathbf{L}| \in [9724,\ 372{,}138]$.

The method of Yasser [10] only uses the AES-256 cipher operator to encrypt the data stream of a shape-file. Thus, the entropy of this method is dependent on the length of the secret key and the length of the data stream. In the work of Bang [12], about 70% of polylines/polygons are selected for the encryption process. Bang used a common secret key to encrypt all DC values in a vector map in the DWT domain. The length of the secret key was 512 bits. Thus, the entropy of this method is dependent on the length of the secret key. We applied the methods of both Yasser and Bang for the GIS vector maps of Scotland for comparison. The entropy of our method was higher than that of Yasser (1.2×10^5 vs. 6.6×10^6 *dB*), while the entropy of Bang's method was fixed at 4608 *dB* for every GIS vector map (Table 2). In conclusion, our method offers more security than previous methods. Figure 10 shows the difference between the entropy of our method and the entropy of the methods of Yasser and Bang according to the number of objects. The entropy of our method is much higher than that of Yasser or Bang.

Table 2. The entropy of the proposed method compared with previous methods.

Scotland Maps	# Objects	Entropy (dB)		
		Our Method	Yasser's Method	Bang's Method
Railway	9724	128,817	9216	4608
Land-use	41,242	632,315	55,296	4608
Waterway	53,389	838,434	124,416	4608
Nature	99,835	1,657,985	235,008	4608
Road	372,138	6,886,591	299,520	4608

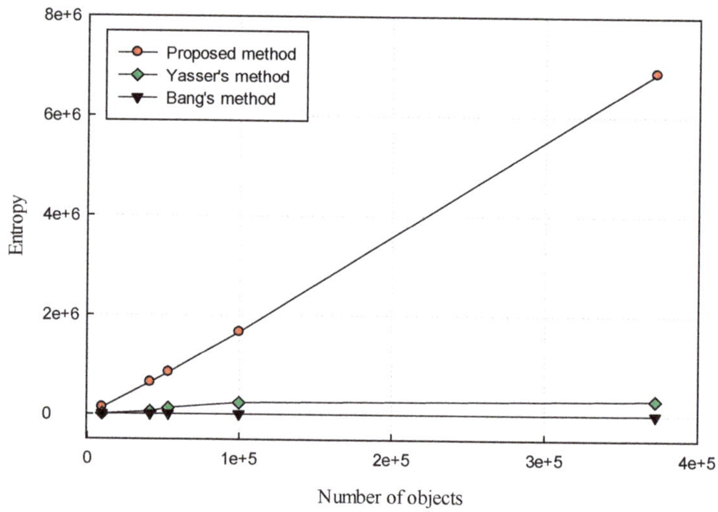

Figure 10. Entropy according to the number of objects.

4.3. Computation Time

We implemented the proposed method using the C# language in Visual Studio 2013. We then conducted our experiments on a PC with Intel Core i7 Quad 3.5 GHz, 8 GB of RAM, and Windows 7 64-bit. Section 3 indicates that the computation time of the proposed method is dependent on the number of objects in the vector maps. With the GIS vector maps of Scotland, the computation time of our method ranged from 77 ms to 1896 ms, with the size of maps ranging from 1951 Kbs to 78,136 Kbs. We also implemented the methods of Yasser and Bang using the C# language in Visual Studio 2013, in a similar environment, to measure and compare the computation time between methods. Yasser performed full encryption for GIS vector maps, which means he encrypted notation, header, text, and all geometric objects. Thus, the computation time of Yasser's method is dependent on the size of the GIS vector maps. Compared with Yasser's method, the computation time of our method was less (40 ms vs. 2870 ms; Table 3). In the method of Bang, the computation time is dependent on the computation time of the selection process, the number of selected objects, and the computation time of the DWT and inverse DWT processes. The computation time of the DWT and inverse DWT processes is dependent on the number of vertices in each object. If an object has many vertices, the computation time of the DWT process is long. Conversely, if the number of selected objects is small, the computation time is short (Table 3). Figure 11 shows the computation times of the proposed method compared to the methods of Yasser and Bang according to the size of the vector maps. The time of the proposed method is significantly shorter than that of previous methods.

Table 3. Computation time of the proposed method compared with previous methods.

Scotland Maps	Size (Kb)	Computation Time (ms)		
		Our Method	Yasser's Method	Bang's Method
Railway	1951	77	117	416
Land-use	12,452	263	759	1916
Waterway	27,954	668	1705	5564
Nature	51,417	1367	3136	9036
Road	78,136	1896	4766	23,563

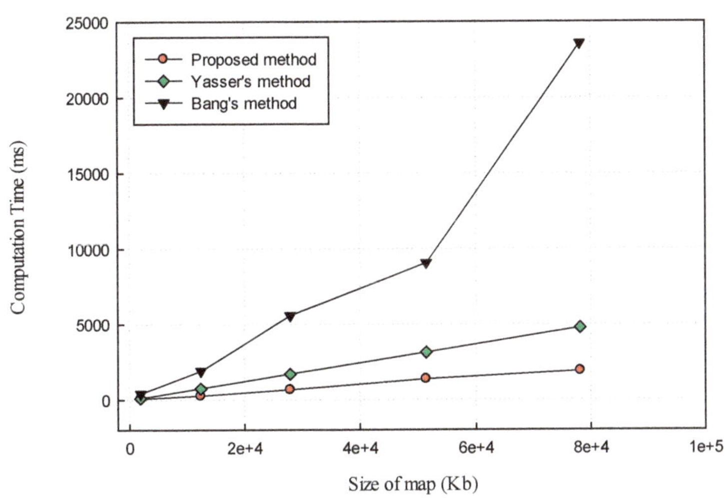

Figure 11. Computation time according to the size of map.

5. Conclusions

In this paper, we proposed a random encryption algorithm based on multi-scale simplification and the Gaussian distribution for GIS vector maps. Experimental results showed that the proposed method is very effective with GIS vector maps that contain many geometric objects. The presented method provides higher security than previous methods. The computation time of the method is significantly shorter than that of previous methods, and it could be used to replace previous methods for secured storage and transmission. By encrypting only geometric objects, the proposed method can be responsive to the various formats of GIS vector maps. In addition, to apply the method presented in this paper, developers only need to extract geometric objects before performing the encryption process. Furthermore, the proposed method can be applied to the security of online and off-line map services [20,21].

Author Contributions: Idea and Validation, S.-H.L.; Supervision and Evaluation, K.-R.K.; Research and Methodology, G.N.P.; Implementation and Software, S.T.N.; Visualization and Verification, A.N.B.; Data and Analysis, D.V.T.

Funding: This research was supported by the FPT University; and the FPT Software, Hanoi, Vietnam; and also supported by the Basic Science Research Program through the National Research Foundation of Korea (NRF) funded by the Ministry of Science and ICT (No. 2016R1D1A3B03931003, No. 2017R1A2B2012456).

Acknowledgments: Thanks for the funding support.

Conflicts of Interest: The authors declare no conflict of interest.

References

1. What Is Geographic Information Systems (GIS)? Available online: https://gisgeography.com/what-gis-geographic-information-systems/ (accessed on 30 October 2019).
2. Wang, C.; Peng, Z.; Peng, Y.; Yu, L.; Wang, J.; Zhao, Q. Watermarking geographical data on spatial topological relations. *Multimed. Tools Appl.* **2012**, *57*, 67–69. [CrossRef]
3. Lee, S.H.; Kwon, K.R. Vector watermarking scheme for GIS vector map management. *Multimed. Tools Appl.* **2013**, *63*, 757–790. [CrossRef]
4. Ohbuchi, R.; Ueda, H.; Endoh, S. Robust watermarking of vector digital maps. In Proceedings of the IEEE International Conference on Multimedia and Expo, Lausanne, Switzerland, 26–29 August 2002; pp. 577–580.
5. Chena, S.C.; Wang, X.; Rishea, N.; Weiss, M.A. A web-based spatial data access system using semantic R-trees. *J. Inf. Sci.* **2003**, *167*, 41–61. [CrossRef]
6. Bertino, E.; Damiani, M.L. A Controlled Access to Spatial Data on Web. In Proceedings of the 7th Conference on Geographic Information Science, Heraklion, Greece, 12–13 November 2004; pp. 369–377.
7. Rybalov, N.B.; Zhukovsky, O.I. Access to the Spatial Data in the Web-Oriented GIS. In Proceedings of the Siberian Conference on Control and Communications, Tomsk, Russia, 20–21 April 2007; pp. 104–107.
8. Bertino, E.; Thuraisingham, B.; Gertz, M.; Damiani, M.L. Security and privacy for geospatial data: Concepts and research directions. In Proceedings of the International Workshop on Security and Privacy in GIS and LBS, Irvine, CA, USA, 4 November 2008; pp. 6–19.
9. Fuguang, M.; Yong, G.; Menglong, Y.; Fuchun, X.; Ding, L. The fine-grained security access control of spatial data. In Proceedings of the 18th International Conference on Geo-informatics, Beijing, China, 18–20 June 2010; pp. 1–4.
10. Wu, F.; Cui, W.; Chen, H. A Compound Chaos-Based Encryption Algorithm for Vector Geographic Data under Network Circumstance. In Proceedings of the Cardholder Information Security Program, Hainan, China, 27–30 May 2008; pp. 254–258.
11. Li, G. Research of Key Technologies on Encrypting Vector Spatial Data in Oracle Spatial. In Proceedings of the International Conference on Industrial Electronics and Computer Science, Wuhan, China, 25–26 December 2010; pp. 1–4.
12. Yasser, D.; Elghafar, I.A.; Tammam, A. Protecting GIS Data Using Cryptography and Digital Watermarking. *Int. J. Comput. Sci. Netw. Secur.* **2010**, *10*, 75–84.
13. Jang, B.J.; Lee, S.H.; Kwon, K.R. Perceptual Encryption with Compression for Secure Vector Map Data Processing. *J. Digit. Signal Process.* **2014**, *25*, 224–243. [CrossRef]
14. Bang, N.V.; Lim, S.H.; Moon, K.S.; Lee, S.H.; Kwon, K.R. Selective Encryption Scheme for Vector Map Data using Chaotic Map. *J. Korea Multimed. Soc.* **2015**, *18*, 818–826. [CrossRef]
15. RSA Laboratories. *Password-Based Cryptography Standard*; RSA Security LLC: Bedford, MA, USA, 2006.
16. Maria, I.R. *Gaussian Probability Density Functions: Properties and Error Characterization*; Institute for Systems and Robotics: Lisbon, Portugal, 2004.
17. Gaussian Distribution. Available online: http://hyperphysics.phy-astr.gsu.edu/hbase/Math/gaufcn.html (accessed on 30 October 2019).
18. United Kingdom Shapefile. Available online: https://www.igismap.com/united-kingdom-shapefile-download/ (accessed on 30 October 2019).

19. *White Paper: ESRI Shape-File Technical Description*; Environmental Systems Research Institute: Redlands, CA, USA, 1998.
20. Open GIS Geography Markup Language (GML) Encoding Standard. Available online: https://www.opengeospatial.org/standards/security (accessed on 30 October 2019).
21. Mapinfor. Available online: http://www.pbinsight.com/welcome/mapinfo/ (accessed on 30 October 2019).

© 2019 by the authors. Licensee MDPI, Basel, Switzerland. This article is an open access article distributed under the terms and conditions of the Creative Commons Attribution (CC BY) license (http://creativecommons.org/licenses/by/4.0/).

applied sciences

MDPI

Article

Evaluation of Effective Cognition for the QGIS Processing Modeler

Zdena Dobesova

Department of Geoinformatics, Faculty of Science, Palacký University, 17. Listopadu 50, 77146 Olomouc, Czech Republic; zdena.dobesova@upol.cz

Received: 28 December 2019; Accepted: 17 February 2020; Published: 20 February 2020

Abstract: This article presents an evaluation of the QGIS Processing Modeler from the point of view of effective cognition. The QGIS Processing Modeler uses visual programming language for workflow design. The functionalities of the visual component and the visual vocabulary (set of symbols and line connectors) are both important. The form of symbols affects how workflow diagrams may be understood. The article discusses the results of assessing the Processing Modeler's visual vocabulary in QGIS according to the Physics of Notations theory. The article evaluates visual vocabularies from the older QGIS 2.x and newer 3.x versions. The paper identifies serious design flaws in the Processing Modeler. Applying the Physics of Notations theory resulted in certain practical recommendations, such as changing the fill colour of symbols, increasing the size and variety of inner icons, removing functional icons, and using a straight connector line instead of a curved line. Another recommendation was to provide a supplemental preview window for the entire model in order to improve user navigation in huge models. Objective eye-tracking measurements validated some results of the evaluation using the Physics of Notations. The respondents read workflows to solve different tasks and their gazes were tracked. Evaluation of the eye-tracking metrics revealed the respondents' reading patterns of the diagram. Evaluation using both Physics of Notation theory and eye-tracking measurements inspired recommendations for improving visual notation. A set of recommendations for users is also given, which can be applied easily in practice using a contemporary visual notation.

Keywords: algorithm; cognition; computer languages; eye-tracking measurement; gaze tracking; human-computer interaction; open source software; symbols; visualisation

---

1. Introduction

Today, open source GIS software competes with commercial GIS software. The user's choice not only depends on the price but also the degree of functionality in parts of the GIS software. Users need to satisfy their requirements. One of the demands is the automatic processing of spatial data as a sequence of steps. Visual programming languages (VPLs) are used to design steps of processes in the form of workflow diagrams. GIS operations are not used in isolation but as a part of a chain of operations to completely process data. An overview and basic description of several visual programming languages in GIS are given in this article [1]. ModelBuilder for ArcGIS, Macro Modeler for IDRISI, Model Maker and Spatial Model Editor for ERDAS IMAGINE and Workflow Designer for AutoCAD Map 3D are mentioned. All systematic description and evaluation of VPLs in GIS is presented as habilitation [2]. VPLs in GIS are data-centric notations that serve to express a process in detail. Only AutoCAD Map uses hybrid symbols where one symbol both for operation and input/output data altogether is. Other VPLs have a unique set of simple symbols for data and unique symbols for operations and control of flow. GIS workflow does not express a generalised conceptual model of processing, and they are more detailed.

Open source software QGIS is competitive with commercial GIS software in designing workflow diagrams using VPL. The accessibility of a visual programming language increases the usability of QGIS. The possibility of designing workflows could be a reason for selecting open source QGIS.

The Processing Modeler is a graphical editor in QGIS software. This editor allows workflows to be designed in graphical form using a visual programming language. Workflow diagrams in QGIS are termed as a model.

VPLs differ in their visual notation generally, and the symbols in GIS software are various. The visual notation consists of graphical symbols (visual vocabulary), a set of compositional rules (visual grammar) and definitions of the meaning of each symbol (visual semantics). The visual notation is important from the point of user perception and cognition. In his theory, the Physics of Notations, D. Moody stated that it is necessary to use cognitively effective visual notations [3]. Cognitively effective means optimised for processing by the human mind.

This article presents an assessment of the visual notation of the QGIS Processing Modeler using the Physics of Notations theory in combination with eye-tracking measurement. The presented research started with version QGIS 2 in 2014 and has been continued with version 3 up to now. The long-term release (LTR) version QGIS 3.4 Madeira and partly version 3.6 Noosa was used for assessment. Some features of visual notation were empirically tested using the eye-tracking method on version QGIS 2. Finally, some improvements to visual notations are suggested in this article.

The research question was "What is the level of effective cognition in QGIS Processing Modeler." This research aimed to evaluate and improve cognition of visual notation in QGIS. The results bring new and innovative ideas that improve the usability of and satisfaction with QGIS software.

These tasks fall under investigation in Human-Computer Interaction (HCI) research. Standard ISO 9241-210:2019 Ergonomics of human-system interaction—Part 210: Human-centred design for interactive systems [4] provides requirements and recommendations for human-centred design principles and activities of computer-based interactive systems. In the center of HCI research and UX research (User Experience), is the understanding and design of interactive digital systems and their human users [5]. Their common aim is to innovate novel computing user interfaces to satisfy usefulness, ergonomics, and efficiency of using digital systems [6,7]. The improvement is based on theories and both on empirical testing in laboratories [8] e.g., the eye-tracking measurement presented in this article.

2. QGIS Processing Modeler and its Graphical Notation

2.1. History of QGIS Processing Modeler

The Processing Modeler was implemented in version QGIS 2.0 Dufour, released in 2013. The next development of the Processing Modeler aimed to increase the functionality of the editor. The author of the graphical editor was Victor Olaya from Spain. In version QGIS 2.6 Brighton, released in 2014, the Processing Modeler was rewritten and provided extra functionality, such as allowing nested models with no depth limit [9].

Furthermore, adding Python script to the model was supported in version 2.x. Python script could be downloaded from an online external collection of scripts created by different users and adopted to a newly created user model. The software architecture and features of the QGIS processing framework are described in this article [10].

In 2018, the new series of version QGIS 3.x began with version 3.0 QGIS Girona. The Processing Modeler underwent extensive changes and included additional and changed input parameters and algorithms. Specifically, the colours of basic symbols were changed, and the interface and degree of functionality were redone. For example, zoom in and zoom out functions [11,12]. The two Input and Algorithm panels can be positioned differently in the interface and now float above the processing window [11,12]. The storage format of the model was also changed [13]. File extension model 3 was used instead of extension model.

2.2. Description of Interface and Graphical Notation

A graphical editor Processing Modeler is embedded in QGIS and runs in a separate window. The interface is divided into two areas [14]. Two switchable panels are on the left side. The "Inputs" panel is the source for different types of input data. The "Algorithms" panel is the source of operations that can be added to the model (workflow diagram). The large window at right is a canvas for designing the model (Figure 1).

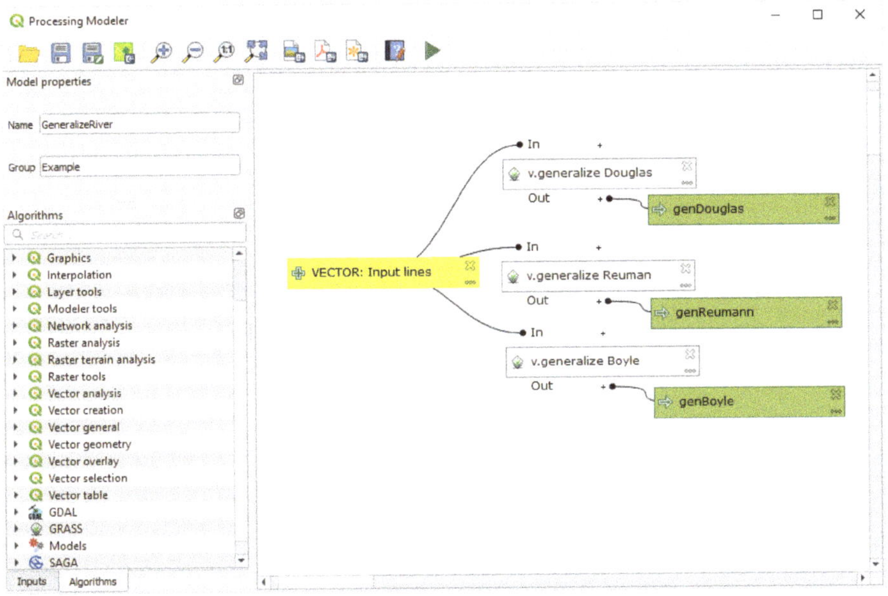

Figure 1. The interface of the QGIS Processing Modeler version 3 and an example model in visual programing language.

Selected inputs and algorithms can be added to the model by dragging and dropping it into the modeler canvas. Being movable, the position of the symbols is the user's choice. When input data is added to the model, the type of data and name of data are set. The input data are considered input parameters. Inputs are not assigned to particular existing data at the directory or values of variables. Their names could, therefore, be more descriptive than the data's real name. This would be an advantage because the names of parametrical inputs could be more general. It improves comprehension of the model for other users. The Algorithms panel provides GIS operations (processing algorithms) from several types of open source software apart from QGIS—these are GDAL, GRASS and SAGA. Previously, created QGIS models are also displayed. Python scripts and operations from the ORFEO library were accessible in the older version 2.

Grey connector lines are automatically drawn immediately after adding operations and assigning the existing inputs to the operation in the model. The lines connect symbols of input data with the symbol of operations. The output data symbol is also automatically linked to the model after naming outputs from the operation. The connector lines are then automatically drawn between operation and output data. The user cannot draw the connector lines manually with a mouse or reconnect the symbols. The shape of the connector lines is curved and ends with a black point. When the positions of the symbols change, the shape of lines is automatically redrawn with a different curvature.

Modeler's visual vocabulary contains three rectangular symbols (Figure 2). The size of the symbols is the same and cannot be changed. Originally, the violet rectangle represented input data, the blue rectangle represented output data and the white rectangle represented the operation. The fill colours were changed in version 3. The symbol for input data is now yellow, the symbol for output data is green and the symbol for operations remains white. At first glance, this was perhaps to emphasise that this is a model from the new Processing Modeler version. The user can subsequently very clearly distinguish between existing models from older version 2 and the latest version 3. Comparing the brightness of symbols (compare the greyscale between versions in Figure 3), the input symbol is lighter in tone and the output symbol is darker than in version 3. The difference in brightness is important for colour-blind people or people with perception limitations. For these people, only the difference in brightness is helpful in distinguishing objects. The colour setting of any computer application, design of web pages respects the colour-blind people to apply different brightness of menus, text boxes and other graphical objects of interfaces. From that point, the new colours of symbols in version 3 are better due to different brightness. The difference in brightness maybe not made intentionally by QGIS designers but it is valuable.

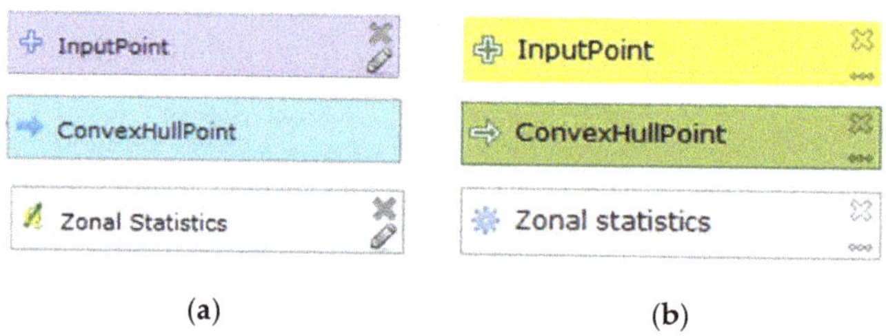

Figure 2. Graphic symbols (from top, input data, output data and operation); (**a**) symbols from version 2 at left, (**b**) symbols from version 3 at right.

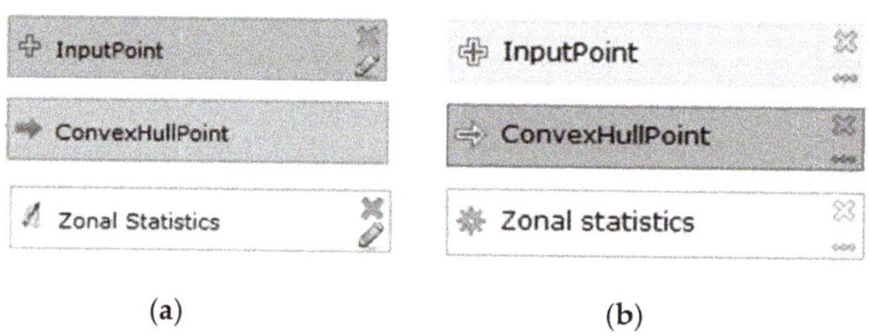

Figure 3. Greyscale symbols (**a**) symbols from version 2 at left, (**b**) symbols from version 3 at right.

The rectangular symbols contain some inner icons. The input data symbols are indicated with a *plus sign* icon. The output data symbol is indicated by an *arrow*. Both icons are on the left side. The operation symbols have different icons according to the source library of operation or type of operation. For example, the QGIS 2 icon for Zonal Statistics is shown in Figure 2. The input data symbols and operation symbols have two icons on the right side: *across* and *pencil* in version 2. These icons depict the delete and edit functions. They can be considered *operational icons*. In version 3, the *cross* icon remained, and the icon for editing is *three dots*. The green output symbol was also assigned these two operational icons. It means that the label of the output data symbol is editable. The option to assign a default name and path for the output data is provided in the editing dialogue. When the output symbol is deleted, the output data is automatically assigned as a temporary output in operation. No symbol in the model indicates a temporary output.

Generally, the use of icons is very helpful. According to Szczepanek, icons can be divided into three groups of icons in software interfaces [15]. The first group is *universal icons*, which can be understood without explanation (e.g., a floppy disk for the save operation). The second group is *domain-specific icons* (e.g., for any GIS software), and the third group is *application-specific icons* (e.g., for

QGIS software). In the case of the QGIS Processing Modeler, the icons can be sorted as follows: a pencil, three dots and cross icons are universal icons. The plus icons for data are midway between universal and domain-specific icons. The plus icon is frequently used in some GIS interfaces and means adding a layer to the current project. The icons of source libraries (which are in the white operation symbol) belong to the application-specific group of icons. All icons can be understood very well.

3. Theory and Measurement Methods

3.1. Theory Physics of Notations

Physics of Notations is an objective theory for evaluating visual notation [3,16]. This theory is widely used in all areas of software engineering, not only GIS software because creating diagrams is frequently required in information technology (IT). The first work was an assessment of the Esri ModelBuilder in the area of VPL in GIS applications [17]. The Physics of Notations theory can be used not only for evaluating existing notation but also improving graphical notation or designing new ones. It means that the visual notation in QGIS can be assessed and improved if any drawbacks are identified. Exploring this theory for the new design is very beneficial in design new graphical vocabulary for any purpose. This paper presents the opportunity to make new suggestions according to the theory for QGIS Processing Modeler.

The Physics of Notations theory states nine principles that recommend fulfilling cognitively effective notation. *Cognitive effectiveness* is defined as the speed, ease and accuracy with which a representation can be processed by the human mind [18]. The aim is to read the diagram quickly, without mistakes, and comprehend it accurately.

The nine principles are organised as connected ideas where the first central principle is the Principle of Semiotic Clarity. The modular structure of Physics of Notations is designed to make it easy to add or remove principles, emphasising that they are not fixed or immutable but can be modified or extended by future research [19].

The principles are [3]:

- **Principle of Semiotic Clarity**

The principle of Semiotic Clarity expresses a one to one correspondence between the syntactic model and semantic features. According to this principle, symbol redundancy, symbol overload, symbol deficit or symbol excess is not permissible. The principle reflects the ontological analysis.

- **Principle of Perceptual Discriminability**

The second principle of Perceptual Discriminability states that different symbols should be clearly distinguishable from each other by visual variables.

- **Principle of Visual Expressiveness**

The principle of Visual Expressiveness states that the full range of visual variables and their full capacity should be used to represent the notational elements. Colour is one of the most effective visual variables. The human visual system is very sensitive to differences in colour and can quickly and accurately distinguish them. Differences in colour are found three times faster than shape and are also easy to remember [20]. The level of expressiveness is measured from level 1 (lowest) to 8 (highest).

- **Principle of Graphic Economy**

The principle states that the number of symbols in a graphical vocabulary must be manageable by human working memory. The choice of symbol affects the ease of memorizing and recalling visual diagrams. The magic number seven express a suitable number of symbols. The range for an of 7 ± 2 symbols is suitable. More different symbols in basic graphical vocabulary than nine are demanding for comprehension.

- **Principle of Dual Coding**

The principle suggests using the text to support the meanings of symbols and clarity. Two methods (graphics and text) provide the user with information and improve comprehensibility. The base is on the duality of mental representation [21].

- **Principle of Semantic Transparency**

This principle evaluates how symbols associate the real meaning of an element. Here, associations are sought between the shape or other visual symbol variables and their real properties, and the form implies content.

- **Principle of Complexity Management**

This principle recommends producing hierarchical levels of the diagram and dividing it into separate modules and create hierarchical structures. It is suitable for large models when comprehension exceeds human working memory capacity. Modularity means scaling information into separate chunks. Modularisation is the division of large systems into smaller parts or separate subsystems. Practice shows that one subsystem should be only large enough to fit on one sheet of paper or one screen. This subsystem is then represented at a higher level by one symbol. Hierarchical structuring allows systems to be represented at different levels of detail (levelled diagram) with the ability to control complexity at each level. This promotes understanding of the diagram from the highest level to the lowest, which improves the overall understanding of the diagram. Both mechanisms can be combined into the principle of recursive decomposition.

- **Principle of Cognitive Interaction**

The principle recommends increasing the options for navigating in the model. The reader must be able to follow the chain of operations easily. The connector lines affect navigation.

- **Principle of Cognitive Fit**

The principle proposes to realize different sets of graphical vocabularies for the same type of semantics, where information is represented, for different tasks and different groups of users in different ways. It recommends the use of multiple visual dialects, each of which is suitable for different types of tasks and different user spectrums (according to experience).

3.2. *Eye-Tracking Measurement and Experiment*

The eye-tracking equipment was used to evaluate the comprehensibility and discriminability of visual symbols in models. This method was assumed as a combination and extension of Physics of Notations results as an experimental method.

Testing was conducted at an eye-tracking laboratory in the Department of Geoinformatics, Palacký University in Olomouc (Czech Republic). The eye-tracker SMI RED 250 with software SMI Experiment Suite 360° was used for the experiment. The test was designed using the SMI Experiment Center program. The results were visualised using SMI BeGaze. An evaluation was also conducted using software Ogama 4.5 and V-Analytics. For statistical evaluation, the STATISTICA software was used. The size of the monitor to record eye movements and display models was 1920 × 1080 pixels. The sampling frequency of the eye-tracker SMI RED was 250 Hz [22].

The complex eye-tracking experiment consisted of 22 workflow diagrams from Processing Modeler version 2. Several models with different sizes, functions, and arrangements of flow orientation (vertical, horizontal, and diagonal directions) were tested. The workflow diagrams were displayed individually on the screen in random order to prevent a learning effect [23]. Shuffling ensured that each respondent saw the models in a different order.

The respondents were first-year students at the end of the semester in a master's programme of Geoinformatics. They had attended lectures where the design of models in Processing Modeler version 2 was practised. They created various examples of models with different functionalities and sizes. The group of respondents was assumed to be advanced users. A total of 22 respondents participated in the eye-tracking testing, aged 22–25.

The term stimulus is applied in the process of eye-tracking testing [24]. The stimuli, in this case, were the models (workflow diagrams). Each model was associated with a comprehension task to record the cognitive process. Response time and the correctness of user answers were measured for each comprehension task as in other research [25–27]. The set of models or maps and comprehension tasks are often used to evaluate the usability of visualisation methods in cartography and GIS [28,29].

Research in the area of workflow diagrams has also been organised at our eye-tracking laboratory for other GIS VPLs. The reading patterns are described for models in ArcGIS ModelBuilder [30]. The significant effect of the orientation of connector lines is mentioned. Also, the influence of bends on connector lines was tested for ModelBuilder [22]. Being able to change colour helps to discriminate graphical symbols in ModelBuilder. This was also demonstrated using eye-tracking experiments at our university laboratory [31].

The eye-tracking experiment consisted of two parts for the QGIS Processing Modeler. The first part only displayed models without any task. This part is called free viewing. The second part contained 22 models that were introduced with comprehension tasks. The respondents solved the tasks by clicking on the stimulus at the right location all answer the question. All tested diagrams are in Appendix A. Clicks were recorded as an answer. All stimuli were interleaved with a fixation cross in the middle of the screen to provide the same starting point for all respondents. The fixation cross was displayed for 600 milliseconds before each stimulus.

The research combined two above mentioned methods in evaluation. They were very different. The first report findings by application theoretical principles. It produced results in a text form with list of insufficiencies, good features, recommendations and ideas. The second is the experimental method where the objective measurement was constructed using user testing. Both methods could be assumed as cross-validation of results, but mainly eye-tracking is an extension of received results. The research tried to combine both attempt to receive more complexity results such as finding reading patterns. In the phase of preparing the eye-tracking experiment, we considered how to test the principles of Physics of Notations. The task was aimed to receive answers that correspond to the principle definitions. The design of the experiment was done so much coherent to the principles. However, only the set of principles is possible to test in a limited way. It is impossible to design the eye-tracking as one task to one principle. There are more influences on the respondent's perception. Moreover, the last principles, Complexity Management and Principle of Cognitive Interaction, are hard to test because of no sufficient solution present in that visual vocabulary. It was also impossible to test case of the Principle of Cognitive Fit when no visual dialects exist in the Processing Modeler.

Two hypotheses were proposed before eye-tracking testing:

Hypothesis 1 (H1). *Insufficiencies in Semiotic Clarity, Perceptual Discriminability, Visual Expressiveness and Semantic Transparency adversely affect the correctness of user answers.*

Hypothesis 2 (H2). *Insufficiencies in Semiotic Clarity, Perceptual Discriminability, Visual Expressiveness and Semantic Transparency adversely affect the effectiveness of comprehension.*

To evaluate these two hypotheses, the number of correct answers (for H1), the time required to answer, and eye-tracking metrics were measured. Eye-tracking metrics such as the length of the scanpath, number of fixations and average time of fixations were calculated (for H2). All results are presented in Section 4.

4. Evaluation of Processing Modeler

4.1. Evaluation of Effective Cognition by Physics of Notations Method

Systematic application of Physics of Notations theory on Processing Modeler follows.

4.1.1. Principle of Semiotic Clarity

When this principle is applied to symbols in the Processing Modeler, it is evident that both input data and output data symbols are overloaded. In version 2, one symbol represents nine different data types: vector, raster, string, file, table, table field, number, extent and boolean. The newer version 3 offers 22 different types of input data. The user has to assign the data type immediately when an input data symbol is assigned to the model. The data type (for example point, line, polygon) is assigned immediately when a model is designed, despite there being no evidence about data type in the graphical symbol.

Detected symbol overloads could be solved in the following manner: remove the inner *plus icon* at the left in the symbol and replace it with more specific icons that express the type of data. Suggestions for vector and raster data symbols are given in Figure 4. Both icons are adopted from the QGIS interface. The lower pair uses the compound icons from version 3, where the symbols for vector and raster are supplemented by a small plus icon to express input of data. These compound icons are better than simple icons in version 2. The former and larger plus icon is substituted by a plus icon that forms a part of the vector or raster icon. New icons could be suggested for file, folder, string, number, table and field using any universal icon. Data such as extent, CRS, map layer, etc. need domain-specific icons. The same inner icon sets could be used for the output data symbol where a compound symbol can contain bigger icon of data type and small output arrow that is original in output symbol. The suggestions follow Szczpanek icon theory [15]. These suggestions would increase the number of icons in the graphical vocabulary and solve the overload of the two original symbols for input/output data.

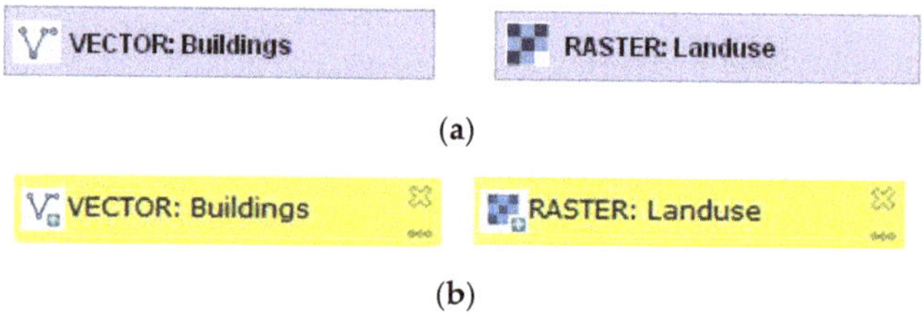

Figure 4. Graphic Examples of data symbols with the addition of inner icons according to data types; (**a**) symbols for version 2 at the top, (**b**) symbols for version 3 at the bottom.

4.1.2. Principle of Perceptual Discriminability

The colour, shape, orientation, brightness and other visual variables are what the user uses for discrimination of symbols in practice. Systematic couple comparison shows the distance between every two symbols. The visual distance is measured by several different characteristics (number of visual variables). Pairwise comparison of the version 3 symbols according to this principle is given in Table 1. In the Processing Modeler, the symbols differ only in colour and brightness, and the rectangular shape is the same for all symbols. The visual distance is two in all pairs. The characteristics are poorer in the symbols of the older version 2. The only differences are in colour, and there the visual distance is one (the difference in brightness is only between the data symbol and operation symbol). Perceptual discriminability of the symbol through colour is almost satisfactory by differing in tone. The Processing Modeler does not have the option for the user to define the colour to express other meanings of symbols, for example, to distinguish the final data and intermediate data in a large model.

Table 1. Pairwise comparison of graphical symbols and their visual distances.

Pair	Symbol 1	Symbol 2	Visual Distance
Pair 1	InputPoint	Zonal statistics	2 (colour, brightness)
Pair 2	ConvexHullPoint	Zonal statistics	2 (colour, brightness)
Pair 3	InputPoint	ConvexHullPoint	2 (colour, brightness)

Considering this principle of perceptual discriminability, the distinctiveness of the white symbol from the canvas is poor. The model's canvas and symbol for operation have the same white colour, from which a new recommendation emerged: change the fill colour for the operation symbol from white to orange-brown (Figure 5). The result of the pairwise comparison of all symbols in the vocabulary remains the same. The discriminability of symbol and canvas is better than with the white symbol.

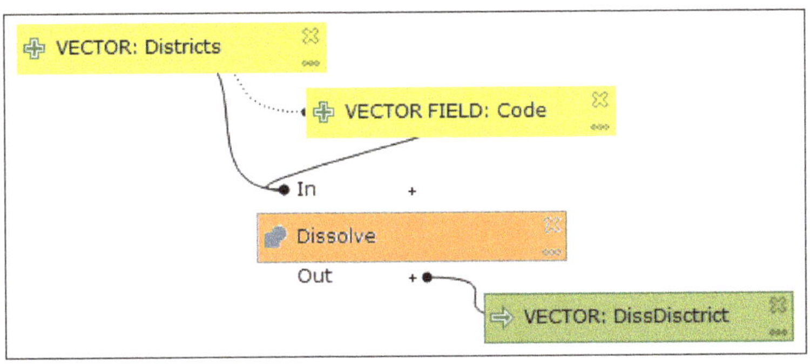

Figure 5. Suggested orange-brown fill for the operation symbol used in the workflow.

4.1.3. Principle of Visual Expressiveness

The recommendation according to this principle is to use maximum visual variables in symbols. Only colour is used as the fill for graphic elements. Other visual variables such as symbol *shape*, *size*, *texture*, *orientation* and *position* are not used in the Processing Modeler. The shape is the same rectangle for all symbols. The size of the symbols does not vary and cannot be changed. Brightness is used in version 3 (Figure 3). The new symbol vocabulary is improved by using a greater variation in brightness between symbols and maybe various shapes of symbols. The visual variable of position is only applied when the output data symbol is automatically placed near the right side of the producing operation (Figure 6). This manner of near-automatic placement of output symbols is the same in version 3 (Figure 13). However, the position of the output data symbol is very often changed by the user, which then moves the operation symbol. The former position of output data remains without following the symbol of the sourcing process. The mutual position linking the symbol is not fixed. The positioning of the output data symbol is only a weak and unstable use of the *position* variable.

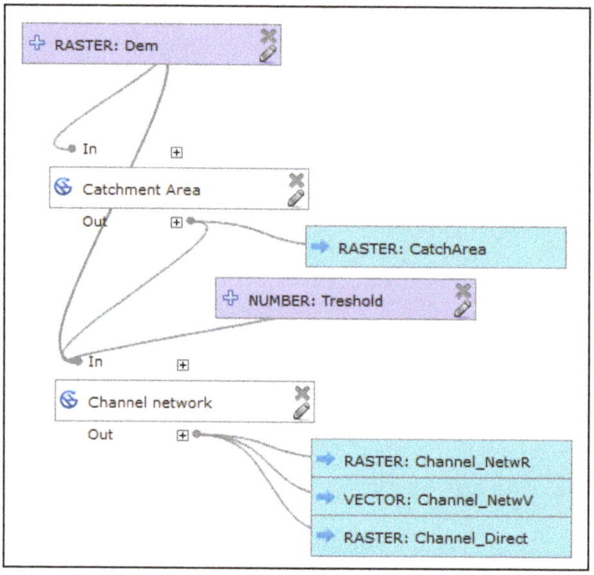

Figure 6. The default placement of blue output data symbols near the producing white operations in version 2.

The graphical vocabulary is at a low level 1 in a maximum scale of 8 in terms of the principle of Visual Expressiveness.

The QGIS Processing Modeler does not offer the *loop* and *condition* functions. To implement these functions in order to control operations, the draft offered in this paper uses the visual variables of shape and colour (Figure 7). The pink rectangle with oblique sides represents the cycle operation, and the light yellow rhombus represents the condition. These symbol shapes correspond to the classic shapes of flowchart symbols. In the vocabulary in version 3, these shapes differ from basic rectangular shapes in vocabulary and colour. By using these new symbols, the number of variables used increases to two. The total number of symbols would be five in the vocabulary. These symbols fulfil the principles of Discriminability and Visual Expressiveness. The principle of Graphic Economy would also be fulfilled (explanation of the principle of Graphic Economy follows).

Figure 7. Draft for loop and condition symbols.

4.1.4. Principle of Graphic Economy

The number of base graphical elements is three, which meets the requirement for cognitive management and the requirement for a range of 7 ± 2 symbols. Even with all the previous suggestions for changes with two symbols for the condition and cycle (under the principle of Visual Expressiveness) and suggestion for a blue symbol for the sub-models (see below in the Complexity Management principle), the total number of symbols is six. Altogether, the requirement of that principle is fulfilled. The vocabulary will be economical.

4.1.5. Principle of Dual Coding

This principle suggests accompanied descriptive text to the symbols. For models in the Processing Modeler, the text completes the data symbol with the data name and the operation symbol with the operation name. The user assigns the data name arbitrarily, which is always an input parameter. The input data symbol is never bound to specific data stored on the storage medium in the model's design mode. The name can be edited as desired. The operation name is added to the symbol automatically according to the selected operation and can also be changed in version 3 (not possible in version 2). The option to edit the operation name is a good improvement in functionality and allows the model to be better understood. Renaming the operation is especially advantageous when the same operation appears multiple times in one model. Therefore, it is possible to describe or specify the meaning of the operation. Figure 8 depicts a diagram where *v.generalize* operations are called three times, but each time with a different generalisation algorithm. The selected algorithm is added manually by the user to the operation name. The operation name's editing option improves the clarity of the model.

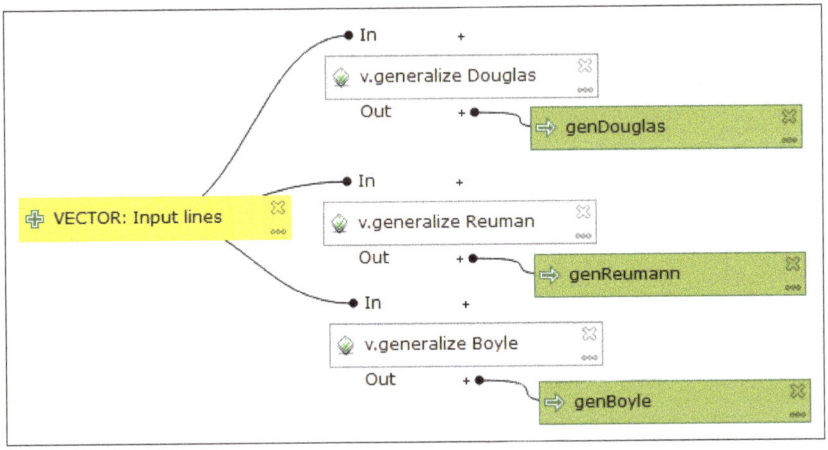

Figure 8. Default label of white symbols of operation supplemented by the user with the generalisation algorithm names: Douglas, Reuman, Boyle.

For long names that do not fit into the rectangle, the name is automatically truncated and completed with an ellipsis (Figure 9). If a Semantic Transparency modification (see below—deletion of functional icons) were implemented, it would increase the space for longer operation and input data names, which would be beneficial.

Figure 9. Automatic truncation of the long operation name ending in an ellipsis.

To follow the principle of Dual Coding, modifying the input data labels is suggested. It would be helpful if labels concerning the data type improved the data symbols by using capitals. Examples are given in Figure 10. If this is added automatically when symbols are added, the user only need arbitrarily select the data name. Additionally, the user's data name (e.g., input lines) emphasises the spatial type. Manually describing a data type is possible in the current stage of notation. There is a space for good use of Dual Coding by users in naming symbols. The Processing Modeler meets the Dual Coding principle; however, comprehensibility could be improved with the proposed modification by specifying the data type with captioning.

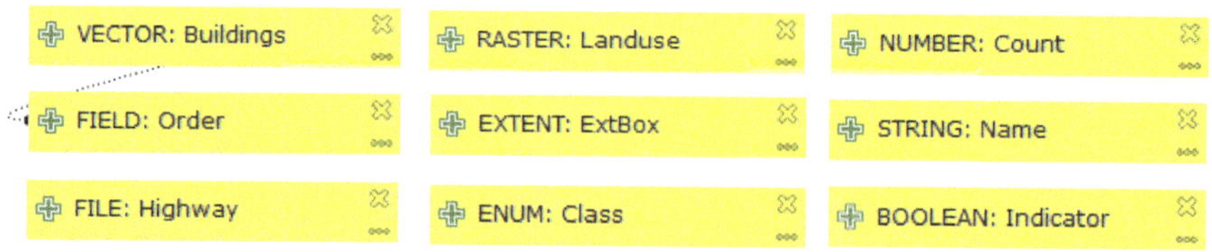

Figure 10. Suggestions for improving the labels of input data symbols with labels for a data type in capitals.

The text is still used in the models to list the operation parameters when the plus symbol is pressed above the operation (Figure 11). After this, the black dot divides itself into several black dots according to the number of join connector lines. The operation parameter list does not contain the values of these parameters and often dumps overlapping lines leading to the rectangle. This is retained in version 3. It would be useful to add a list of specific parameter values here. The current form of textual information is not useful to users. It is perhaps only useful in terms of expressing which symbol assigns concrete parameters to the operation.

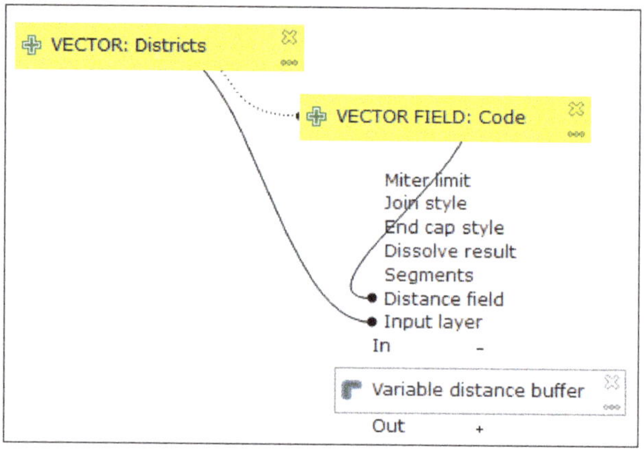

Figure 11. List of the operation parameters of the 'Variable distance buffer' with missing values.

4.1.6. Principle of Semantic Transparency

Symbols could be associated with the real meaning of an element according to this principle. The shape and colour of the symbols do not carry any association; they are semantically general in the Processing Modeler. This is the same in other visual programming languages for the GIS application. In those symbols, the inner icon of the *plus sign* symbol on the input data symbol is used at the left. The output data symbol depicts an inward *arrow* icon. Icons can also carry semantic meaning. These icons can be considered almost semantically immediate. The *plus* icon indicates new data for processing. The *arrow* icon indicates the processing result in a certain direction. However, the previous proposal under the principle of Semiotic Clarity is useful and also improves semantic immediacy. It suggests that each data type has an icon, such as in Figure 4 (the plus icon is replaced or is a part of the compound symbol in version 3). Here, it is clear that the change resulting from applying the Semiotic Clarity principle also leads to an improvement in Semantic Transparency.

For operations, icons are mainly used to represent the source library. Rather, these icons are semantically generic because they do not explain anything about the purpose of the operation. However, these icons are a good guideline for determining the source library. It should be considered that many libraries contain operations with the same name (clip, buffer, etc.). In the Processing Modeler, version 3 sometimes uses an icon that represents the type of operation (namely for QGIS operations). Figure 5 shows the operation Dissolve, which has a specific icon that represents this operation. Another specific icon for the operation 'Merge vector layer' is a model in Figure 13. The size and graphics of the icons are not suitable for improving the association of operation and their meanings. The icons are small and use only grey tones. A good example of large colour and detailed icons that describe the purpose of the operation is demonstrated in the Spatial Model Editor (Figure 12) embedded in the ERDAS IMAGINE software [32]. The icons take up more space than the lower text in the symbol. The icons are prominent. The graphical vocabulary of the Spatial Model Editor has a high Semantic Transparency. The graphical vocabulary of the Spatial Model Editor is inspiring for the redesign of Processing Modeler symbols. The final recommendation is to reshape the rectangle to square to adopt bigger icons and then put the text label bellow icon.

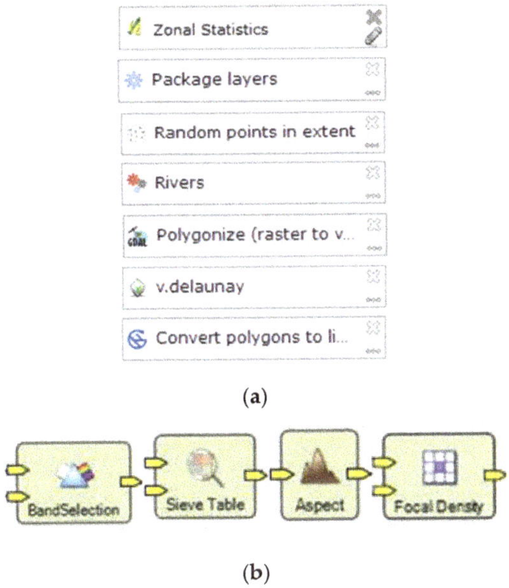

Figure 12. (**a**) Samples of operation symbols with inner icons that represent the source or library (top to bottom): QGIS 2, QGIS 3, QGIS 3—specific icon for operation Random points in extent, existing model Rivers, GDAL, GRASS, SAGA; (**b**) symbols of operations in the Spatial Model Editor (bottom line).

The graphical vocabulary of the QGIS Processing Modeler has semantic opacity, except for some operations, where a greater positive semantic immediacy can be observed (Figure 12—a third symbol from the top: *Random points in extent*).

4.1.7. Principle of Complexity Management

This principle recommends producing hierarchical levels of the diagram and dividing it into separate modules and hierarchy. In textual/visual programming, this is achieved with sub-programs (sub-routines) or sub-models that can be designed and managed separately. The hierarchical model contains only two levels, no more.

The Processing Modeler allows existing models to be added to other models in the interface—panel algorithms (Figure 13). This has the right degree of modularity according to Complexity Management in both versions 2 and 3. The symbol of the model has a three-gear wheel icon (three connected balls in version 2) at the left of the symbol. Otherwise, a white rectangle is used. Since it would be good to differentiate the symbol of the individual operation from the sub-model with an icon, a colour fill other than white would be appropriate. A suggested depiction—blue fill colour for sub-models— is shown in Figure 13. The visual resolution of other symbols is maintained. The number of symbols increases to seven after a new one is added for the sub-model. The final count of seven symbols fulfils the principle of Graphic Economy.

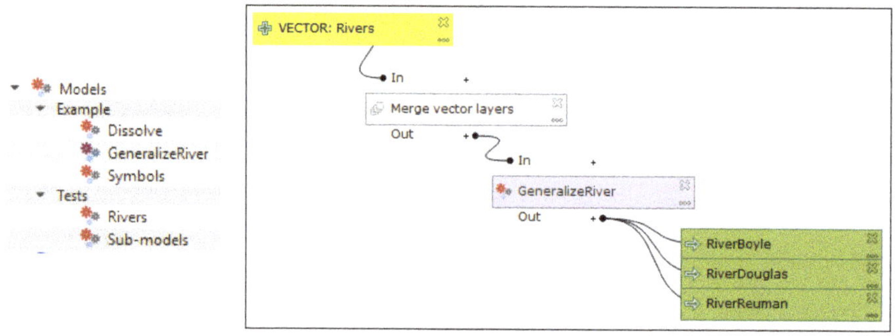

(a) (b)

Figure 13. (**a**) Groups of existing models that can be added to other models; (**b**) suggested blue colour fill for the sub-model.

4.1.8. Principle of Cognitive Interaction

The principle recommends increasing the options for navigating in the model. The connector lines affect navigation. In the Processing Modeler, round connector lines join symbols. The lines are rendered automatically. Symbols very often overlap lines when symbols are manually moved (Figure 6). Lines also sometimes cross each other, and they are not parallel. The user must manually attempt to find the best position for symbols in order to prevent overlapping and perplexing criss-crossing of curved lines. Previous research recommended that the number of edge crossings in drawings should be minimised [33]. For these reasons, curved connector lines do not appear to be the proper solution. It is often difficult to trace the connector's direction. A suggested change is to replace curved lines with straight lines (Figure 14). Straight lines ensure good continuity for reading. Good continuity means minimizing the angular deviation from the straight line of two followed edges connecting two nodes [34]. In this new suggestion for the Processing Modeler, straight lines could be optionally angled at an oblique or right angle when it is necessary to avoid a symbol. An acute angle is not suitable because of its smooth line tracking. If curved connectors remain in notation, there is necessary to add the user control over shaping these connectors to prevent crossing and overlapping.

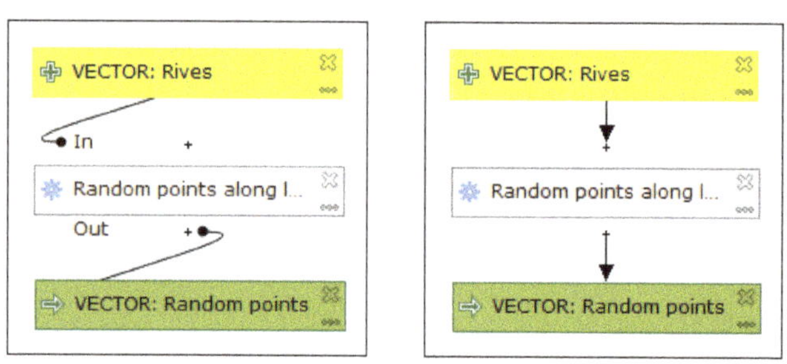

(a) (b)

Figure 14. Replacing curved connector lines (**a**) with straight connector lines (**b**).

Operation symbols linked with lines and a black dot offset from the edge of the symbol unnecessarily occupy space in the model's area. It would be possible to terminate the lines directly on an edge or at the plus sign of the symbol to save space.

Finally, the ability to display a model's thumbnail in the separate preview window helps to navigate the model. The preview window has not yet been implemented. In terms of cognitive interaction, version 3 was supplemented by a zooming function. The functions zoom in and zoom out in the model was absent in version 2.

Aligning the symbols makes reading the model quicker and easier. No automatic function for aligning the model to the grid is implemented. Symbols usually snap to a grid in other graphical software. No snapping function is given in the Processing Modeler. Post alignment to the vertical or horizontal line of the model could, therefore, be beneficial for design. All arrangement of symbol positions depends on user diligence, which is entirely manual work in the Processing Modeler. Manually aligning is a time-consuming activity.

4.2. Evaluation by Eye-Tracking Measurements

The eye-tracking experiment was designed in a complex way to confirm or reject the hypothesis H1 and H2. All tested diagrams are in Appendix A. The design of the test contain more tasks to find maximum information. Some models serve for evaluation repetitively for a different purpose, e.g., find symbol, compare orientation, or read the labels. After testing only reliable answers and correct record by eye-tracker were finally evaluated and presented in the article.

The first evaluation concerned the discriminability of symbols. These tasks required finding input data and output data symbols in the models. The task was: "*Click on the symbol where the input data are*" (task A1, A2, A3 in Appendix A). The number of incorrect answers recorded was zero. The next task was "*Click on the symbol where the output data are*" (task A4, A5, A6). The wrong answers were two times 2 (A4, A5), and 4 for task A6. However, model A6 has a big influence of arrangement to answer. It means that the input and output symbols were nearly high in *Perceptual Discriminability*, but the errors report about space for improvements of symbols such as inner icons that are suggested in Section 4.1. for increasing transparency and using all visual variables.

Besides the number of correct/wrong answers, the time of the first click was recorded. The distribution of times had not normal distribution (tested by Shapiro-Wilk test). Non-parametrical test Kruskal-Wallis tested if the medians of the "first click time" of all tasks (A1–A5) is equal. Kruskal-Wallis tested whether time samples originate from the same distribution. The result of the statistical test revealed the there is no significant difference between finding the symbol of input and output. It means that basic symbols are discriminable and none of them is dominant in perception.

The next task aimed to verify the influence of *Dual Coding* (but the influence of the discriminability present). The task was: "*Click on the symbol where the 'Fixed Distance Buffer' operation is called*" (task A13, A14, A15). Once again, it was necessary to find the white symbol and read the labels in the symbols. A total of 21 correct answers were recorded (one incorrect). The results for 22 respondents were calculated as an attention heat map (Figure 15). The heat map expresses the calculation of places where the peaks of gaze fixations are by all respondents. The figure shows that all white symbols correctly attracted the gaze of respondents. Respondents searched for white operation symbols and then read a particular operation label. The highest attentions were recorded at the two places where the *Fixed Distance Buffer* operation was (top and bottom). Next, the lower peaks of fixations are at another white symbol with the different operation. It is evident that white colours of operation attract the gaze. Both principles of *Visual Expressiveness* (and also *Perceptual Discriminability)*, by using white colour fill, and *Dual Coding* were verified. In fact, this stimulus did not confirm the poor distinguishing of white symbols from a white canvas. The poorest distinguishing result was expected in the theoretical part of this article.

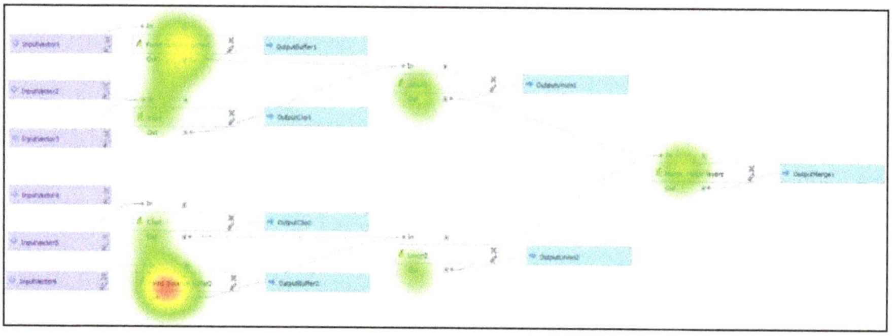

Figure 15. Attention heat map for the task to find the 'Fixed Distance Buffer' operation.

The principle of Semantic Transparency was difficult to test. The transparency of the icons was tested with the task: *"Are all operations from the same source library?"* The tested model is shown in Figure 16 (Appendix A16, 17, 18). Three incorrect answers were recorded from 22 respondents.

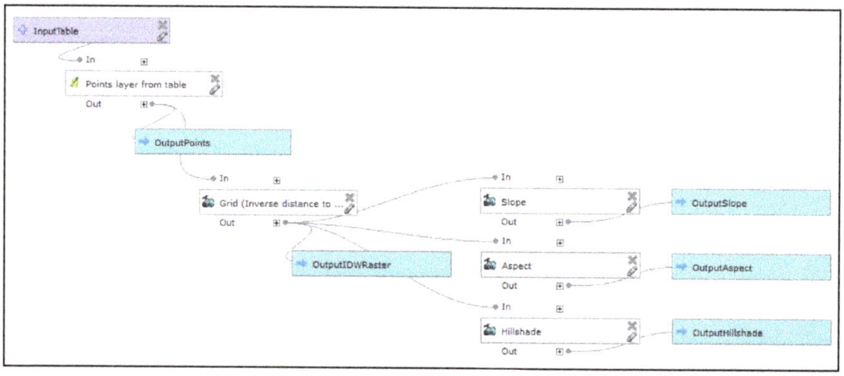

Figure 16. The model with different source libraries of operations in a diagonal arrangement.

Semantic Transparency of data types was only possible to solve in the Processing Modeler with expressive text. This was verified in a model where the task was: *"Does the input data have the same data type as the output data?" (A10, 11, 12)*. The data symbols were labelled with the words "table" and "raster" as a part of the data name in the model. It is "user design help" to the respondents to distinguish the data type. The number of incorrect answers was three for two models and two incorrect answers for A11 task. In these models, the response time was longer than the previously presented models and tasks. The average time of fixation was also longer. It verifies the necessity of reading labels by users. The solution of semantic transparency by text label consume more time for comprehension. The longer times confirms the hypothesis H2 about negative influences of insufficiencies to effective comprehension.

Both of the experiments mentioned above (about source library and the comparison of the data type of input and output data) verified that Semantic Transparency was low in the Processing Modeler.

The results about the number of correct and incorrect answers in all tasks presented in this section report that some insufficiencies adversely affect the cognition as it is stated in hypothesis H1.

From eye-tracking testing, we received not only cross-validation of results by Physics of Notations but other new information. The interesting information was finding the reading patterns and influence of flow orientation to the respondent reading directions. To find the reading pattern of users, gazes were aggregated. A comparison of the same diagram from the free viewing section and a section with tasks is given in Figure 17. Aggregations in both cases revealed that the orientation of data flow expressed by connector lines had a significant influence. Reading began in the middle of the stimulus according to the middle fixation cross in the previous stimulus. Gazes were attracted to the upper left corner and continued horizontally to the right. People's habits of reading lines of text were very strong, especially in free viewing (Figure 17a). The lower part of Figure 17b also depicts strongly followed lines. Only a small number of gazes skips between two main horizontal workflow lines. Free viewing was not as systematic as task-oriented gaze aggregations.

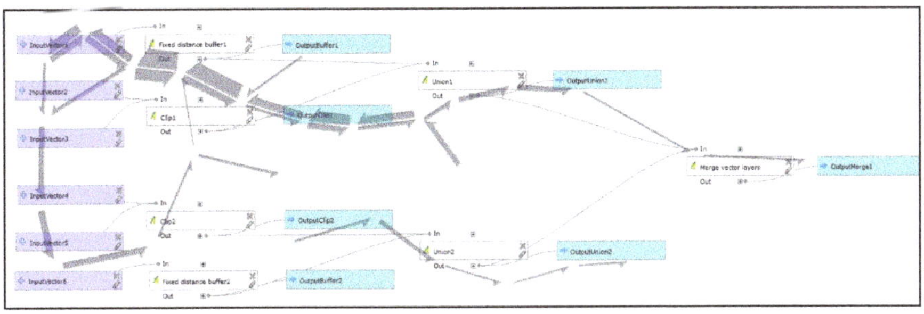

(**a**)

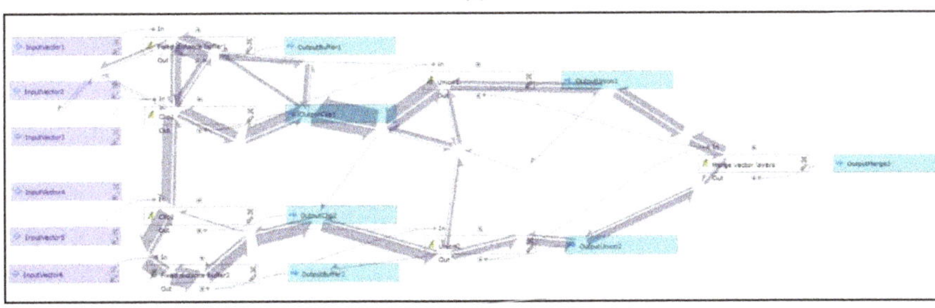

(**b**)

Figure 17. Aggregate directions of scanpaths of respondents in the model during free-oriented reading (**a**) and task-oriented reading (**b**).

Two models tested the effect of *symbol alignment in the model*. This finding can be linked to the principle of *Cognitive Interaction*. The first model had aligned symbols; the second model had no aligned symbols. The functionality was the same. The question was the same for both models: *"How many functions are in the model*?" (task A8, A9). It was enough to count only the white rectangles in large models. The number of the expected correct answers was eight. The first aligned model recorded two incorrect answers and the second recorded seven. The average task time was much shorter in the first tasks. The non-parametrical Kruskal-Wallis test was used for eye-tracking metrics due to non-parametrical distribution of measured values. It tested if medians of groups are equal means they have the same distribution [35]. The significance level for all Kruskal-Wallis tests was set to p-value 0.05. The test was run three times for several fixations, scanpath lengths and number of fixations per second. The test compared tree measured values for the aligned and non-aligned model (A8 and A9). Kruskal-Wallis test found statistically significant differences for all metrics: the number of fixations, scanpath lengths, and number of fixations per second. The model where the symbols were unaligned showed much worse values for all metrics (task A9). Therefore, aligning the symbols in the model made it easier to read and understand the model. This eye-tracking evaluation supports the recommendation for the new function of the automatic alignment of symbols in this graphical editor.

Three groups of models with the same functionality were prepared to test *the orientation of flow* and find if any orientation is better for users. The models in each group differed only by orientation. Three orientations were tested: vertical, horizontal, and diagonal. Comparing the orientation could be also assumed as a contribution to the principle of *Cognitive Interaction*. An example diagonal orientation of flow in a model is shown in Figure 16, horizontal in Figure 17. Variations of these models with three types of orientations were designed. The triplets are [A10, A11, A12], [A13, A14, A15], [A16, A17, A18] in Appendix A. To prevent the bias, the same task was in each group. The aim was to find the best model orientation. The results from a Kruskal-Wallis test were not statistically significant. In some cases, horizontal orientation had the shortest average time of solution. In some models, diagonal orientation was better in average times of fixation, and horizontal models had the shortest scanpath length. The results were completely ambiguous, and the orientation preferences were not the same in all triplets. There is certainly a great deal of effect depending on the given question and model sizes. Eye-tracking did not reveal the best orientation of flow.

5. Results

Research into the QGIS Processing Modeler brought useful results and suggestions. The combination of Physics of Notations theory and eye-tracking measurements determined that Perceptual Discriminability, Dual Coding and Graphic Economy were nearly good with space of improvements. The worst situation is in Semantic Transparency. Some of the recommendations can help improve Semiotic Clarity, Visual Expressiveness and Semantic Transparency.

All recommendations can be divided into two groups. The first group is for developers of the Processing Modeler and the second for users in practice. Suggestions for the first group for larger sizes and colours for inner meaning icons increased the Semantic Transparency. This solution also increased the Semiotic Clarity of symbols. Another suggestion for improvement was adding colour fill to the operation symbol of sub-models. Straight connector lines are better than curved lines, optionally users shaped lines are more suitable. New symbols for IF and loop FOR commands were based on new shapes and different colours. The readability of models improved the automatic alignment function of the symbols to the grid.

Users can benefit from some recommendations in practice. Correct labelling of symbols and expressing data types in capitals (VECTOR, RASTER, STRING, NUMBER, etc.) is very useful. Aligning symbols, preventing overlapping, and crossing of lines improved the effective comprehensibility of a model. Design and using of sub-model fulfil the Complexity Management principle. There is a space for user broader use of sub-models. Reading speed increased in one type of orientation (horizontal or diagonal) without any changes to one of the flow direction. These user recommendations were presented to students attending lectures at the Geoinformatics department at the Palacký University in Olomouc every year. The author of the article has had a positive experience in applying the knowledge acquired by the teacher in research and solving practical problems. This positive teacher experience is described in an article about the database design for the university's botanical gardens (BotanGIS project) [36].

The presented evaluation and list of suggestions could assist by inspiring designers of visual programming languages in GIS software. Some recommendations could also be useful for the broader community of users to increase effective cognition of any graphical depiction.

Table 2 reports all findings and recommendation in summarised form, and concrete graphical improvements are in the figures of the article.

Table 2. List of principles, level of satisfaction by Physics of Notations, eye-tracking findings, and recommendations for Processing Modeler v.3.

Principle	Physics of Notations Evaluation	Eye-Tracking Results	Recommendations
Semiotic Clarity	Symbols of data are overloaded.	Some wrong answers indicate an overload.	• Add various icons in symbols of data types. • Add all icons for spatial functions.
Perceptual Discriminability	Visual distance is 2.	No dominant symbols in perception.	• Change the colour of the operation to orange.
Visual Expressiveness	Level 1, the only colour is used as visual variables.	Some wrong answers indicate weak expressiveness.	• The new pink symbol for loop and light yellow for the condition symbol. It increases expressiveness to level 2.
Graphic Economy	3 symbols fulfils the economy.	Only some wrong answers.	• With the addition of new symbols, a total number of 7 fulfil better the economy.
Dual Coding	Good possibility to change the text.	The text helps users find the proper symbols.	• User renaming to express the data type. User supplement the operation name with some other information about parameters.
Semantic Transparency	Semantically general	Low	• Remove functional icons on the right side. • Add larger sizes inner domain-specific colour icons like in the Spatial Model Editor. • Reshape the rectangle to adopt bigger icons and put the text label bellow icon.
Complexity Management	Modularisation to sub-models is possible.Only one level in the hierarchy.	Not tested	• Change the colour of a sub-model to blue.
Cognitive Interaction	Unmanageable crossing and overlapping of lines worsen interaction.	The arranged model has lower time, scanpath,	• Straight lines or curved user lines • Add ending arrow to lines. • Add preview window. • Add the function of the automatic alignment.
Cognitive Fit	Dialects are missing	Not tested	• In the area of GIS users, dialects have not to sense to make them.

Funding: This research received no external funding.

Acknowledgments: This article was created with the support of the Education for Competitiveness Operational Programme – European Social Fund (project CZ.1.07/2.3.00/20.0170 Ministry of Education, Youth, and Sports of the Czech Republic).

Conflicts of Interest: The authors declare no conflict of interest.

Appendix A

Eye-tracking experiment: List of tasks and models from QGIS Processing Modeler

The appendix presents the list of models and assigned tasks that were used in the eye-tracking experiment. The order of models was random in testing.

Note 1: All models were also used in the first part of testing – free viewing part

1. Symbols for data and operations (testing of discriminability)

Task A1. *Click on the symbol where the input data are.*

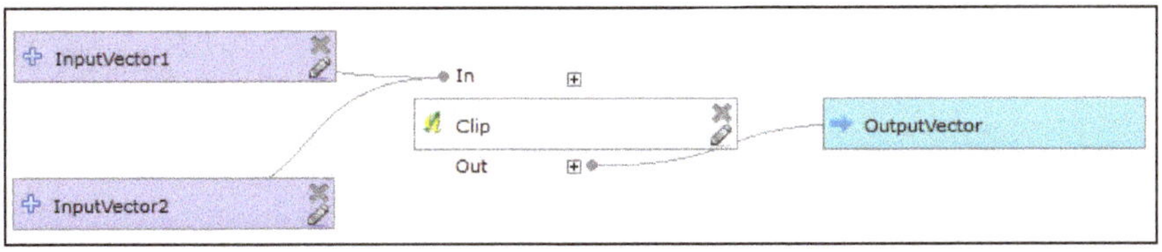

Task A2. *Click on the symbol where the input data are.*

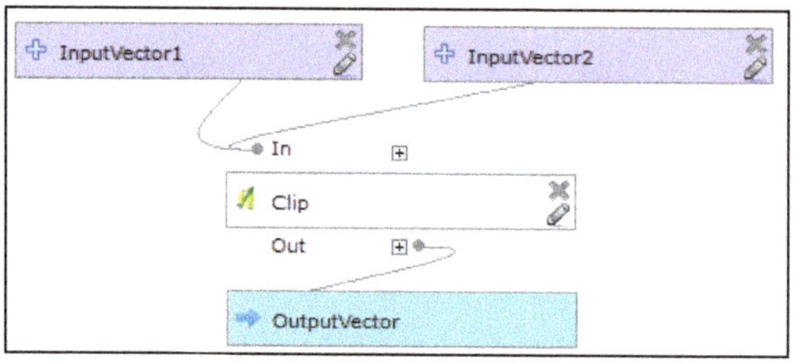

Task A3. *Click on the symbol where the input data are.*

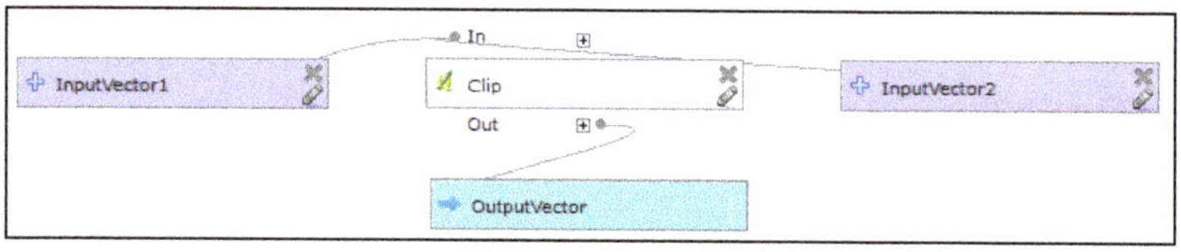

Task A4. *Click on the symbol of output data.*

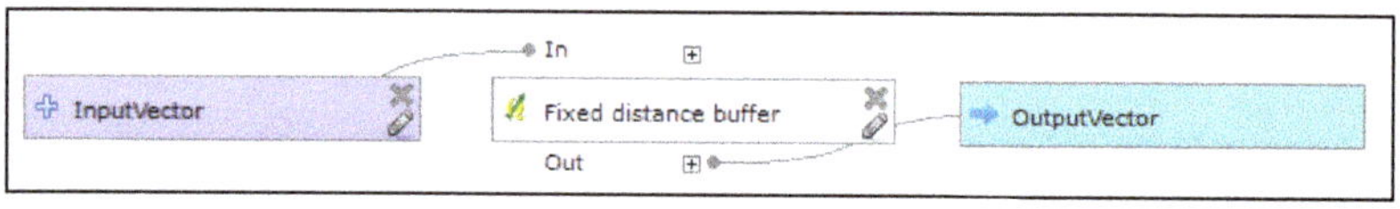

Task A5. *Click on the symbol of output data.*

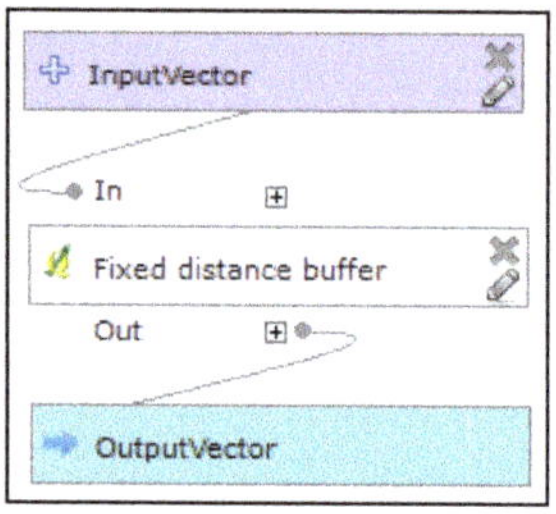

Task A6. *Click on the symbol of output data.*

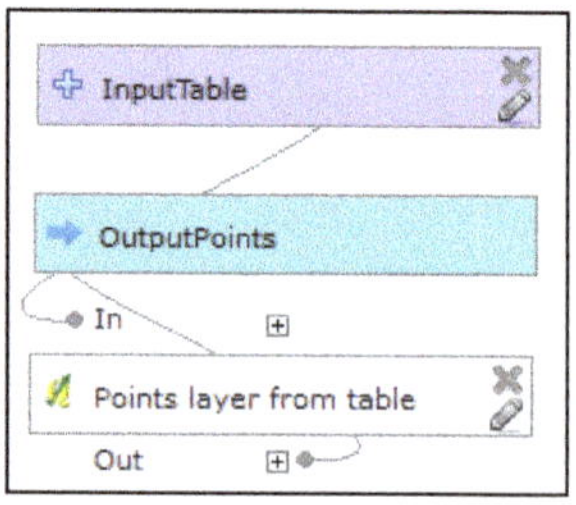

Task A7. *How many functions are in the model? (Correct answer* **8***)*

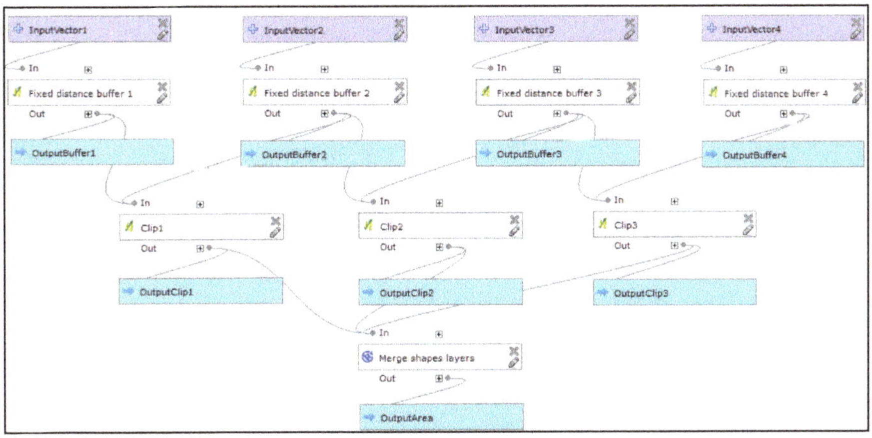

Task A8. *How many functions are in the model? (Correct answer* **8***)*

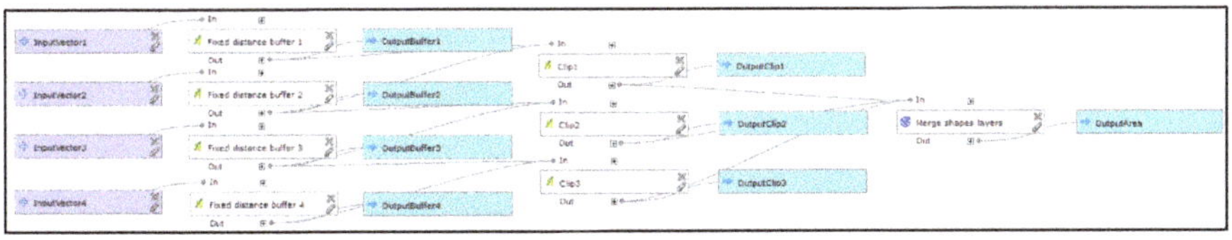

Task A9. *How many functions are in the model? (Correct answer* **8***)*

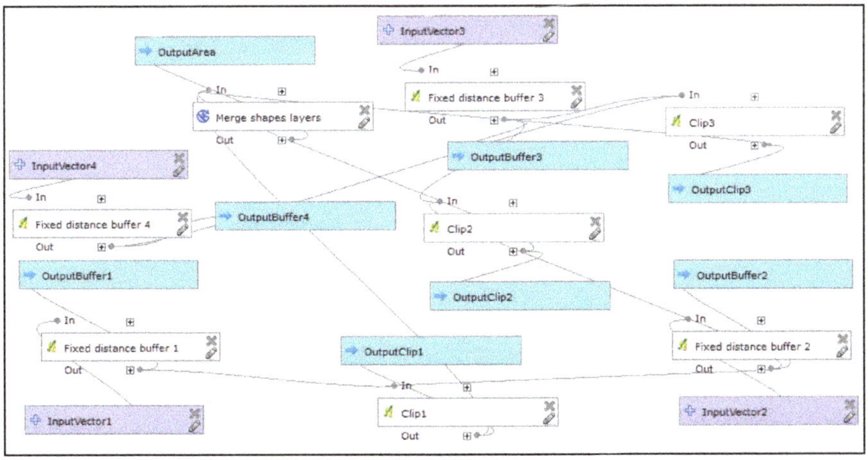

2. Test of dual codding

Task A10. *Are the input data the same type as output data?*

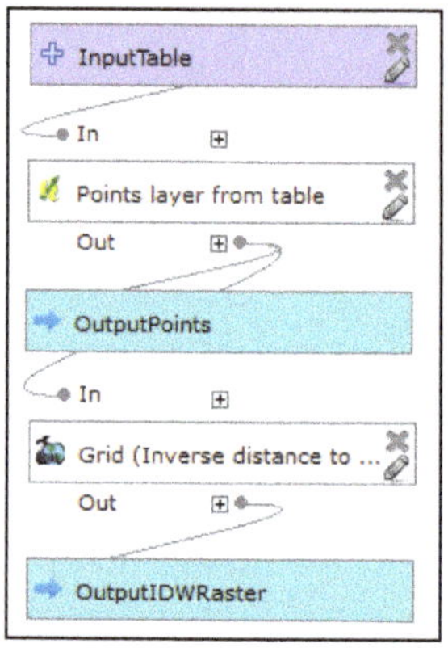

Task A11. *Are the input data the same type as output data?*

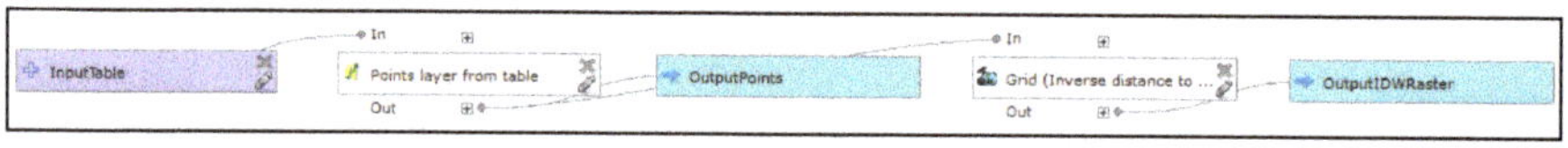

Task A12. *Are the input data the same type as output data?*

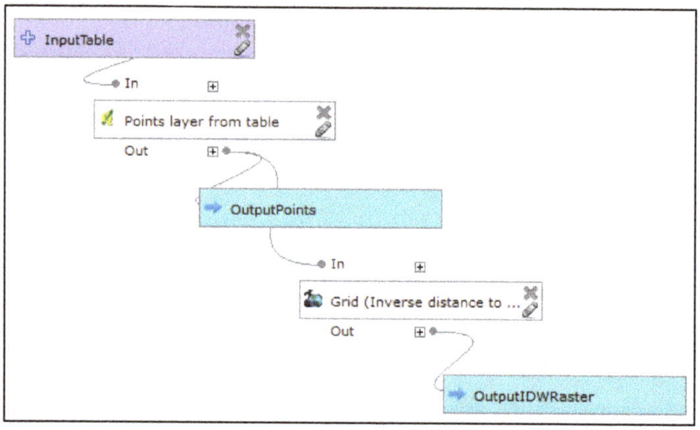

Task A13. *Click on the symbol where the 'Fixed Distance Buffer' operation is called.*

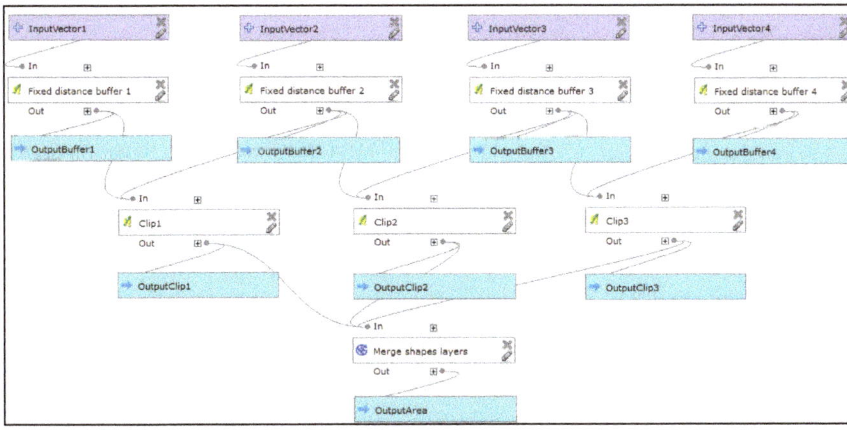

Task A14. *Click on the symbol where the 'Fixed Distance Buffer' operation is called.*

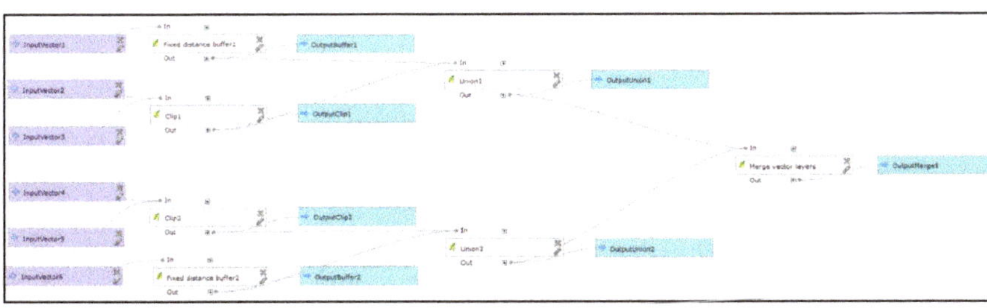

Task A15. *Click on the symbol where the 'Fixed Distance Buffer' operation is called.*

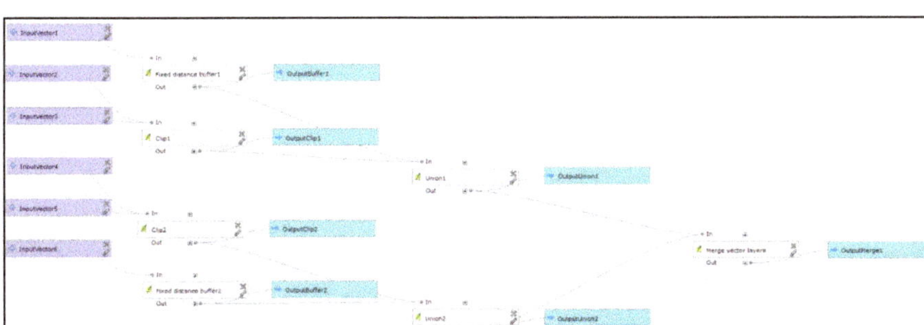

3. Test of inner icons (semantic transparency)

Task A16. *Are all operations from the same source library? (Correct answer* ***No****)*

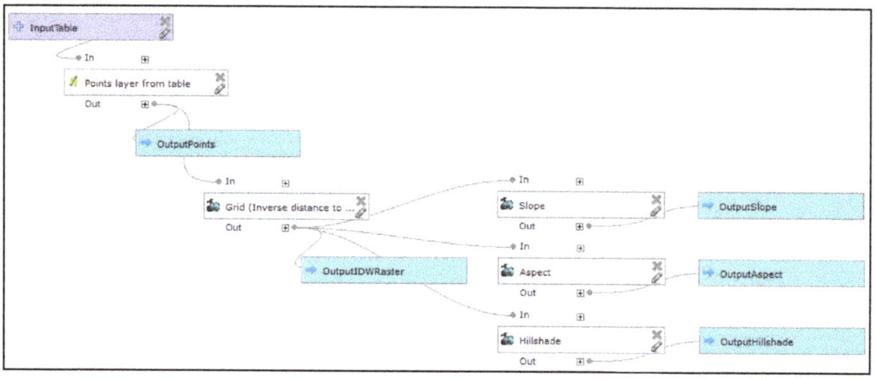

Task A17. *Are all operations from the same source library? (Correct answer* ***No****)*

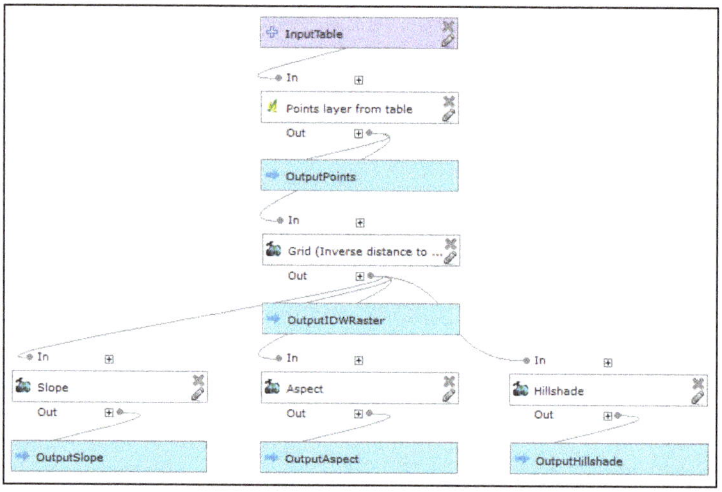

Task A18. *Are all operations from the same source library? (Correct answer* ***No****)*

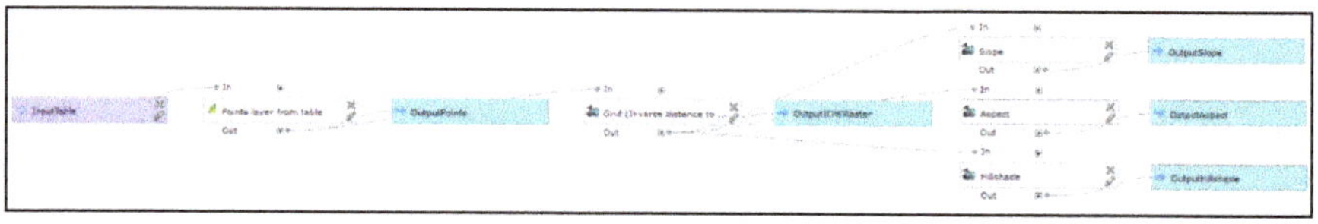

4. Other tested models and tasks

Task A19. *What operation is used? (Correct answer: Clip)*

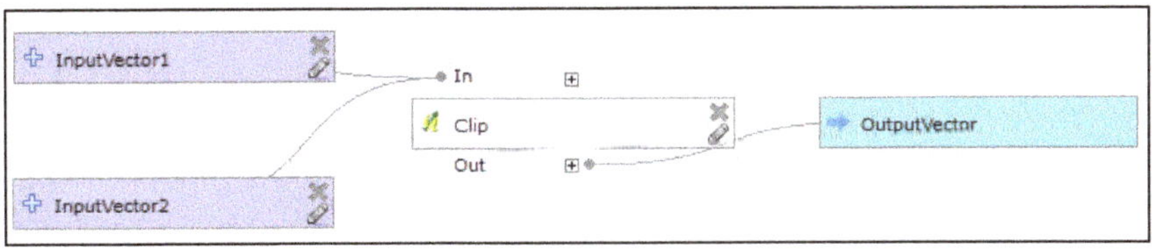

Task A20 *What operation is used? (Correct answer: Fixed Distance Buffer)*

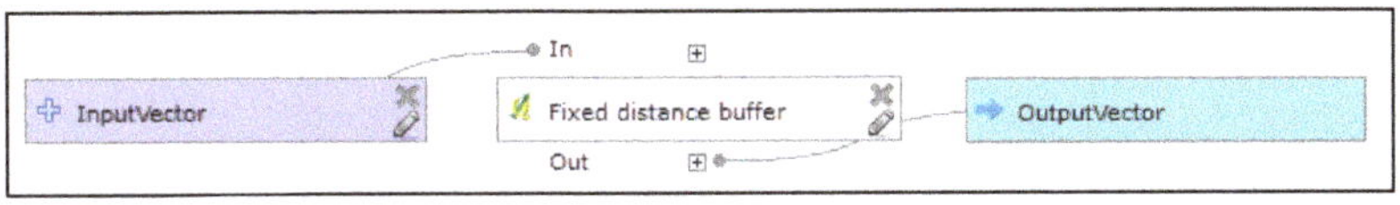

Task A21: *Click on the symbol of operation.*

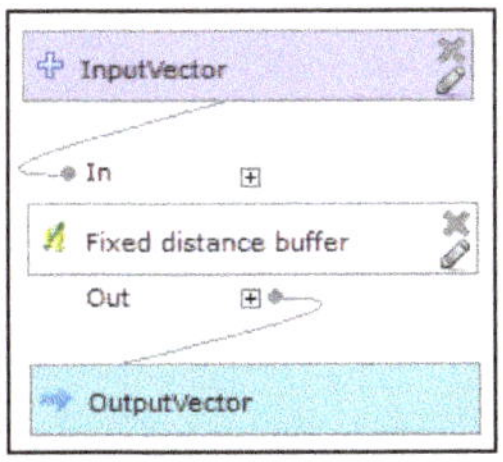

Task A22: *Click on the symbol of operation.*

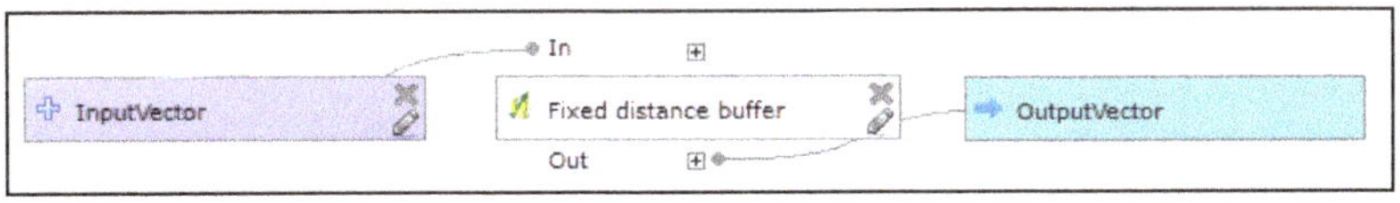

References

1. Dobesova, Z. Data flow diagrams in geographic information systems: A survey. In Proceedings of the 14th SGEM GeoConference on Informatics, Geoinformatics and Remote Sensing, Albena, Bulgaria, 19–25 June 2014; pp. 705–712. [CrossRef]
2. Dobesova, Z. Graphical notation of visual programming languages in GIS. Habilitation Thesis, University of Pardubice, Pardubice, Czech Republic, 2017.
3. Moody, D.L. The "Physics" of Notations: Toward a Scientific Basis for Constructing Visual Notations in Software Engineering. *IEEE Trans. Softw. Eng.* **2009**, *35*, 756–779. [CrossRef]
4. ISO. ISO 9241-210:2019 Ergonomics of human-system interaction. In *Part 210: Human-Centred Design for Interactive Systems*; ISO: Geneva, Switzerland, 2019.
5. Marsh, J. *UX for Beginners*; O'Reilly Media, Inc.: Sebastopol, CA, USA, 2016.
6. Komarkova, J.; Machova, R.; Bednarcikova, I. Users requirements on quality of Web pages of municipal authority. *E M Ekon. Manag.* **2008**, *11*, 116–126.
7. Reeves, S. What Is the Relationship Between HCI Research and UX Practice? Available online: https://www.uxmatters.com (accessed on 4 February 2020).
8. Komárková, J. *Quality of web geographical information systems*; University of Pardubice: Pardubice, Czech Republic, 2008; ISBN 978-80 7395-056-9.

9. QGIS. Changelog for QGIS 2.6. Available online: http://www.qgis.org/en/site/forusers/visualchangelog260/index.html?highlight=processing%20modeler#processing (accessed on 5 June 2016).
10. Graser, A.; Olaya, V. Processing: A Python Framework for the Seamless Integration of Geoprocessing Tools in QGIS. *Isprs Int. J. Geo-Inf.* **2015**, *4*, 2219–2245. [CrossRef]
11. GIS_Geography. The Hidden Powers of QGIS 3: 33 Truly Underappreciated Features and Plugins. Available online: https://gisgeography.com/qgis-3/ (accessed on 4 May 2019).
12. KARTOZA. Changelog for qgis Version 3.0.0. Available online: http://changelog.qgis.org/en/qgis/version/3.0.0/ (accessed on 5 April 2019).
13. Schwarze, J. Using QGIS 2.x Processing Model in QGIS 3.0? Available online: https://gis.stackexchange.com/questions/277373/using-qgis-2-x-processing-model-in-qgis-3-0 (accessed on 15 May 2019).
14. QGIS. Documentation for QGIS 2.6, QGIS User Guide, QGIS Processing Framework, the Graphical Modeler. Available online: https://docs.qgis.org/2.6/en/docs/user_manual/processing/modeler.html (accessed on 20 August 2016).
15. Szczepanek, R. Toolbar icons for applications. *Geoinform. Fce Ctu* **2008**, *3*, 79–89. [CrossRef]
16. Moody, D.L. The "Physics" of Notations: A scientific Approach to Designing Visual Notations in Software Engineering. In Proceedings of the 32nd ACM/IEEE International Conference on Software Engineering—Volume 2, Cape Town, South Africa, 8–10 May 2010; pp. 485–486. [CrossRef]
17. Dobesova, Z. Using the Physics of Notations to analyse ModelBuilder diagrams. In Proceedings of the 13th International Multidisciplinary Scientific GeoConference, Albena, Bulgaria, 28 June–6 July 2019; pp. 595–602. [CrossRef]
18. Larkin, J.H.; Simon, H.A. Why a Diagram is (Sometimes) Worth Ten Thousand Words. *Cogn. Sci.* **1987**, *11*, 65–100. [CrossRef]
19. Moody, D. Theory development in visual language research: Beyond the cognitive dimensions of notations. In Proceedings of the Visual Languages and Human-Centric Computing, Corvallis, OR, USA, 20–24 September 2009; pp. 151–154. [CrossRef]
20. Mackinlay, J. Automating the design of graphical presentations of relational information. *Acm Trans. Graph.* **1986**, *5*, 110–141. [CrossRef]
21. Paivio, A. *Mental Representations: A Dual Coding Approach*; Oxford University Press: London, UK, 1986; p. 322.
22. Dobesova, Z. Empirical testing of bends in workflow diagrams by eye-tracking method. *Adv. Intell. Syst. Comput.* **2017**, *3*, 158–167.
23. Martin, D.W. *Doing psychology experiments*; Wadsworth Cengage Learning: Belmont, TN, USA, 2008; p. 363.
24. Holmqvist, K.; Nyström, M.; Andersson, R.; Dewhurst, R.; Jarodzka, H.; Van de Weijer, J. *Eye tracking: A Comprehensive Guide to Methods and Measures*; Oxford University Press: Oxford, UK, 2011; p. 537.
25. Figl, K.; Mendling, J.; Strembeck, M. The influence of notational deficiencies on process model comprehension. *J. Assoc. Inf. Syst.* **2013**, *14*, 312–338. [CrossRef]
26. Störrle, H.; Baltsen, N.; Christoffersen, H.; Maier, A.M. On the Impact of Diagram Layout: How Are Models Actually Read? In Proceedings of the 2014ACM/IEEE International Conference on Model Driven Engineering Languages and Systems, Valencia, Spain, 28 September–3 October 2014.
27. Störrle, H. On the impact of layout quality to understanding UML diagrams: Diagram type and expertise. In Proceedings of the IEEE Symposium on Visual Languages and Human-Centric Computing (VL/HCC), Innsbruck, Austria, 30 September–4 October 2012; pp. 49–56. [CrossRef]
28. Pődör, A. Usability study on different visualization methods of crime maps. *Int. J. Geoinform.* **2015**, *11*, 15–22.
29. Sedlák, P.; Komárková, J.; Hub, M.; Struška, S.; Pásler, M. Usability evaluation methods for spatial information visualisation case study: Evaluation of tourist maps. In Proceedings of the ICSOFT-EA 2015—10th International Conference on Software Engineering and Applications, Proceedings; Part of 10th International Joint Conference on Software Technologies, Colmar, France, 20–22 July 2015; pp. 419–425.
30. Dobesova, Z. Student reading strategies of GIS workflow diagrams. *J. Adv. Soc. Sci. Educ. Humanit. Res.* **2016**, *70*, 319–325.
31. Dobesova, Z. Testing of Perceptual Discriminability in Workflow Diagrams by Eye-Tracking Method. In *Advances in Intelligent Systems and Computing*; Springer: Cham, Switzerland, 2018; Volume 661, pp. 328–335. [CrossRef]
32. ASPRS. *American Society for Photogrammetry and Remote Sensing*; ERDAS IMAGINE: Bethesda, MD, USA, 1993.
33. Reingold, E.M.; Tilford, J.S. Tidier Drawings of Trees. *IEEE Trans. Softw. Eng.* **1981**, *7*, 223–228. [CrossRef]

34. Ware, C.; Purchase, H.; Colpoys, L.; McGill, M. Cognitive measurements of graph aesthetics. *Inf. Vis.* **2002**, *1*, 103–110. [CrossRef]
35. Hendl, J. *Overview of Statistical Methods: Data Analysis and Meta Analysis*; Portál: Prague, Czech Republic, 2009; p. 695.
36. Dobesova, Z. Teaching database systems using a practical example. *Earth Sci. Inform.* **2016**, *9*, 215–224. [CrossRef]

© 2020 by the author. Licensee MDPI, Basel, Switzerland. This article is an open access article distributed under the terms and conditions of the Creative Commons Attribution (CC BY) license (http://creativecommons.org/licenses/by/4.0/).

applied sciences

MDPI

Article

Geo-Sensor Framework and Composition Toolbox for Efficient Deployment of Multiple Spatial Context Service Platforms in Sensor Networks

Sehrish Malik and DoHyeun Kim *

Department of Computer Engineering, Jeju National University, Jeju 63243, Korea; serrym29@gmail.com
* Correspondence: kimdh@jejunu.ac.kr

Received: 11 October 2019; Accepted: 6 November 2019; Published: 20 November 2019

Abstract: Geo-sensor is the term used for the deployment of a wireless sensor network (WSN) in a real environment, which can be a hideous task due to many influential variables in a given environment. The spatial context of a sensor in a smart environment can be of huge significance and can also play an important role in improving the smart services provision. In this work, we propose a DIY geo-sensor framework and data composition toolbox for the deployment of sensors data in smart IoT environments along with geographical context. A geo-sensor framework is deployed, which enables the registration of multiple geo-sensor networks by providing multiple geo-sensor platforms. The framework's logic is based on the combination of a geo-sensor service registry, geo-sensor composition toolbox, and geo-sensor platforms. A geo-sensor platform provides the content to the toolbox, enabling relaxed real-time geo-sensor data management. Our proposed work is two-fold. Firstly, we propose the design details for the geo-sensor framework and composition toolbox. The proposed design for the geo-sensor framework aims to provide a DIY platform for multiple geo-sensor networks and services deployment, giving access to multiple users resulting in reuse of resources and reduction in deployment costs by avoiding duplicate deployments. Secondly, we implement the proposed design based on RESTful web services and SOAP web services. Performance comparison analysis is then performed among the two web services to find the best suited implementation for given scenarios. The results of the performance analysis prove that RESTful web services are the better choice for ease of implementation, access, and light-weightiness.

Keywords: geo-sensor framework; geo-sensor platform (GIS); sensor networks; do-it-yourself (DIY)

1. Introduction

A system designed to integrate and display the spatial data of "things" is called a geographical information system (GIS) [1]. GIS also enables the user to store, edit, analyze, and share the geographical data of a given entity. GIS represents data in a useful and meaningful way for users to understand and utilize, also enabling the users to process large amounts of data. Wireless sensors networks (WSN) produce large amounts of data, which is processed for different purposes and intelligent decision making. Integrating the WSN with GIS can make the task of analyzing the data gathered by WSN more evocative for the users to process and analyze [1].

Geo-sensor is the deployment of a wireless sensor network (WSN) in a real environment, which can be a hideous task due to many influential variables in a given environment. A wireless sensor network is a combination of spatially distributed wireless devices that are deployed to monitor physical or environmental conditions such as temperature, light, sound, and humidity. These wireless devices have sensors to collect data from the surrounding environment and pass the data onto some main location. Each wireless device also has a transmitter and a receiver, which are used to communicate with other wireless devices or gateways within a network. The gateway, also known as middleware, transfers the sensed data to the main location, which logs in the data, and the data is then available to view, process, and analyze via the user interface [2].

Physical sensors from sensor networks are uploaded as virtual sensor objects with sensor profiles created and stored in the databases. These sensor profiles are then used to extract the sensors' types, values, and environmental scenario contexts along with the geographical contexts based on the physical locations and location based constraints and parameters. The geo-sensing services are then created by combining the sensors' contextual information and geographical contextual information; based on each systems requirements and scenarios. Figure 1 shows the conceptual overview of the geo-sensor networks' services formation.

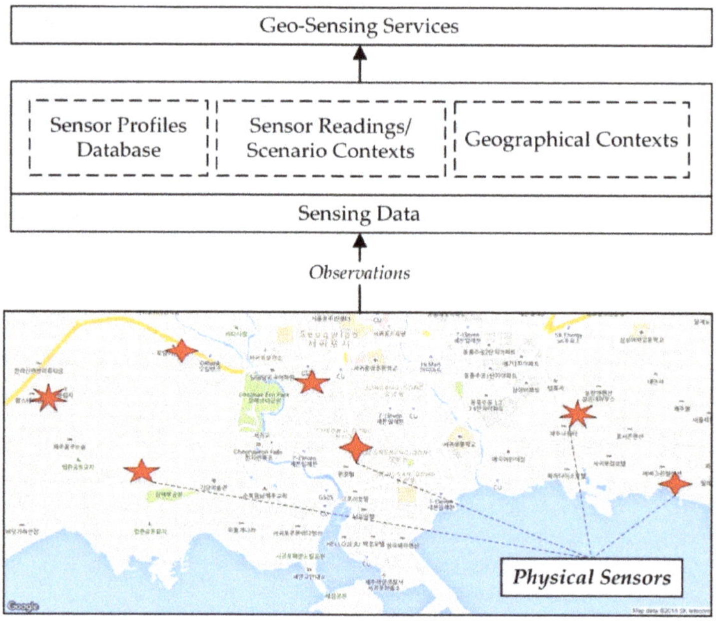

Figure 1. Conceptual overview of geo-sensor networks and services.

Geo-sensor applications and services are of huge significance and are applied to multiple scenarios and environments based on the contextual information. We can conclude from our literature review study, presented in Section 2, that little effort is made in proposing standard designs for geo-sensor frameworks for multiple geo-sensor network deployment and service creation; where the user gets the freedom of uploading their networks based on multiple locations and contextual information and creating and using the services accordingly.

Such systems where more freedom is given to the users are termed as DIY (do-it-yourself) systems. Recently, DIY approaches for IoT (Internet of Things) applications have gained much hype. The concept of IoT can be simply put as connecting "things", such as objects and entities, to the internet and making them smart. The concept of DIY was first proposed in 2011 [3], and multiple approaches have been proposed in recent years for DIY virtual services in IoT networks [3–10]. Specifically in the geo-sensing category, not much work has been done for DIY implementation except for one recently started project named as community-centered urban sensing (CCUS) and DIY sensing devices [11], which collects environmental sensing data, visualizes the data and aims to develop customized digital infrastructure for empowering communities for better sensing their environments, and voicing their opinions by taking part in feedback process. The project is still limited in visualization and analysis of sensing data and in its preliminary stages.

Hence, a DIY geo-sensor framework for the deployment of geo-sensor networks, where multiple users could upload their geo-sensor platforms and extend access of use to other users, can pave ways for unlimited potential in geo-sensor applications and research fields.

In this work, we propose a geo-sensor framework that can be used by multiple clients to deploy their own geo-sensor networks, bind their sensor objects to desired locations based on DIY, generate geo-sensor services for the uploaded networks, and manage the services with a geo-sensor composite toolbox. The availability of such a platform will allow the ease of deployment and management for the geo-sensor networks and also the reuse of geo-sensor services. The deployed geo-sensor platform and its services can be reused by authorized users in different scenarios, with the online availability of the geo-sensor framework. This also reduces the deployment cost in scenarios where such deployed networks and services can be reused, as an alternative to additional deployments. It will also reduce the deployment redundancy. The proposed work offers the following novelty points as

- Ease of deployment for geo-sensor networks based on DIY,
- Multi-platform geo-sensing services generation,
- Independent geo-sensor composite toolbox for ease of management,
- Visualization tool for users,
- Removing redundancy.

The multiple spatial context service platforms use the sensor profile information and geographical context of the service to create geo-sensor services. We present the detailed system design for the deployment of geo-sensor framework, geo-sensor composition toolbox, and geo-sensor multiple spatial services platforms for the wireless sensor networks. The proposed system is implemented using both the RESTful web services and SOAP web services. RESTful web services allocate URI (Uniform resource identifier) to each resource and use basic CRUD operations over HTTP for making the interface uniform throughout and are simpler to implement.

The rest of the paper is divided as follows; Section 2 gives an overview of related work. Section 3 presents the system design for a geo-sensor framework and multiple platforms deployment based on the geo-sensor framework. In Section 4, the implementation of the proposed framework is presented. In Section 5, we give a performance analysis of a system based on system queries and access times, and Section 6 concludes the paper.

2. Literature Review

The proposed work falls under the category of contextual sensing. All physical objects (e.g., sensors, actuators, things) or human beings have some context attached to them in a given scenario. The context can be defined as "any information that can be used to characterize the situation of an entity [12]. The context of an entity can be either external or internal; external contexts are perceived from the physical environment, such as sensing values from sensors or the location from GPS, whereas internal contexts are customized individual level contexts within a system [13]. In our proposed work, we focus on the external context as our scope is limited to geo-sensing services.

One of the subcategories for applications of geo-sensors is the spatiotemporal contexts based services. The term spatiotemporal means belonging to both space and time; hence, the application of spatiotemporal is not just limited to the extraction and use of physical locations contexts [14,15]. Many studies have been presented based on the applications of spatiotemporal with other attributes added to the given physical locations based on the specific time; such as effect of weather conditions at a given time at a location, such as rainfall on the mobility pattern in a city [16,17], or the effect of environmental conditions, such as air quality on the traffic flow within a city [18]. The work presented in [19], discusses and analyzes the contextual sensing based on integrating contextual information with human and technical geo-sensor information for smart cities. The authors highlight the significance of contextual information and discuss the challenges regarding spatiotemporal contextual information. Three groups of sensors are discussed along with the three dimensions of sensing as data generation, geographic phenomena, and type of sensing. The authors aim to explore the use of geo-sensing capabilities and contextual information integration for the future development of smart cities.

The sensors can be of various types, and, hence, their applications vary depending on the type of sensor and context of the sensor. Common terms used for different sub-types of sensors are mentioned as environmental sensors, mobile sensors, and pervasive sensors and people as sensors [19]. Multiple works based on sensing categories and contextual information has been presented. In environmental sensors, some contextual studies include flood monitoring [20,21], air quality monitoring [22–26], and weather [27,28]. In mobile sensors, some of the contextual studies presented include wearable ambient sensors applications [29–34]. In pervasive sensing, some contextual studies include smart-aware environments/homes, assisted living [35–42], healthcare [43–45]. In people as sensors, some of the contextual studies include flood monitoring [46,47], sensing platforms for smart cities [48–54], healthcare [55–59].

Very few studies have presented and proposed the design and implementation of a GIS toolbox. The work presented in [60] proposes an adaptive GIS toolbox for hydrological modeling. In order to increase reusability and portability, the modules are programmed in an object-oriented fashion. The tasks of modeling elements of the hydrological cycle can be done using the toolbox, and it also supports different temporal and spatial scales. The presentation of the spatial, temporal, and geographical data is done using GRASS. The work presented in [61] proposes a landscape genetics GIS toolbox. It maps the genetic landscapes and summarized multiple genetic landscapes. Genetic diversity can be visualized using these tools. Together, these tools create genetic landscape surfaces directly from tables containing genetic distance or diversity data and sample location coordinates, greatly reducing the complexity of building and analyzing these raster surfaces in a Geographic Information System. The work in [62] presents the principal strategies of object-oriented analysis, discusses how the combination with fuzzy methods allows implementing expert knowledge, and describes a representative example for the proposed workflow from remote sensing imagery to GIS. The strategies are demonstrated using the first object-oriented image analysis software on the market, e-Cognition, which provides an appropriate link between remote sensing imagery and GIS. The study in [63], demonstrates that the integration of satellite remote sensing and GIS was an effective approach for analyzing the direction, rate, and spatial pattern of land-use change. The further integration of these two technologies with Markov modeling was found to be beneficial in describing and analyzing land-use change processes. These wireless devices have sensors to collect data from the surrounding environment and pass the data onto some main location. Each wireless device also has a transmitter and receiver, which are used to communicate with other wireless devices or gateway within a network. The gateway, also known as middleware, transfers the sensed data to the main location, which logs in the data, and the data is then available to view, process, and analyze via a user interface. The work in [64], reports an investigation into the application of the integration of remote sensing and geographic information systems (GIS) for detecting urban growth and assessing its impact on surface temperature in the region. Remote sensing techniques were used to carry out land use/cover change detection by using multi-temporal Landsat Thematic Mapper data. Urban growth patterns were analyzed by using a GIS-based modeling approach. The integration of remote sensing and GIS was further applied to examine the impact of urban growth on surface temperatures.

3. Geo-Sensor Framework

In this section, we present our proposed geo-sensor framework's design and implementation modules along with configurations.

Figure 2 shows the conceptual diagram for the geo-sensor framework for the deployment of multiple platforms. The data collected from the sensor networks is first passed onto the sensor middleware, where the sensor network's mapping to a dedicated sensor platform is done. Then, the data is passed onto the sensor platform via sensor middleware. Each sensor network's spatial context is managed at a Geo platform. At the geo-sensor composite toolbox, the assistance for the DIY geo-sensor network's deployment is provided. Geo-sensor framework integrates the services provided by the geo-sensor composition toolbox and geo-sensor platform to enable the geo-sensor services provision and the network's visualization for the client end.

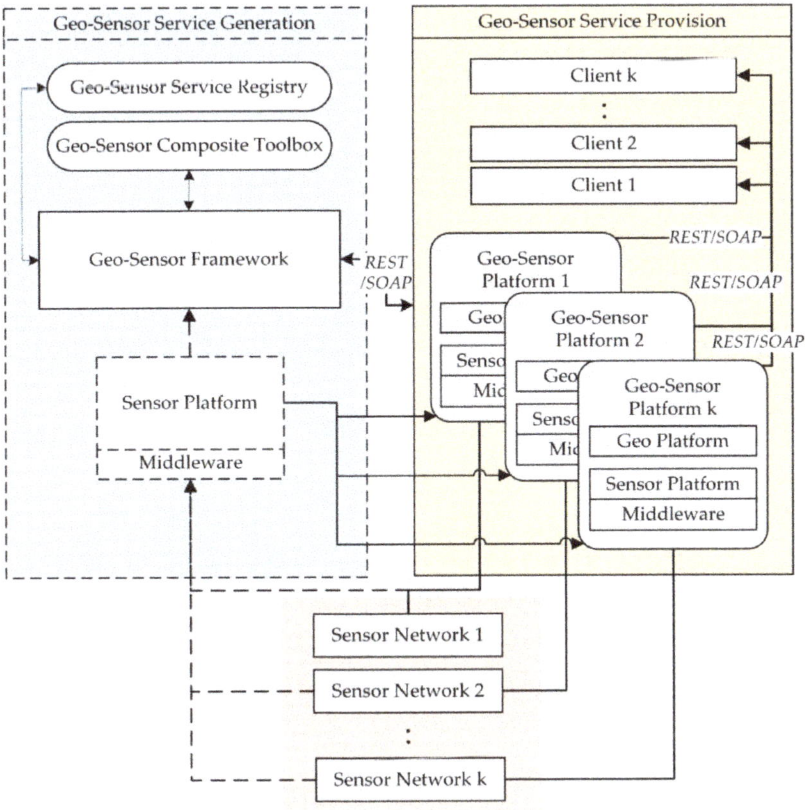

Figure 2. Geo-sensor framework conceptual diagram for multi-platforms.

3.1. *Geo-Sensor Framework and Geo-Sensor Platform Generation*

In this sub-section, we present the system design for our proposed geo-sensor framework and multiple geo-sensor platform generation.

In Figure 3, we present the detailed configurations diagram for the proposed geo-sensor framework module interactions for multiple platform deployment. A geo-sensor framework offers a main service of geo-sensor content service. The geo-sensor content service has logical implementation for sensors involved in the network and geographical context of the sensors provided by sensor content service and geo content service, respectively. The sensor content service offers a sensor information manager and sensor's middleware configuration manager. The geo-sensor framework has a virtual sensor platform at one end, from where the sensor network's information is passed to the framework. It creates the sensors networks' profiles using the geo-sensor composite toolbox, which is later also visualized to the client end for client-level updates. Each time a service is created at the geo-sensor framework, it has to be registered at the geo-sensor service registry to maintain each sensor network's detailed history log. The geo-sensor framework eventually deploys the multiple geo-sensor platforms for multiple networks by mapping each of the network's information to its respected platform using the geo and sensor content services, geo-sensor composite toolbox, and geo-sensor service registry.

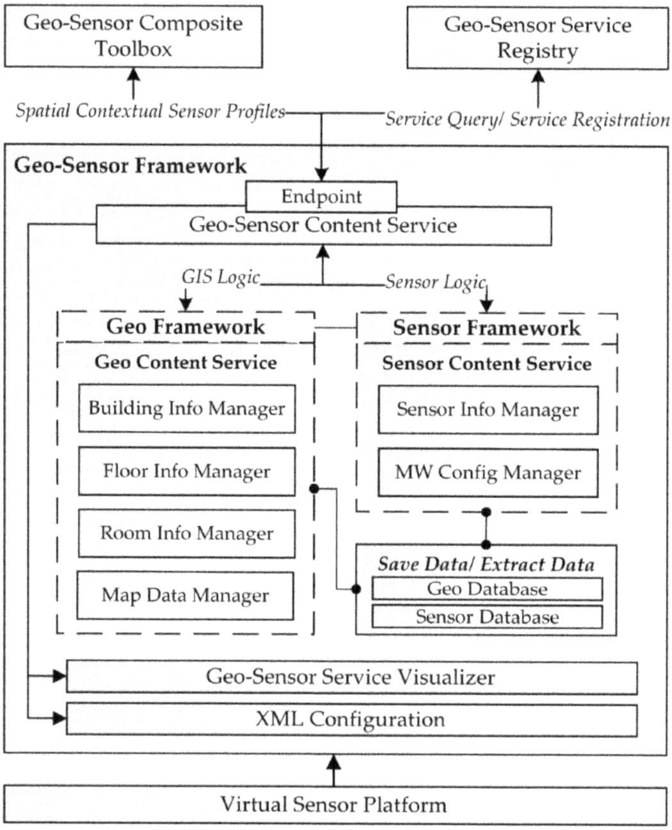

Figure 3. Proposed geo-sensor framework configuration.

Geo-sensor content service handles the deployment logic for the spatial context of the sensors in the sensor network. It then maps the spatial contextual information at the geo platform module in and sensor information at the sensor platform in order to generate an integrated geo-sensor platform for the given sensor network. The mapped information at the geo-sensor platform is processed at the geo-sensor composite toolbox for maintaining the spatial context-based sensor profile for the sensor networks. Once the spatial context-based profiles are completed, the information mapped onto the geo-sensor platform is ready for the client visualization client services' provision. The geo-sensor provider service at the geo-sensor platform handles the GIS service provision deployed at the geo platform sub-module and sensor service provision deployed at the sensor platform sub-module. The integrated logic of both the sub-modules enables the geo-sensor profile visualization and service provision to the client; for the spatial context-based sensor networks' profile queries (Figure 4).

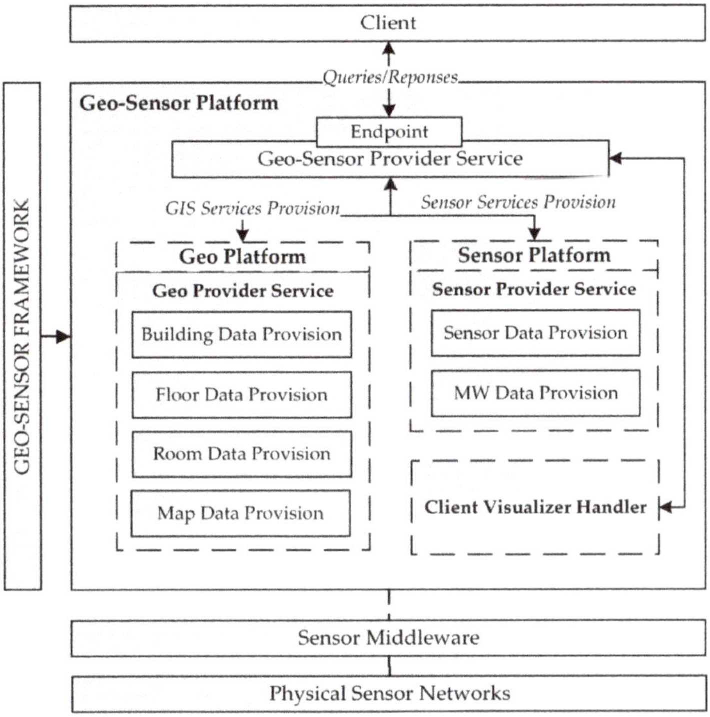

Figure 4. Proposed geo-sensor platform configurations.

Figure 5 depicts the flowchart for the generation of the geo-sensor platform. Firstly, the sensor network addition request is made by the client. Once the sensor addition request is made, the middleware allocation to the sensors involved in the network is done. The physical sensors ID mappings are done to the virtually generated IPs. Next, the spatial context profile for each sensor involved in the network is created. The spatial context-based profiles include the sensor model and type information as well as the sensor's geographical information. After creating the profiles, the content services are registered at the service registry, and, finally, the geo-sensor platform for the added sensor network is deployed, along with the services provision, at the client view.

Figure 6 shows the sequence diagram for geo-sensor service provision based geo-content service, sensor content service, geo provision service, and content provision service. The sensor data is passed onto the content service via sensor middleware. First, the middleware configurations are performed, and then sensor readings, after a certain interval, are sent to the sensor platform. The client query regarding sensor network is passed onto the sensor provision service under a geo-sensor platform, and the client query regarding the spatial context of the sensor is passed onto the geo provision service in the geo-sensor platform. The edit queries from the client are passed onto the content service, which interacts with the geo-sensor toolbox and performs the edit operations and sends a response back.

3.2. Geo-Sensor Composite Toolbox

In this sub-section, we present the geo-sensor composite toolbox module of the proposed system.

The geo-sensor composite toolbox has a vital role in the overall system. It provides the basic functionalities, which enable the client to add a sensor network along with its' geo-spatial context and also manages the network remotely. It has three functioning sub-modules as geo-sensor information manager, middleware information manager, and service information publishing (Figure 7).

The geo-sensor information manager generates the spatial contextual based sensor profile for the sensor network and creates the geo-sensor profile handler for the content services access in the geo-sensor framework. The middleware manager generates the middleware mapping information for each sensor platform and creates the access handler for the geo-sensor framework. Service information publishing generates the service scripts for each added sensor network.

A geo-sensor profile is composed of building information, floor information, and room information with marks on a map image representing the geographical location of the physically installed sensors. It also contains the physical sensor mappings for the virtual platforms and detailed sensor information. The geo-sensor profile provides the geo-sensor query functionalities as add geo-sensor network data, access geo-sensor network data, update geo-sensor network data, or delete geo-sensor network data.

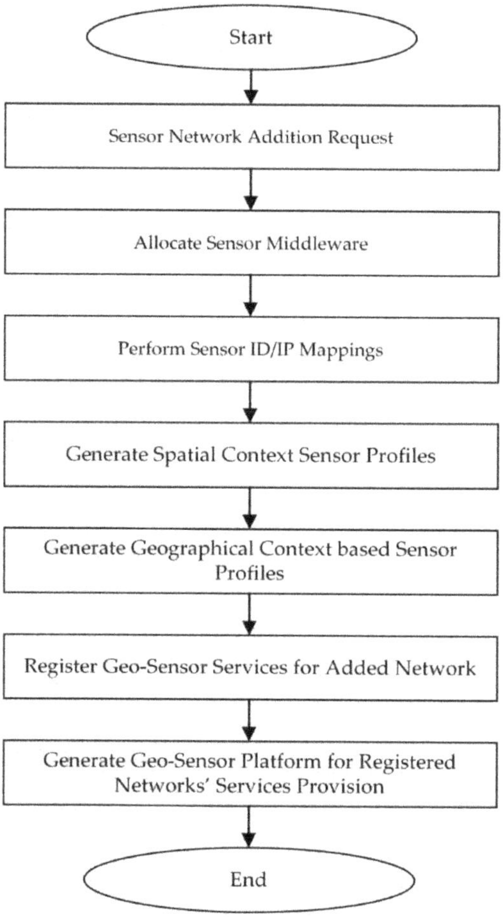

Figure 5. Geo-sensor platform's generation procedure flowchart.

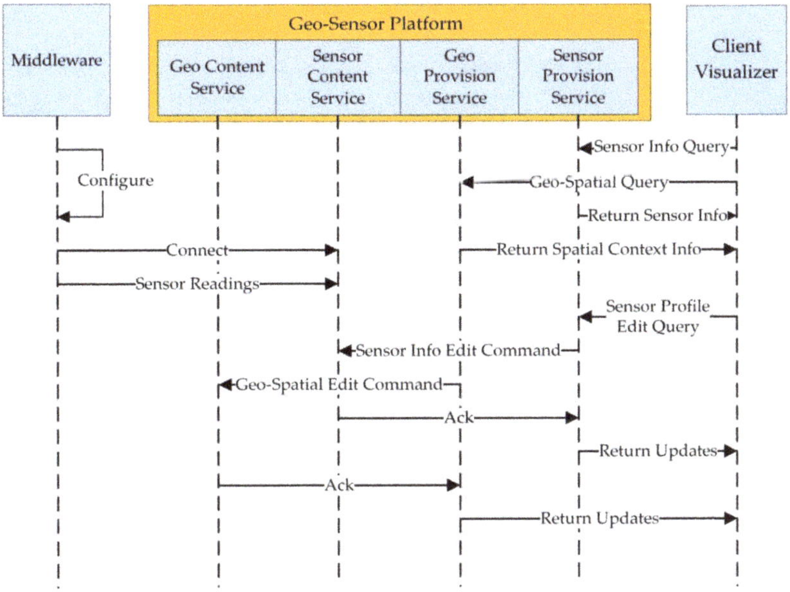

Figure 6. Geo-Sensor Service Provision Service Sequence Diagram.

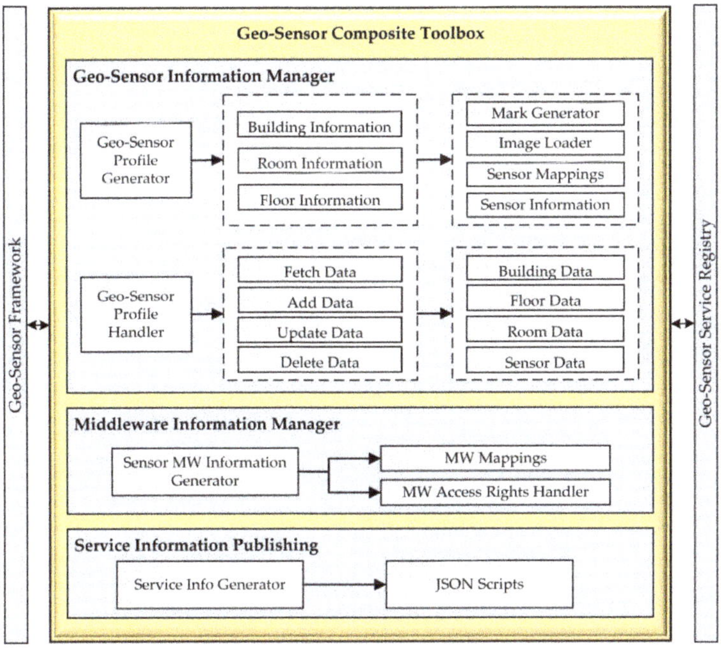

Figure 7. Geo-sensor composite toolbox configuration.

Figure 8 shows the sensor database entity relationship diagram. In the sensor database, we have sensor information, sensor type information, sensing data, sensing list, sensing category, and middleware information. Sensor_Information table saves sensor information. Sensing_List table saves the sensor's sensing type. Sensing_Category table saves detailed information about sensing type. Type_Information table saves the sensor's type. Sensing_Data table saves sensing data.

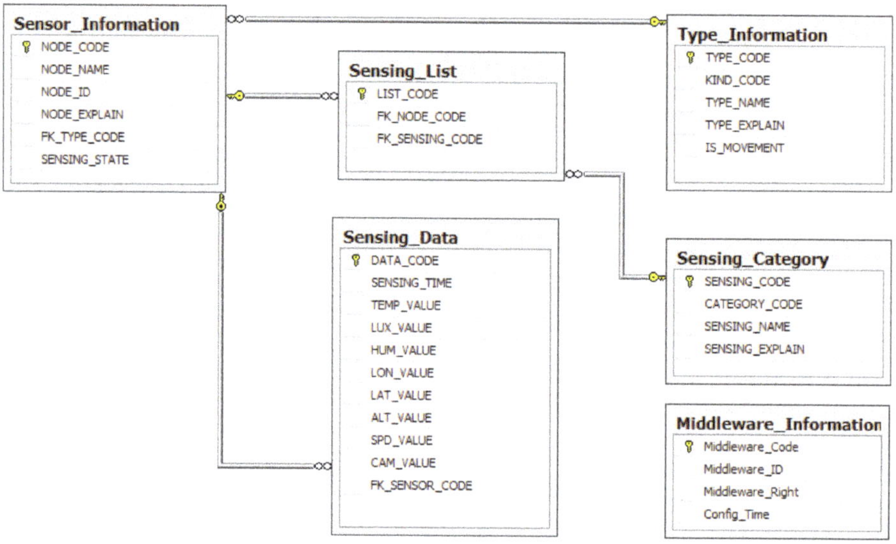

Figure 8. Sensor database relationship diagram.

Figure 9 shows the geo database entity relationship diagram. In the geo database, we have map information, map percent information, map data, building information, building mark, floor information, floor percent, room information, and room mark. Map_Informaiton table saves total map information (map size, minimap, minimap size, etc.). Percent_Information table saves the percent image information for the total map image (10%–100% reduced total map image size information). Map_Data table saves the total map image data. Building_Information table saves building information. Building_Mark table saves building area information. Floor_Information table saves floor map (map size, minimap, minimap size, etc.). Floor_Percent table saves the percent image information for floor map image (10%–100% reduced floor map image size information). Floor_Data table saves floor map data. Room_Information table saves room information. Room_Mark table saves room area information.

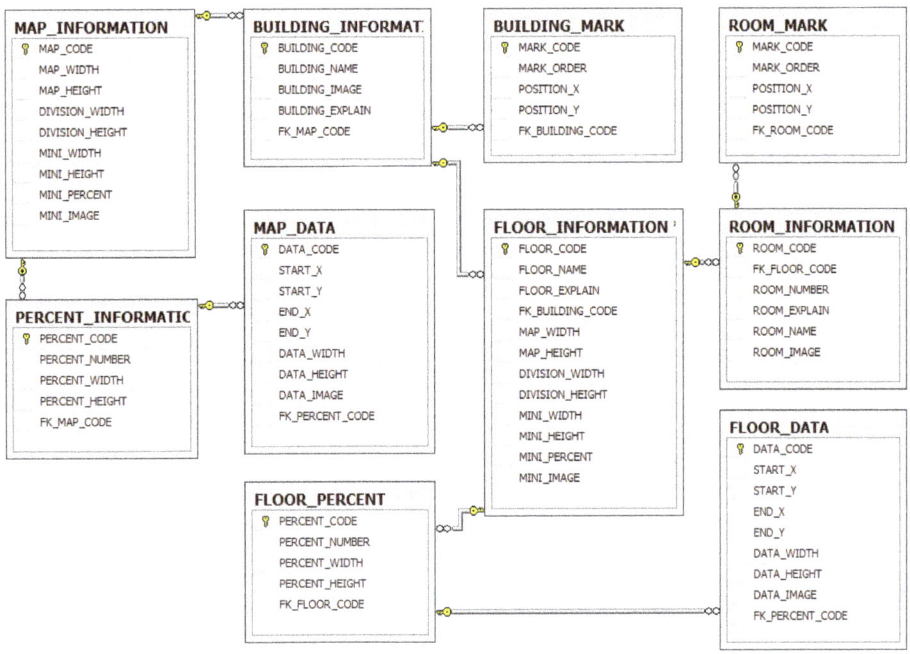

Figure 9. Geo database relationship diagram.

4. Implementation

In this section, we implement the proposed architecture for the geo-sensor framework and multi-platform deployment. The sub-sections below illustrate the implementation environment, APIs (Application Program Interface) developed for the platforms, system implementation output, and benchmarking environment.

4.1. Implementation Environment

The implementation environment is shown below in Table 1. We have implemented RESTful based web-services on the windows net framework and web-based application client.

Table 1. Implementation environment of this system.

Components	Version
Operation System	Microsoft Windows 10 pro (×64)
Microsoft Visual Studio	2015
Microsoft SQL Server Management Studio	2016
Intel System Studio IoT Edition	2017
Putty.exe	2014
Yocto Linux	2014
Intel Edison with Kit for Arduino	2014

4.2. Implemented System Output

In this sub-section, we present the implementation results of our proposed geo-sensor framework and platform deployment based on the geo-sensor composite toolbox.

In Figure 10a, the execution screen for the sensor platform's sub-module is presented, while in Figure 10b, the execution screen for the geo platform's sub-module is presented. In Figure 10c, the geo-sensor framework's deployed available geo-sensor platforms' list is given. Figure 10d shows the geo-sensor platform screen after the selection of the "computer department platform" in Figure 10c. The geo-sensor platform screen shown in Figure 10d has a full map view of the Jeju National University with a selected department highlighted to navigate through its floor for accessing to geo-sensor provision services.

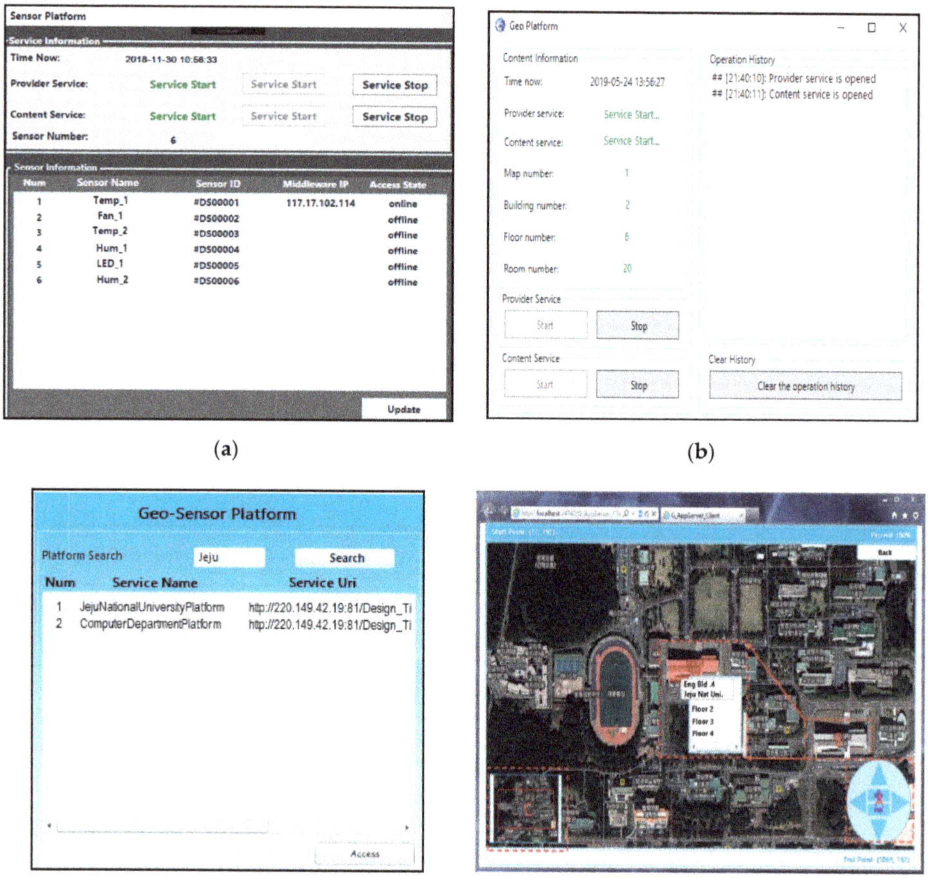

(c) (d)

Figure 10. Geo-sensor platform service initiation deployed by the geo-sensor framework: (**a**) sensor platform sub-module; (**b**) geo platform sub-module; (**c**) geo-sensor platforms view; (**d**) selected geo-sensor platform.

Once a geo-sensor platform is selected, as shown in Figure 10c, the client can either view its geographical based sensor details or head to the composition toolbox. Figure 11 shows the spatial context-based visualization of installed sensors at the selected geographical location. Once the client selects onto a specific sensor placed at a specific location onto the map, the client can view detailed sensor information, such as sensor name, sensor key, sensor location, sensor work state, and the sensor's current readings.

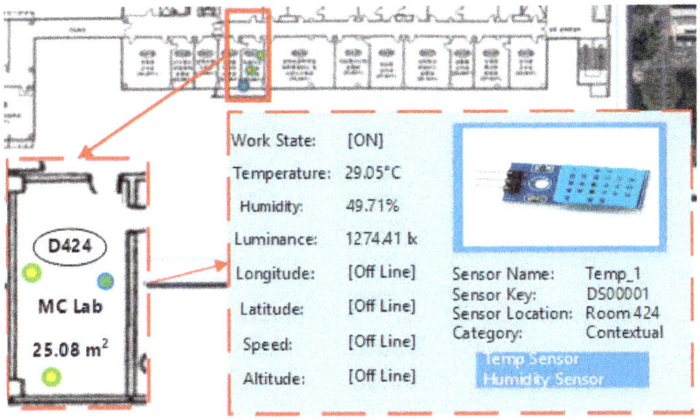

Figure 11. Geo-sensor visualization for Internet of Things (IoT) environments.

Figure 12 shows the geo-sensor composite toolbox execution screen for building information management (Figure 12a) and floor information management (Figure 12b). The composition toolbox enables services for making changes, such as the addition of new sensors network, deletion of sensor or map data, or updating of sensor or map data. Figure 12a shows the building information management screen where the client can get overall building information, such as building map information or the building's floor information. In order to navigate to a particular floor, the client would select the floor number from the building map, as shown in the highlighted region A in Figure 12a. After selecting the floor number, the client would navigate to the floor information management, as shown in Figure 12b. Within floor information management is detailed room-wise floor information along with available sensors in each room. Upon clicking a room, the client would get the option to bind a sensor onto the selected location. On clicking the binding sensor option, the client would navigate to the execution screen shown in Figure 12c. The client can bind a new sensor at the selected spot or delete/update an existing one.

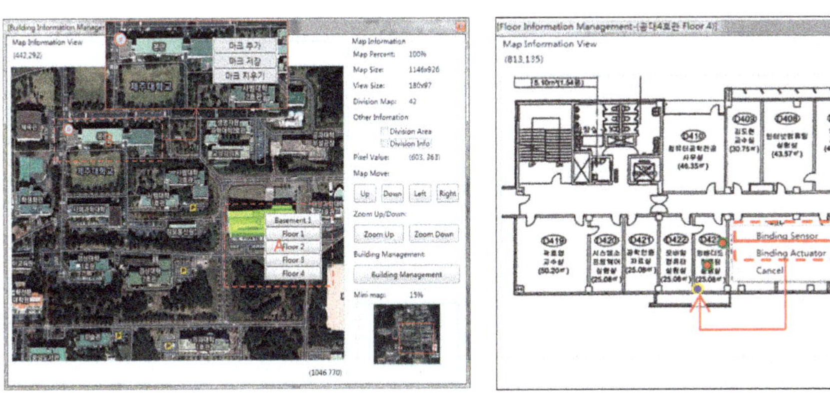

(a) (b)

Figure 12. *Cont.*

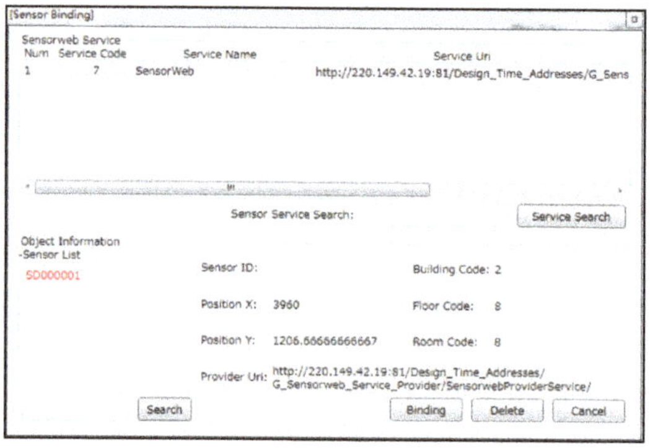

(c)

Figure 12. Geo-Sensor composite toolbox execution screen: (**a**) building information management; (**b**) floor information management; (**c**) sensor binding.

5. Analysis Results

In this section, we present the performance analysis and comparisons for our proposed geo-sensor framework for IoT environments. In Section 5.1, we present the analysis for the geo-sensor framework and geo-sensor toolbox performance in terms of response times, upload execution times for virtual objects (VOs) of the sensor, and geo-sensor services access response time.

We have implemented the proposed system based on both SOAP and RESTful web services, and in Section 5.2, we present the comparisons among the RESTful and SOAP web-services based implementations.

5.1. Geo-Sensor Framework Performance

In Figure 13, we evaluate the connection request response time from the geo-sensor framework for selected geo-sensor service visualization. The service visualization will load the geo-sensor network with a loading site map along with the sensors list, details, and available controls. In order to test the geo-sensor service access and visualization response time, multiple access attempts are made to the service. It takes around 231.74 ms to access the service visualization, which seems to be a reasonable amount of time for the proposed scenario.

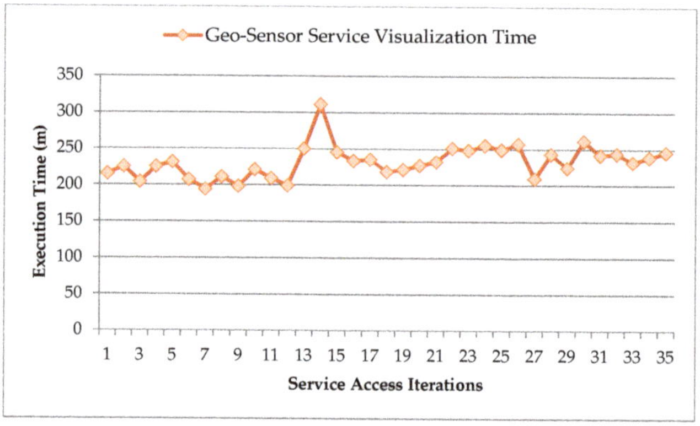

Figure 13. Execution time for geo-sensor service visualization and access request.

However, the tested services are for loading the mini-map visualization, sensors involved along with the details of each sensor, sensor images if available, and sensors' states. For viewing the total map, a separate request is made further.

In Figure 14, we compare the execution time results for the total map size loading process in comparison to the mini-map size loading process.

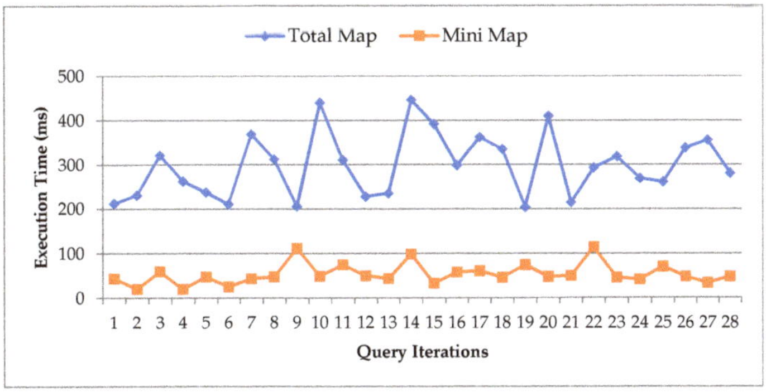

Figure 14. Execution time comparisons for loading of total map vs. mini-map.

Now, we test the response time for multiple services access by multiple users at the same time to evaluate the system performance under-load scenario. Figure 15 presents the comparison of the connection request response time made from multiple users to access multiple services at the same time. The service access and visualization have to be loaded for multiple requests simultaneously in this scenario. We have evaluated multiple geo-sensor service access with a varying number of requests and varying number of active services. In the graph below, we can clearly observe that as the load of access requests and active sessions increases, the delay in responses the system witnesses, and, hence, the response time also increases. In our performed evaluation scenarios, the maximum execution time under 51 requests and 10 active users turn out to be around 3880.63 ms, and the average response time for the system under varying load scenarios turn out to be 1558.16 ms. Hence, the system response time under the worst load scenarios have delays up to a maximum of around a little less than 4 s, and on average, with varying load scenarios, remains around 1.5 s. We can safely say that the overall system performance under varying loads is tolerable, considering system service load requirements.

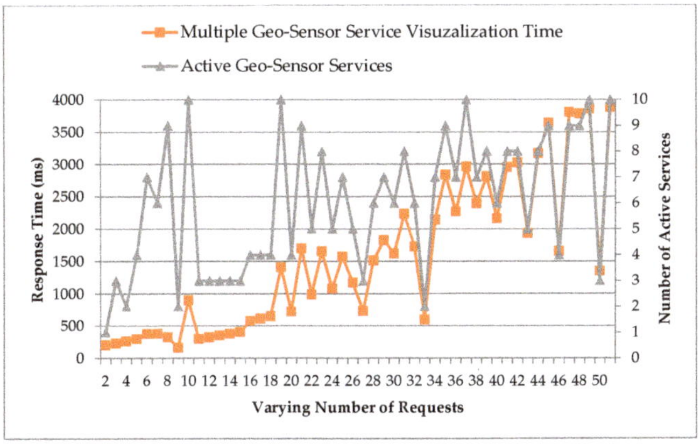

Figure 15. Multiple geo-sensor service access analysis with varying number of requests and varying number of active services.

5.2. Performance Comparisons for RESTful and SOAP Based Web Services Implementation

In this sub-section, we evaluate the performance of system operations executed based on RESTful web services in comparison to the execution based on SOAP web services.

In Figure 16, we present the comparison results for geo context-based sensor binding process between RESTful and SOAP implementations. It shows the time taken in order to bind a sensor object onto a geographical location in the specific position of the given map. We can observe a clear difference and decrease in the execution time taken by the RESTful based services as compared to SOAP, which proves the RESTful based implementation to be better in the given scenario.

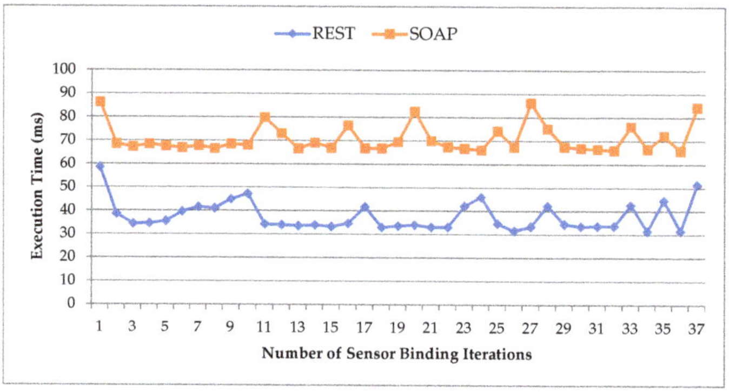

Figure 16. Execution time comparisons for geo-context based sensor binding.

In Figure 17, we present the map navigation query's response time comparisons based on pixel movement between the REST and SOAP implementations of the proposed framework. Each time the client navigates the map view, new blocks of maps are loaded depending on the move. It shows the comparison of the experiment results for loading the map move data from the geo sub-module and sensor data onto the moved location from the sensor sub-module. The data from the geo sub-module and the sensor sub-module are loaded via the geo-sensor platform and deployed by the geo-sensor framework. Every time, both the protocols, SOAP and REST, are tested with the same interval of move operation as a move of 100 pixels, 150 pixels, 200 pixels, 250 pixels, 300 pixels, 350 pixels, and 400 pixels. The results in response to each move operation for both the web services implementations are recorded in terms of the response time by the geo-sensor platform. The results clearly show that REST based web services outperform the SOAP based web services.

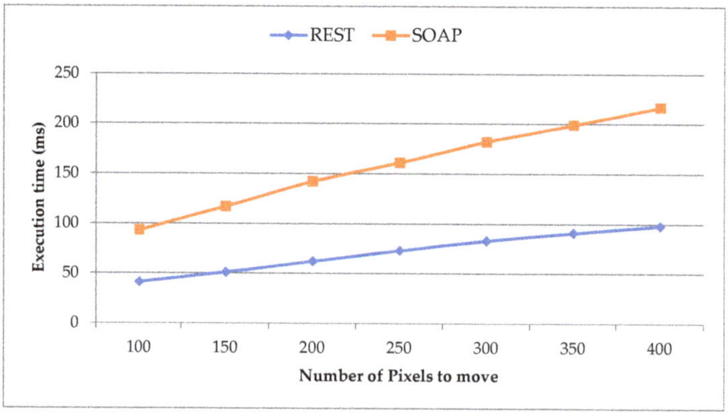

Figure 17. Map navigation query based on pixels move request

In this sub-section, we can conclude that REST is better in performance in comparison to SOAP for our application. The REST web services perform better through caching the information, and REST also allows the use of JSON, which provides faster data parsing. With JSON, REST offers better service to the browser uses. Also, our proposed framework is web-based and uses JSON. Table 2 provides a comparison between REST and SOAP web services.

Table 2. SOAP web services vs. REST web services.

Work	SOAP	REST
Style	Protocol	Architectural style
Function	Function driven	Data driven
Data Format	XML	Diverse data formats (e.g., JSON, plain text, HTML, XML)
Security	WS-Security and SSL	SSL and HTTPS
Bandwidth	More resources	Fewer resources
Data Cache	No	Yes

6. Discussion and Conclusions

In this work, we first study the geo-sensor networks and geo-sensor services and present a user flexible DIY geo-sensor framework, which can be used for virtually uploading the geo-sensor networks' objects and services in the cyber world. The proposed work first focuses on the design of a geo-sensor framework based on multiple sensor platforms and geo platforms instances for the deployment of multiple geo-sensor services to the geo-sensor framework. Every time a user wishes to upload a geo-sensor network to the geo-sensor framework, the user gets dedicated instances of the sensor platform and geo platform for uploading of the geo-sensor network and geo-sensor services. Once uploaded, the authorized users can access the available geo-sensor services via the geo-sensor platform. The proposed framework provides real-time context data management using the geo-sensor composite toolbox. Using such a setup, it is possible to easily add other systems providing sensor nodes and map the information to the geo-sensor platform in order to make the geo-sensor networks easily accessible and manageable.

Many works and projects have been done in geo-sensing, mainly focusing on the goal of collecting real-time spatial and environmental data. Most of these studies relied on volunteers for the data collection process [65,66]. The people-centric sensing involved the data collection done by people carrying mobile sensors to various locations [67]. Many of the proposed approaches were one way or the other, solely for the benefit of specific entities, companies, or organizations trying to carry out research of data collection for their specific purposes. Zhang et al. [11] pointed out this fact and started a DIY project (CUSS) with an aim to involve people in not only the data collection process but also in the decision-making process based on the collected data. The project CUSS is in its initial stages and has limitations in its definitions and applications. The users in CUSS are also not provided with the facilities of creating their own geo-sensor services and managing them via a geo-sensing composite toolbox, instead, the CUSS is currently more focused on data collection by DIY VOs created by users. It provides a sensing visualization for a smart city's overall environmental picture and smart city decisions. Whereas, we provide a framework that users can use to create their own scenario-based virtual geo-sensor networks by binding the objects as VOs and service creation. Table 3 shows the comparisons of our deployed framework and CUSS.

Table 3. Operations and services provision comparison from user-level.

Work	Bind Geo-Sensor VOs	Geo-Sensor Service Creation	Geo-Sensor Service Management	Geo-Sensor Service Sharing	DIY
Proposed Geo-Sensor Framework	Yes	Yes	Yes	Yes	Yes
CCUS [12]	Limited	Limited	No	Limited	Yes

Further, in this work, we have implemented the geo-sensor services based on the RESTful protocol and SOAP protocol and drawn comparisons among both the web services performance in terms of various operation response times. In Table 4, we present the average performance result comparisons between the REST web services and SOAP web services, and we also take out the average difference between the performances of both services in terms of execution time for various geo-sensor operations evaluated above, such as sensor binding, map navigation, map viewer, and map management. Map management includes the operations of map loading, building map actions, floor map actions, map zoom in, and map zoom out. We can clearly observe from the results that the RESTful web services are better suited and efficient for the deployment of such frameworks with multiple platforms.

Table 4. Average performance results comparisons between REST and SOAP.

Geo-Sensor Operation	Avg. REST Execution (ms)	Avg. SOAP Execution (ms)	Avg. Difference (ms)
Sensor Binding	37.71	70.76	33.05
Map Navigation	71.28	158.71	87.42
Map Viewer	98.1	167.72	69.62
Map Management	65.07	115.08	50.01

Author Contributions: S.M. and D.K. designed the geo-sensor system. S.M. implemented the system and performed experiments and testing. S.M. and D.K. wrote this paper.

Funding: This research was supported by the MSIT(Ministry of Science and ICT), Korea, under the ITRC(Information Technology Research Center) support program(IITP-2019-2014-1-00743) supervised by the IITP(Institute for Information & communications Technology Planning & Evaluation), and This work was supported by Institute for Information & communications Technology Planning & Evaluation(IITP) grant funded by the Korea government(MSIT) (No.2019-0-01456, AutoMaTa: Autonomous Management framework based on artificial intelligent Technology for adaptive and disposable IoT). Any correspondence related to this paper should be addressed to DoHyeun Kim.

Conflicts of Interest: The authors declare no conflict of interest.

References

1. Ayday, C.; Safak, S. Application of wireless sensor networks with GIS on the soil moisture distribution mapping. In Proceedings of the Symposium GIS Ostrava, Ostrava, Czech Republic, 25–28 January 2009.
2. Aldasouqi, I.; Atoum, J. Stream Processing Environmental Applications in Jordan Valley. *Int. J. Comput. Sci. Secur. (IJCSS)* **2011**, *5*, 1.
3. Eddy, N. Gartner: 21 Billion IoT Devices to Invade by 2020—InformationWeek. Available online: http://www.informationweek.com/mobile/mobile-devices/gartner-21-billion-iot-devices-toinvade-by-2020/d/d-id/1323081 (accessed on 1 October 2019).
4. De Roeck, D.; Slegers, K.; Criel, J.; Godon, M.; Claeys, L.; Kilpi, K.; Jacobs, A. I would DiYSE for it!: A manifesto for do-it-yourself internet-of-things creation. In Proceedings of the 7th Nordic Conference on Human-Computer Interaction: Making Sense through Design, Copenhagen, Denmark, 4–17 October 2012; pp. 170–179.

5. Heath, N. How IBM's Node-RED Is Hacking Together the Internet of Things. Available online: http://www.techrepublic.com/article/node-red/ (accessed on 1 October 2019).
6. Quoc, H.N.M.; Serrano, M.; Le-Phuoc, D.; Hauswirth, M. Super stream collider–linked stream mashups for everyone. In Proceedings of the SemanticWeb Challenge Co-Located with ISWC2012, Boston, MA, USA, 11–15 November 2012.
7. Mazzei, D.; Montelisciani, G.; Fantoni, G.; Baldi, G. Internet of Things for designing smart objects. In Proceedings of the 2014 IEEE World Forum on Internet of Things (WF-IoT), Seoul, Korea, 6–8 March 2014; pp. 293–297.
8. Pachube—The Internet of Things Real-TimeWeb Service and Applications. Available online: http://www.appropedia.org/Pachube (accessed on 1 October 2019).
9. Kefalakis, N.; Soldatos, J.; Anagnostopoulos, A.; Dimitropoulos, P. A Visual Paradigm for IoT Solutions Development. In *Model and Data Engineering*; Springer Science and Business Media LLC: Berlin, Germany, 2015; Volume 9001, pp. 26–45.
10. Lee, H.; Sin, D.; Park, E.; Hwang, I.; Hong, G.; Shin, D. Open software platform for companion IoT devices. In Proceedings of the 2017 IEEE International Conference on Consumer Electronics (ICCE), Las Vegas, NV, USA, 8–10 January 2017; pp. 394–395.
11. Zhang, Z.; Mondschein, A.; El Khafif, M. Human-Tool Assemblage: Designers in the Big Data World. *J. Digit. Landsc. Archit.* **2018**, *3*, 397–405. [CrossRef]
12. Dey, A.K. Understanding and using context. *Pers. Ubiquitous Comput.* **2001**, *5*, 4–7. [CrossRef]
13. Hong, J.; Suh, E.H.; Kim, J.; Kim, S. Context-aware system for proactive personalized service based on context history. *Expert Syst. Appl.* **2009**, *36*, 7448–7457. [CrossRef]
14. Lukowicz, P.; Choudhury, T.; Gellersen, H. Beyond context awareness. *IEEE Pervasive Comput.* **2011**, *10*, 15–17. [CrossRef]
15. Schmidt, A.; Beigl, M.; Gellersen, H.W. There is more to context than location. *Comput. Graph.* **1999**, *23*, 893–901. [CrossRef]
16. Sagl, G.; Blaschke, T.; Beinat, E.; Resch, B. Ubiquitous Geo-Sensing for Context-Aware Analysis: Exploring Relationships between Environmental and Human Dynamics. *Sensors* **2012**, *12*, 9800–9822. [CrossRef]
17. Cools, M.; Moons, E.; Creemers, L.; Wets, G. Changes in Travel Behavior in Response to Weather Conditions. *Transp. Res. Rec. J. Transp. Res. Board* **2010**, *2157*, 22–28. [CrossRef]
18. Merbitz, H.; Buttstädt, M.; Michael, S.; Dott, W.; Schneider, C. GIS-based identification of spatial variables enhancing heat and poor air quality in urban areas. *Appl. Geogr.* **2012**, *33*, 94–106. [CrossRef]
19. Sagl, G.; Resch, B.; Blaschke, T. Contextual Sensing: Integrating Contextual Information with Human and Technical Geo-Sensor Information for Smart Cities. *Sensors* **2015**, *15*, 17013–17035. [CrossRef]
20. Chang, N.B.; Guo, D.H. Urban flash flood monitoring, mapping and forecasting via a tailored sensor network system. In Proceedings of the IEEE Interantional Conference on Networking, Sensing and Control 2006 (ICNSC'06), Ft. Lauderdale, FL, USA, 23–25 April 2006; pp. 757–761.
21. Horita, F.E.; De Albuquerque, J.P.; Degrossi, L.C.; Mendiondo, E.M.; Ueyama, J. Development of a spatial decision support system for flood risk management in Brazil that combines volunteered geographic information with wireless sensor networks. *Comput. Geosci.* **2015**, *80*, 84–94. [CrossRef]
22. Mead, M.; Popoola, O.; Stewart, G.; Landshoff, P.; Calleja, M.; Hayes, M.; Baldovi, J.; McLeod, M.; Hodgson, T.; Dicks, J.; et al. The use of electrochemical sensors for monitoring urban air quality in low-cost, high-density networks. *Atmos. Environ.* **2013**, *70*, 186–203. [CrossRef]
23. Resch, B.; Blaschke, T.; Mittlboeck, M. Live geography: Interoperable geo-sensor webs facilitating the vision of digital earth. *Int. J. Adv. Netw. Serv.* **2010**, *3*, 323–332.
24. Resch, B.; Mittlboeck, M.; Girardin, F.; Britter, R.; Ratti, C. Live geography—Embedded sensing for standardised urban environmental monitoring. *Int. J. Adv. Syst. Meas.* **2009**, *2*, 15–167.
25. De Nazelle, A.; Seto, E.; Donaire-González, D.; Mendez, M.; Matamala, J.; Nieuwenhuijsen, M.J.; Jerrett, M. Improving estimates of air pollution exposure through ubiquitous sensing technologies. *Environ. Pollut.* **2013**, *176*, 92–99. [CrossRef]
26. Brienza, S.; Galli, A.; Anastasi, G.; Bruschi, P. A Low-Cost Sensing System for Cooperative Air Quality Monitoring in Urban Areas. *Sensors* **2015**, *15*, 12242–12259. [CrossRef]

27. Devaraju, A.; Kauppinen, T. Sensors Tell More than They Sense: Modeling and Reasoning about Sensor Observations for Understanding Weather Events. *Int. J. Sens. Wirel. Commun. Control* **2012**, *2*, 14–26. [CrossRef]
28. Muller, C.L.; Chapman, L.; Grimmond, C.S.B.; Young, D.T.; Cai, X. Sensors and the city: A review of urban meteorological networks. *Int. J. Clim.* **2013**, *33*, 1585–1600. [CrossRef]
29. Elen, B.; Peters, J.; Poppel, M.; Bleux, N.; Theunis, J.; Reggente, M.; Standaert, A.; Van Poppel, M. The Aeroflex: A Bicycle for Mobile Air Quality Measurements. *Sensors* **2012**, *13*, 221–240. [CrossRef]
30. Nakamura, T.; Kikuya, Y.; Arakawa, Y.; Nakamura, M.; Higashijima, Y.; Maruo, Y.Y.; Nakamura, M. Proposal of web framework for ubiquitous sensor network and its trial application using NO2 sensor mounted on bicycle. In Proceedings of the 2012 IEEE/IPSJ 12th International Symposium, Izmir, Turkey, 16–20 July 2012; pp. 83–90.
31. Bossche, J.V.D.; Peters, J.; Verwaeren, J.; Botteldooren, D.; Theunis, J.; De Baets, B. Mobile monitoring for mapping spatial variation in urban air quality: Development and validation of a methodology based on an extensive dataset. *Atmos. Environ.* **2015**, *105*, 148–161. [CrossRef]
32. Peters, J.; Bossche, J.V.D.; Reggente, M.; Van Poppel, M.; De Baets, B.; Theunis, J. Cyclist exposure to UFP and BC on urban routes in Antwerp, Belgium. *Atmos. Environ.* **2014**, *92*, 31–43. [CrossRef]
33. Thuro, K.; Singer, J.; Festl, J. *A Geosensor Network Based Monitoring and Early Warning System for Landslides*; Springer Science and Business Media LLC: Berlin, Germany, 2013; pp. 79–86.
34. Chuli, H.; Nengcheng, C. Geospatial sensor web for smart disaster emergency processing. In Proceedings of the 19th International Conference on Geoinformatics 2011, Shanghai, China, 24–26 June 2011; pp. 1–5.
35. Essa, I. Ubiquitous sensing for smart and aware environments. *IEEE Wirel. Commun.* **2000**, *7*, 47–49. [CrossRef]
36. Mann, W.; El-Zabadani, H.; Kaddoura, Y.; Helal, S.; King, J.; Jansen, E. The Gator Tech Smart House: A programmable pervasive space. *Computer* **2005**, *38*, 50–60.
37. Munoz, A.; Serrano, E.; Villa, A.; Valdés, M.; Botia, J.A. An Approach for Representing Sensor Data to Validate Alerts in Ambient Assisted Living. *Sensors* **2012**, *12*, 6282–6306. [CrossRef]
38. Kim, S.C.; Jeong, Y.S.; Park, S.O. Rfid-based indoor location tracking to ensure the safety of the elderly in smart home environments. *Pers. Ubiquitous Comput.* **2013**, *17*, 1699–1707. [CrossRef]
39. Palumbo, F.; Ullberg, J.; Štimec, A.; Furfari, F.; Karlsson, L.; Coradeschi, S. Sensor Network Infrastructure for a Home Care Monitoring System. *Sensors* **2014**, *14*, 3833–3860. [CrossRef]
40. Forkan, A.; Khalil, I.; Tari, Z. Cocamaal: A cloud-oriented context-aware middleware in ambient assisted living. *Future Gener. Comput. Syst.* **2014**, *35*, 114–127. [CrossRef]
41. Garcia, N.M.; Rodrigues, J.J.P. *Ambient Assisted Living*; CRC Press: Boca Raton, FL, USA, 2015.
42. Bisio, I.; Lavagetto, F.; Marchese, M.; Sciarrone, A. Smartphone-centric ambient assisted living platform for patients suffering from co-morbidities monitoring. *IEEE Commun. Mag.* **2015**, *53*, 34–41. [CrossRef]
43. Arnrich, B.; Mayora, O.; Bardram, J.; Tröster, G. Pervasive healthcare. *Methods Inf. Med.* **2010**, *49*, 67–73. [CrossRef]
44. Keh, H.C.; Shih, C.C.; Chou, K.Y.; Cheng, Y.C.; Ho, H.K.; Yu, P.Y.; Huang, N.C. Integrating unified communications and internet of m-health things with micro wireless physiological sensors. *J. Appl. Sci. Eng.* **2014**, *17*, 319–328.
45. Mukhopadhyay, S.; Postolache, O.A. *Pervasive and Mobile Sensing and Computing for Healthcare*; Springer: Berlin/Heidelberg, Germany, 2014.
46. Schade, S.; Díaz, L.; Ostermann, F.; Spinsanti, L.; Luraschi, G.; Cox, S.; Nuñez, M.; de Longueville, B. Citizen-based sensing of crisis events: Sensor web enablement for volunteered geographic information. *Appl. Geomat.* **2013**, *5*, 3–18. [CrossRef]
47. Poser, K.; Dransch, D. Volunteered geographic information for disaster management with application to rapid flood damage estimation. *Geomatica* **2010**, *64*, 89–98.
48. Hayes, J.; Stephenson, M. Bridging the social and physical sensing worlds: Detecting coverage gaps and improving sensor networks. In Proceedings of the First Workshop on Pervasive Urban Applications (PURBA) in Conjunction with the Ninth International Conference on Pervasive Computing, San Francisco, CA, USA, 12–15 June 2011.

49. Tamilin, A.; Carreras, I.; Ssebaggala, E.; Opira, A.; Conci, N. Context-aware mobile crowdsourcing. In Proceedings of the 2012 ACM Conference on Ubiquitous Computing, Pittsburgh, PA, USA, 5–8 September 2012; pp. 717–720.
50. Roitman, H.; Mamou, J.; Mehta, S.; Satt, A.; Subramaniam, L.V. Harnessing the crowds for smart city sensing. In Proceedings of the 1st International Workshop on Multimodal Crowd Sensing, Maui, HI, USA, 29 October–2 November 2012; pp. 17–18.
51. Koch, F.; Cardonha, C.; Gentil, J.M.; Borger, S. A Platform for Citizen Sensing in Sentient Cities. In *Computer Vision—ECCV 2012*; Springer Science and Business Media LLC: Berlin, Germany, 2013; Volume 7685, pp. 57–66.
52. Doran, D.; Gokhale, S.; Dagnino, A. Human sensing for smart cities. In Proceedings of the 2013 IEEE/Acm International Conference on Advances in Social Networks Analysis and Mining, Niagara, ON, Canada, 25–28 August 2013; pp. 1323–1330.
53. Liao, C.C.; Hou, T.F.; Lin, T.Y.; Cheng, Y.J.; Erbad, A.; Hsu, C.H.; Venkatasubramania, N. Sais: Smartphone augmented infrastructure sensing for public safety and sustainability in smart cities. In Proceedings of the 1st International Workshop on Emerging Multimedia Applications and Services for Smart Cities, Orlando, FL, USA, 3–7 November 2014; pp. 3–8.
54. Evchina, Y.; Puttonen, J.; Dvoryanchikova, A.; Lastra, J.L.M. Context-aware knowledge-based middleware for selective information delivery in data-intensive monitoring systems. *Eng. Appl. Artif. Intell.* **2015**, *43*, 111–126. [CrossRef]
55. Pantelopoulos, A.; Bourbakis, N.G. A survey on wearable sensor-based systems for health monitoring and prognosis. *IEEE Trans. Syst. Man Cybern. Part C Appl. Rev.* **2010**, *40*, 1–12. [CrossRef]
56. Swan, M. Sensor mania! The internet of things, wearable computing, objective metrics, and the quantified self 2.0. *J. Sens. Actuator Netw.* **2012**, *1*, 217–253. [CrossRef]
57. Resch, B.; Summa, A.; Sagl, G.; Zeile, P.; Exner, J.P. Urban emotions—Geo-semantic emotion extraction from technical sensors, human sensors and crowdsourced data. In *Progress in Location-Based Services 2014*; Springer: Cham, Germany, 2015; pp. 199–212.
58. Garbarino, M.; Lai, M.; Bender, D.; Picard, R.W.; Tognetti, S. Empatica e3—A wearable wireless multi-sensor device for real-time computerized biofeedback and data acquisition. In Proceedings of the 4th International Conference on Wireless Mobile Communication and Healthcare (Mobihealth), Athens, Greece, 3–5 November 2014; pp. 39–42.
59. Resch, B.; Sudmanns, M.; Sagl, G.; Summa, A.; Zeile, P.; Exner, J.P. Crowdsourcing Physiological Conditions and Subjective Emotions by Coupling Technical and Human Mobile Sensors. *GI_Forum* **2015**, *1*, 514–524. [CrossRef]
60. Batelaan, O.; Wang, Z.M.; De Smedt, F. An adaptive GIS toolbox for hydrological modelling. *IAHS Publ. Ser. Proc. Rep. Intern. Assoc. Hydrol. Sci.* **1996**, *235*, 3–10.
61. Vandergast, A.G.; Perry, W.M.; Lugo, R.V.; Hathaway, S.A. Genetic landscapes GIS Toolbox: Tools to map patterns of genetic divergence and diversity. *Mol. Ecol. Resour.* **2011**, *11*, 158–161. [CrossRef]
62. Benz, U.C.; Hofmann, P.; Willhauck, G.; Lingenfelder, I.; Heynen, M. Multi-resolution, object-oriented fuzzy analysis of remote sensing data for GIS-ready information. *ISPRS J. Photogramm. Remote Sens.* **2004**, *58*, 239–258. [CrossRef]
63. Weng, Q. Land use change analysis in the Zhujiang Delta of China using satellite remote sensing, GIS and stochastic modelling. *J. Environ. Manag.* **2002**, *64*, 273–284. [CrossRef]
64. Weng, Q. A remote sensing–GIS evaluation of urban expansion and its impact on surface temperature in the Zhujiang Delta, China. *Int. J. Remote Sens.* **2001**, *22*, 1999–2014.
65. Cuff, D.; Hansen, M.; Kang, J. Urban Sensing: Out of the Woods. *Commun. ACM* **2008**, *51*, 24. [CrossRef]
66. Martino, M.; Britter, R.; Outram, C.; Zacharias, C.; Biderman, A.; Ratti, C. Senseable City. Available online: http://www.mamartino.com/images/senseable_city.pdf (accessed on 19 November 2019).
67. Gabrys, J. Programming Environments: Environmentality and Citizen Sensing in the Smart City. *Environ. Plan. D Soc. Space* **2014**, *32*, 30–48. [CrossRef]

© 2019 by the authors. Licensee MDPI, Basel, Switzerland. This article is an open access article distributed under the terms and conditions of the Creative Commons Attribution (CC BY) license (http://creativecommons.org/licenses/by/4.0/).

applied sciences

MDPI

Review

Review of GIS-Based Applications for Mining: Planning, Operation, and Environmental Management

Yosoon Choi *, Jieun Baek and Sebeom Park

Department of Energy Resources Engineering, Pukyong National University, Busan 48513, Korea; bje0511@gmail.com (J.B.); sebumi1v@gmail.com (S.P.)
* Correspondence: energy@pknu.ac.kr; Tel.: +82-51-629-6562; Fax: +82-51-629-6553

Received: 28 February 2020; Accepted: 20 March 2020; Published: 26 March 2020

Abstract: In this study, geographic information system (GIS)-based methods and applications utilized for mine development were reviewed. Three types of GIS-based studies, namely studies on mine planning, operation, and environmental management, were examined to describe the role of GIS as a decision-making support tool in mine development. This review was conducted by classifying previous GIS-based studies into several subtopics that pertain to mine development activities and the range of environments to be managed. Because the use of GIS is appropriate for spatial data management related to ore deposits and mine environment conditions at various scales, the applications of GIS-based methods in mine development could be expanded further.

Keywords: Geographic information systems; mine planning; mine development; mine operation; environmental management; mine reclamation

---

1. Introduction

With global concerns regarding the exhaustion of high-quality ore deposits near the surface, the risks related to mine development are increasing in terms of economic, technical, and environmental considerations [1]. In detail, a drop in both mineral production and profits for mining companies has resulted in the introduction of large heavy machines and advanced mining technologies to mine low-quality ore deposits at greater depths. Moreover, as mine-induced disasters and hazards have become difficult to properly monitor and control, a negative impact on work safety and the surrounding environment has been evident. Consequently, the necessity for systematic and strategic mine planning, operation, and environmental management based on optimization techniques or decision-making systems are emerging to improve mineral productivity, operational efficiency, and stability in the mine environment.

To accomplish these objectives, geographic information system (GIS)-based methodologies have been effectively used to design and optimize the mine development process. GIS is a comprehensive framework that is used to capture, store, query, analyze, and display spatial data in various applications [2]. In the field of geoscience, various applications have been proposed to manage and control natural phenomena using GIS, and literature reviews on GIS applications have been actively conducted [3–8]. However, in the mining sector, although GIS has been widely applied for mine development, no attention has been paid to evaluating the GIS-based methods and applications pertaining to mine planning, operation, or environmental management.

Thus, the purpose of this study was to investigate and review the GIS-based methods and applications proposed for designing and optimizing mine planning, operations, and environmental management. The scope of this study was confined to relevant published studies in these areas. Keywords (i.e., GIS, mining, planning or design or operation, or environmental management) were input into the Google Scholar website, and 68 articles were identified for the literature review. Finally, 58 articles were selected after consideration of the type of article, similarity with other articles, relevance to scope, and content details.

This paper is organized into six sections. Section 1 introduces mine activities according to the mine development phase. Section 2 describes the GIS-based methods and applications for mine planning. Section 3 presents the literature covering various mine operation systems using GIS. Section 4 reviews the GIS-based methods for mine environmental management and reclamation planning. Section 5 presents the advantages and disadvantages of using GIS in mining and suggests future directions for GIS. Finally, the conclusions are presented in Section 6.

2. Mine Development Phase

Generally, the mine development phase consists of mine planning, construction, operation, and reclamation [9]. In the planning phase, the mineralized target is found through geophysical and geochemical exploration. In the construction and operation phases, infrastructure is built for mineral production, and ore materials are extracted using heavy equipment. Finally, in the reclamation phase, all undesirable materials, such as waste, tailings, and contaminated topsoils, are removed from the mining area, and the underground area is properly filled with rock materials.

To facilitate investment decisions, feasibility studies are performed by considering the technical, economic, and environmental aspects of mine development in the planning phase [10]. The scale and grade of mineral deposits are estimated using the results of computer-aided modeling. The appropriate mining method is selected, and the open-pit boundary is optimized to quantify the volume and amount of ore and waste. Additionally, the mine infrastructure is designed, and potential conflicts are identified to prevent political, legal, and environmental problems. Finally, ore resources and reserves, which refer to the amount and grade of the minable ore body, can be determined.

When the mine development plan is approved, an open-pit or underground mine is constructed, and haulage roads, benches, and processing plants are built in the mining area. Various optimization techniques and decision-making systems are applied to optimize the mineral production and haulage operations. Furthermore, wireless communication systems can play a major role in tracking personnel and equipment and in monitoring the working environment from a remote office. Various management systems related to the underground ventilation network, safety management, and drainage system can be implemented to prevent disasters caused by noxious gases, dust, fires, or water; they can also alert workers about the need to escape, if necessary.

The environmental impacts on the mining area should be monitored and controlled in both the mine operation and reclamation phases [11]. In the mine operation phase, dust, noise, and vibration mainly occurred as a result of drilling and blasting activities. Moreover, slope failure and erosion problems are a frequent occurrence in mining sites. After the mining operation, various mining-induced hazards, including subsidence, water pollution, soil contamination, and deforestation, can occur in mining areas [12]. Therefore, it is crucial to continuously monitor the movement of drainage flows, metal leaching, and sediments, and establish a database framework for effective hazard management. Furthermore, systematic mine rehabilitation planning may prevent and minimize risks to the mining environment and human health.

3. Mine Planning Using GISs

GIS-based research on mine planning is generally classified into four topics: ore reserve estimation, open pit optimization, mine infrastructure design, and potential conflict region analysis.

3.1. Ore Reserve Estimation

In the ore reserve estimation phase, GIS-based technologies, such as spatial analysis and targeting, exposition, 3D data visualization, and processing, can be utilized to identify and assess the location, size, and geometry of mineral deposits buried underground.

Sprague et al. [13] introduced a query framework for spatial targeting within a 3D GIS software environment. They developed query tools that can be used as a GIS-3D plug-in, which included a proximity query (to a "probe" object), property query (numeric attribute), shell query (containment within a closed surface), meta-data query, feature query (dome, depression, curvature), trend query (dip plane, vector), and intersection query (with a "probe" object). The developed plug-in also demonstrated the usefulness of software extensions to GOCAD©. The 3D GIS capabilities of GOCAD© facilitate queries and easy attribute representation, and support spatial targeting to meet mining and mineral exploration needs.

Uygucgil and Konuk [14] integrated geostatistics with GIS capabilities, such as spatial analysis, exposition, and 3D data visualization and processing. In this study, GIS-based visual and interactive analyses were used to examine the relationship between the data in the study area and the total data in the integrated spatial database. A case study was conducted in Turkey to evaluate the accuracy of kriging, a geostatistical method for estimating the shape, spatial location, and volume of open mines. The accuracy of each of the three normal co-kriging estimations was assessed to determine which method was the most accurate. Consequently, the estimate with three variables appears to be more accurate than that calculated with the two-variable models. These results show that the additional information obtained from additional auxiliary variables improves the accuracy of the estimates in the mine data.

Hosseinali and Alesheikh [15] identified and classified weighting methods used in the mine exploration process. Weighting methods can be classified into two main groups: data-driven and knowledge-driven methods. In this study, six weighting methods were identified and scientifically assessed, namely ratio estimation, analytical hierarchy process (AHP), Delphi, weight of evidence, logistic regression, and artificial neural networks (ANNs). The first three are examples of knowledge-driven methods, whereas the last three are classified as data-driven methods. The evaluation of each weighting method on copper deposits in Iran showed that the ANN method is the most accurate because it can accurately predict all borehole characteristics.

Kim et al. [16] developed a new outlier top-cut method to quantify statistically significant data and presented the results of an analysis of borehole data from Au deposits. Generally, a geostatistical interpolation method, such as kriging, is used to estimate the amount of resources. However, if the data used in such an analysis deviate significantly from the mean value, this data should be excluded because it can significantly affect the results. In this process, the existing top-cut method can cause errors by excluding significant data, whereas the new methodology developed in this study can minimize the loss of data and analyze statistically significant outliers. The Getis–Ord Gi* statistic was calculated for each Au sample value in a borehole dataset by considering the 3D spatial coordinates of the borehole data. The Getis–Ord Gi* statistic is calculated by comparing the local sum for a feature and its neighbors to the total sum as follows:

$$G_i^* = \frac{\sum_{j=1}^{n} w_{i,j}x_j - \overline{X}\sum_{j=1}^{n} w_{i,j}}{S\sqrt{\frac{\left[n\sum_{j=1}^{n} w_{i,j}^2 - \left(\sum_{j=1}^{n} w_{i,j}\right)^2\right]}{n-1}}}, \quad (1)$$

where x_j is the attribute value for feature j, $w_{i,j}$ is the spatial weight between feature i and j, n is equal to the total number of features, and:

$$\overline{X} = \frac{\sum_{j=1}^{n} x_j}{n}, \quad (2)$$

$$S = \sqrt{\frac{\sum_{j=1}^{n} x_j^2}{n} - \left(\overline{X}\right)^2}. \quad (3)$$

The resultant Getis–Ord Gi* statistic is a z-score. For statistically significant positive z-scores, the larger the z-score, the more intense the clustering of high values (i.e., a hot spot). When the local sum is very different from the expected local sum, and that difference is too large to be the result of random chance, a z-score is statistically significant.

Figure 1 shows a schematic diagram of a 3D hotspot analysis for borehole data. Figure 2 shows the results of kriging using different top-cut methods for Au deposits. The proposed method can alleviate the overestimation or underestimation that can occur when applying existing methods.

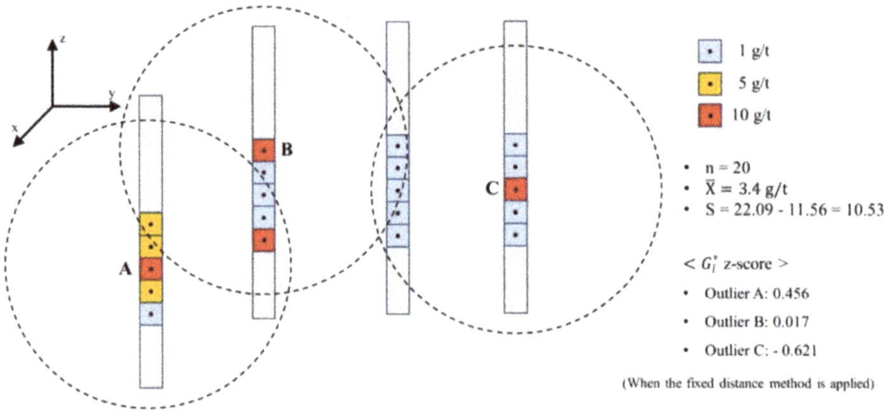

Figure 1. Schematic diagram of 3D hot-spot analysis for borehole data [16].

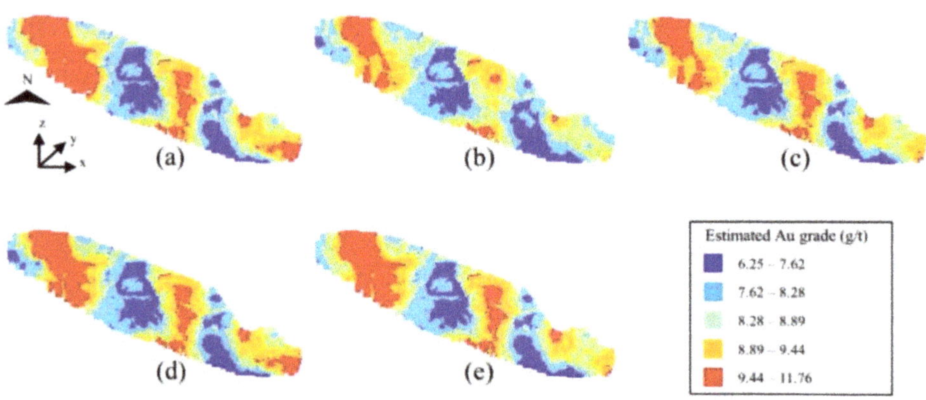

Figure 2. Kriging results of Au grade estimation using (**a**) all data or capped data with a (**b**) probability top-cut, (**c**) 95% percentile top-cut, (**d**) +2 standard deviation top-cut, and (**e**) hot-spot top-cut [16].

3.2. Open-Pit Optimization

Baek et al. [17] proposed a method to quantify the uncertainty that exists in the open-pit optimization results as a result of fluctuations in mineral prices due to changes in the world economy and mineral reserves. In this study, Monte Carlo simulations were used to generate multiple sets of mineral prices based on historical mineral price data, and the boundary optimization was repeatedly performed using this price. The results of the open-pit optimal boundary analysis are examined; 1 is assigned if a block exists within the optimal boundary, and 0 is assigned otherwise. This binarization is performed on all results of optimal boundary analysis. By integrating several binarized block models, a probability model was generated that quantitatively represented the uncertainty derived from changes in the mineral prices.

Figure 3 shows an example of a probability model formation representing uncertainty. The results of a case study applying the proposed method to copper-zinc mixed deposits showed that significant uncertainty existed in the open-pit optimization results. Figure 4 shows the probability model representing the uncertainty in the pit optimization results analyzed through the case study. The method developed in this study was able to quantify the probability that each block representing deposits is included within the open pit optimal boundary when copper and zinc prices increased or decreased from the current reference prices (Figure 5) [17].

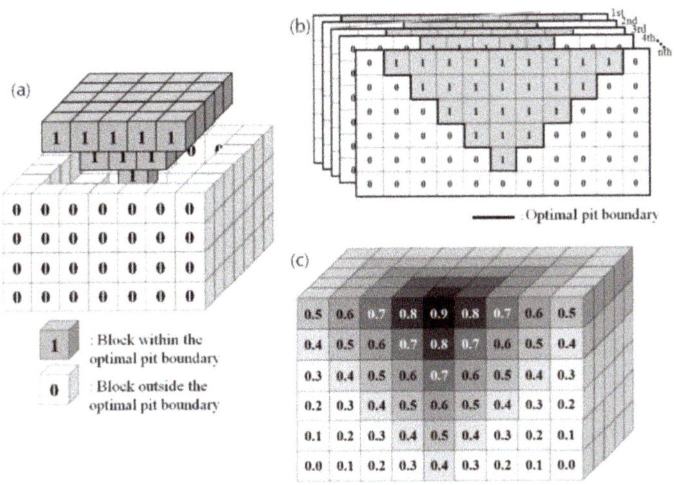

Figure 3. Example of a probability model formation representing uncertainty: (**a**) binarized block model, (**b**) probability model formation using overlay analysis, and (**c**) sectional view of the probability model [17].

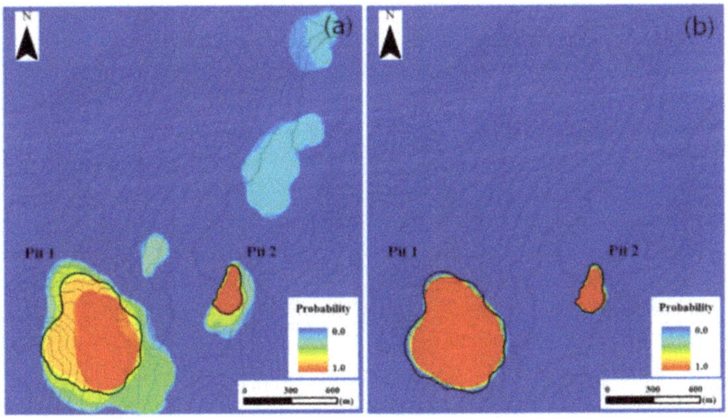

Figure 4. Probability model representing uncertainty in pit optimization results. (**a**) Case 1 considers the variation in copper prices only. (**b**) Case 2 considers the variation in zinc prices only [17].

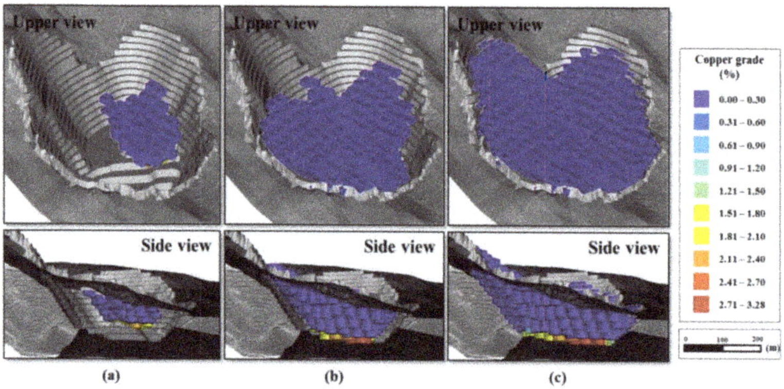

Figure 5. Blocks representing ore reserves above the optimal pit boundary. (**a**) Case 1: probability ≥ 90%. (**b**) Case 1: probability ≥ 80%. (**c**) Case 1: probability ≥ 70% [17].

Sinha et al. [18] used GIS-based technology to develop overburden dumps and backfilled and mined-out zones as existing hilly topography; they also assessed soil erosion rates to identify dam locations that require monitoring. Soil erosion rates from the developed mining landscapes were evaluated using an empirical E30 model. A unique parametric change study was performed by altering the original normalized difference vegetation index values for estimating the soil erosion rate with different vegetation densities [18]. The case studies show that erosion rates increase in barren mining land, while the year-wise plantation decreases. The GIS-based approach presented could help assess the potential for soil erosion and taking the necessary steps before initiating the operation of the mining.

Grenon and Hadjigeorgiou [19] developed a slope stability analysis tool that can be used to integrate slope stability analysis into the data management, ore reserve, and pit optimization processes of an open-pit mine. The module advanced in this study was successfully applied to the analysis of a surface mine. The case study included both kinematic and limit equilibrium stability analyses for bench and inter-ramp designs. In addition, this module utilized GIS-provided visualization tools and designated stability susceptibility zones along the pit. This approach can improve the optimized 3D pit configuration and contribute to a better understanding of the economic impact of various slope and pit designs.

Grenon and Laflamme [20] proposed a methodology that can rigorously determine the inter-ramp and bench-face slope orientations in a digital elevation model (DEM) of an open pit. They conducted a case study on Portage Pit in Canada and analyzed the GIS slope algorithm commonly used to evaluate the slope orientations in the DEM. Consequently, planar regression algorithms based on principal component analysis provided the best results at the inter-ramp and bench-face levels. In addition, the optimal sampling window was 21 × 21 cells for the inter-ramp and 9 × 9 cells for the bench. Subsequent slope stability analysis relying on those assessed slope orientations would provide a more realistic geometry for potential slope instabilities in the design pit [20]. The methodology presented in this study is flexible and can be adopted according to the block size and pit shape of the mine.

3.3. Mine Infrastructure Design

In the mine infrastructure design phase, the optimal positioning of facilities (e.g., crusher, storage, waste dump), power lines, and road design features must be carried out. Because GIS includes functions pertaining to the processing and targeting of spatial data and path analysis, it can be useful at this stage.

Lechner et al. [21] suggested how to optimally design linear infrastructure, such as roads in mines, by considering environmental and social factors. First, environmental and social factors were identified through structured interviews with stakeholders, and the weightings of spatial data inputs were calculated using these factors. These spatial layers were then used to construct a cost surface for the least-cost path analysis (LCPA). In addition, they conducted a case study with the South Konawe Regency, located in southeastern Sulawesi Province. Figure 6 shows a cost surface map for each subfactor based on GIS spatial data. Figure 7 shows the results of the least-cost path and corridor analysis for national and local groups. This figure also shows the differences in the paths identified by different groups. This study demonstrates the capability of GIS with LCPA and social survey methods to identify the infrastructure corridors that have the least impact on social and environmental values. This tool has the potential for use and refinement in more effective multiple-criteria decision-making by communities and government planning authorities [21].

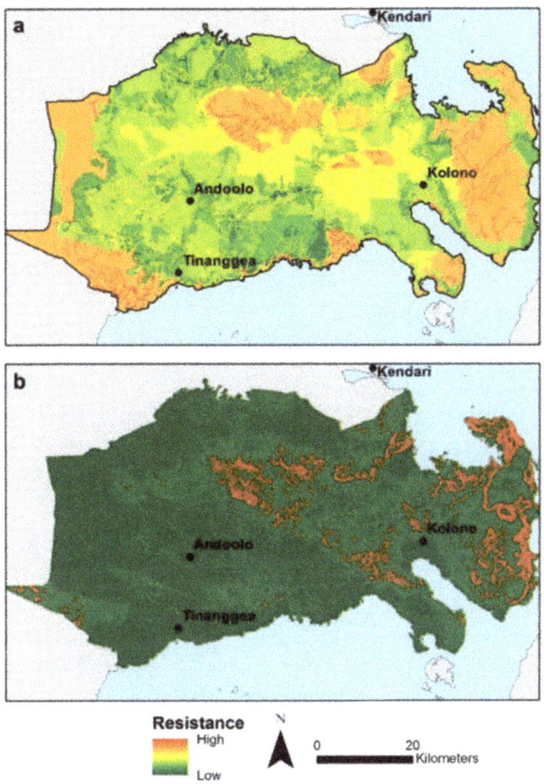

Figure 6. Example of the cost surface produced through (**a**) combining all the weighted GIS layers representing each of the subfactors where spatial data was available for local regions and (**b**) using only topography where high-resistance areas have higher slope values [21].

Blachowski [22] assessed the following aspects of the mining and transport of rock minerals (aggregates) in the context of regional development: the spatial and temporal changes in the distribution and intensity of mining, availability of economic reserves in active mines, magnitude and distribution of road transport flows of aggregates, potential of railways as an alternative means of transport, and valorization of undeveloped aggregate deposits to assess their suitability for future use. In this investigation, a case study was conducted in Lower Silesia, Poland, and cartographic models were developed using GIS to facilitate the analysis of the mining and transport of mineral resources. This model could be used to comprehensively aid with advising on the current state of mineral resources, mining, and transport in the region and to provide insight into future extraction activities.

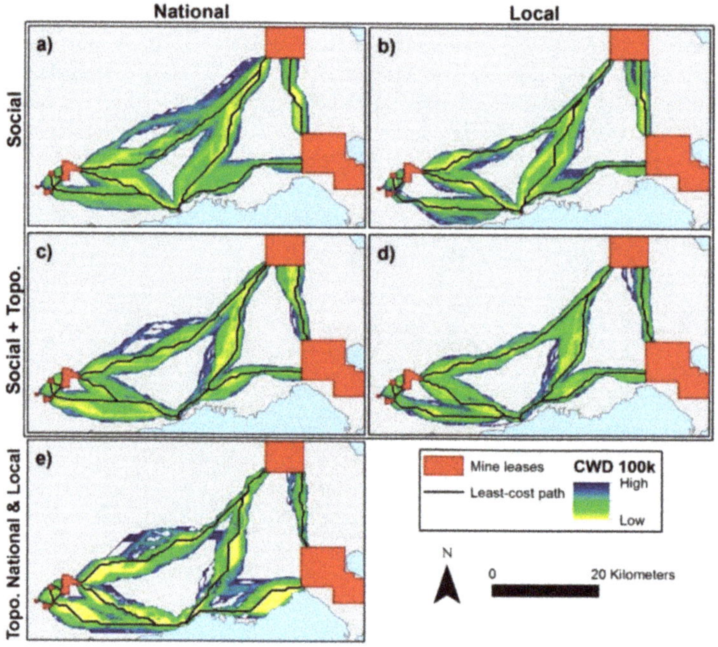

Figure 7. Infrastructure corridors identified with least-cost paths and corridor analysis for national and local groups (**a–d**). (**a**,**b**) Pathways and corridors identified where the resistance surface is based only on social values. (**c**,**d**) Pathways and corridors identified where the resistance surface is based on a combination of topography and social values. (**e**) Pathways and corridors identified only where the resistance surface is based on topography. Corridor width is based on a cost-weighted distance (CWD) threshold of 100,000 m [21].

Baek and Choi [23] proposed a new method for designing roads to efficiently carry out truck haulage operations in open-pit mines that are in the planning or design phase. In this study, terrain data, including the final designs of the pit and bench, obtained using the LCPA method were processed to analyze the haul road layout (Figure 8a). The zigzag haul road layout was determined through an initial analysis simplified by the application of the Douglas–Peucker algorithm (Figure 8b). In addition, the haul road layout was modified according to the radius of curvature constraints presented in the existing guides (Figure 8c). The terrain data and haul road layouts of the mines were then entered into AutoCAD Civil 3D to visualize the haul road of the open-pit mine in three dimensions. These processes are shown in Figure 9. Figure 10 shows the design result of the haul road of the open-pit mine and the 3D visualization that was obtained by combining the mine terrain model designed with benches and the result of the 3D haul road layout design. The proposed method can be used in the planning and design stages as a tool to design the layout of haul roads at civil engineering and construction sites, that is, where no road exists, as well as at mining sites [23].

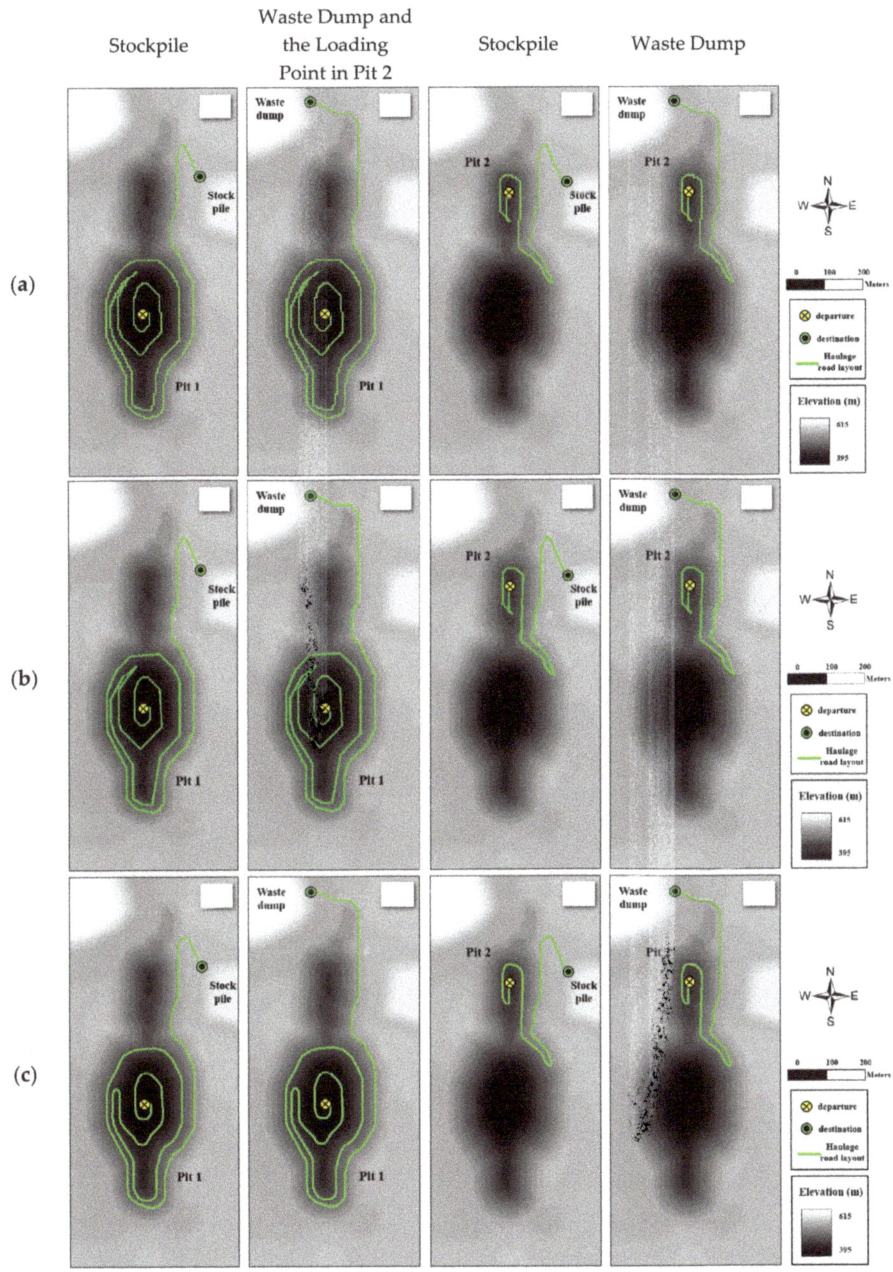

Figure 8. Results of analyzing the 2D haul road. (**a**) Haul road layout in the pit 1 and pit 2 areas analyzed using least-cost path analysis (LCPA). (**b**) Simplification of the haul road layout of pit 1 and pit 2. (**c**) Results of modifying the road layout in pit 1 and pit 2 by considering the radius of curvature (modified from Baek et al. [23]).

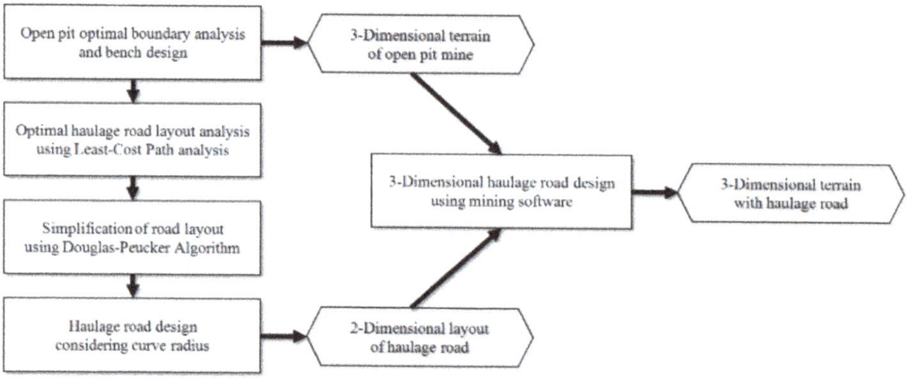

Figure 9. Study process of haul road design for open-pit mines using the proposed method [23].

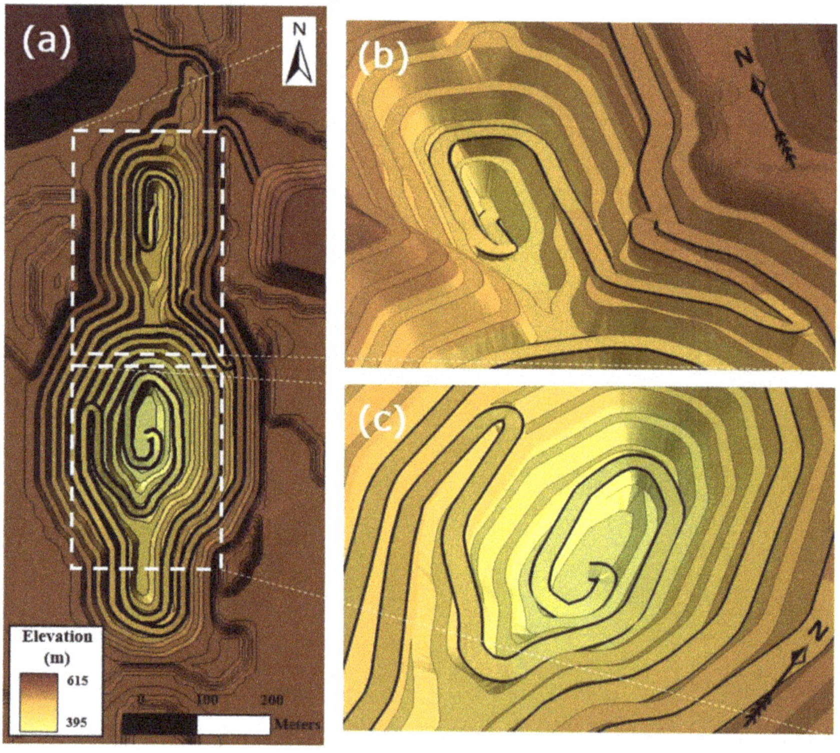

Figure 10. Result of the haul road design. (**a**) Open-pit mine with bench and haul road. 3D views: (**b**) pit 2 area and (**c**) pit 1 area [23].

3.4. Potential Conflict Region Analysis

In the potential conflict region analysis phase, GIS geoprocessing and hydrologic modeling can be utilized.

Jeronimo et al. [24] conducted a case study of the ecological and economic zoning of Cajamarca in Peru, which is rich in gold, to analyze the land-use planning process in the face of conflicting interests regarding future land and water use. GIS techniques were used to evaluate the conflicting policies on land and water use between a powerful pro-mining association led by the central government and a conservation coalition (anti-mining) led by the local government. This study suggests how GIS technology can be used if there is a conflict between stakeholders concerning mine development, such as in Cajamarca, Peru.

Craynon et al. [25] carried out a GIS-based spatial analysis considering social, environmental, and economic factors for the sustainable development of mineral and energy resources. In addition, a case study was conducted to identify potential sustainability conflict zones in mining design. They identified key parameters for consideration in decision-making through the GIS analysis of on-the-ground issues related to sustainability. The case study suggests that the use of this approach could have promoted better communication and planning. The adoption of such a process could assist in the transition to a new regulatory framework that is based on and promotes sustainable development principles [25].

4. Mine Operation Using GIS

The aspects of mine operation can be classified into ore haulage operation optimization, wireless communication system design and management, decision support systems, and mine ventilation/safety/drainage system design and management.

4.1. Ore Haulage Operation Optimization

Several studies have been performed to monitor the ore haulage system of trucks and shovels using spatial visualization functions and optimize the ore haulage routes of dump trucks using raster-based and vector-based LCPA.

Gu et al. [26] designed a GIS monitoring dispatch client to monitor the haulage system of trucks and shovels in an open-pit mine. This system allows users to manage, dispatch, and monitor trucks and shovels through vehicle location displays based on Global Positioning System (GPS) and General Packet Radio Service (GPRS) data, playback of historical paths, and inquiries concerning truck locations.

Choi et al. [27] proposed a raster-based GIS model to determine the optimal haulage routes of dump trucks in large-scale open-pit mines using multi-criteria evaluation and LCPA. In the process of weighting multiple adverse factor scores, pairwise comparisons were applied to quantitatively analyze the different resistances of five factors and three constraints, such as truck speed, proximity to water and ore bodies, existence of curves, visibility, and haul road maintenance. The model was applied to the Pasir open-pit mine, and the result showed that the optimal haulage routes of dump trucks could be determined by considering both the haulage distance and multiple criteria.

Choi and Nieto [28] presented a modified raster-based least-cost path algorithm to account for the effects of terrain relief and curves along a route in the optimal haulage route planning of an open-pit mine. This study determined the optimal haulage routes between loaders and dumps that ensure the least travel time or fuel consumption of dump trucks in the working areas without paved roads. The software, named the Google Earth-based Optimal Haulage Routing System (GEOHARTS), was developed to provide project settings, information visualization, and route optimization functions. Figure 11 presents interfaces of the GEOHARTS software for optimal haulage route planning.

Previous studies on the haulage route optimization of dump trucks in open-pit mines mostly used raster data [29]. However, the raster data can cause several problems, including zigzag-shaped travel paths due to cell resolution and the neglect of the topological relationship among haulage roads. To overcome these problems, Park et al. [29] demonstrated that vector-based network analysis is effective in determining the optimal haulage routes of dump trucks by considering the terrain gradient in open-pit mines. The Dijkstra algorithm [30] was adopted as a basic concept of network analysis, and this study presented an overall method for creating vector network data of an open-pit mine using ArcGIS software. The method consists of four procedures: (a) creating a DEM, (b) digitizing the haulage road network, (c) calculating the terrain gradient of the haulage roads, and (d) calculating the average speed and travel time of the dump truck along haulage roads. It was observed that the proposed method could effectively determine the optimal haulage routes of dump trucks, taking into account changes in truck speed, which depend on the slope gradient of the open-pit mine. Park et al. [31] analyzed the optimal routes of vehicles for auxiliary operations in open-pit mines using a heuristic algorithm for the traveling salesman problem (TSP) within a GIS environment. To compare the quantitative difference of travel routes determined by a heuristic TSP algorithm and intuitive judgment, the visiting points for auxiliary vehicles in the study area were set to 5, 15, and 25. Consequently, this study demonstrated that travel time could be shortened by 20 min when the auxiliary vehicle traveled through 25 working points along the haulage route analyzed by the heuristic TSP algorithm.

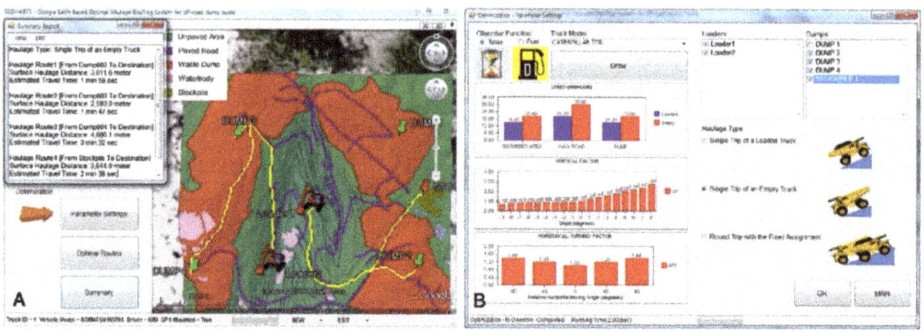

Figure 11. Visualization of the Google Earth-based Optimal Haulage Routing System (GEOHARTS) software for optimal haulage planning in an open-pit mine. (**a**) Optimal haulage routes of empty trucks that ensure the least travel time. (**b**) Parameter settings for haulage route analysis [28].

Park et al. [32] measured the carbon dioxide emissions of diesel vehicles operated in an underground mine using GIS. A GIS database was established for underground haulage road networks; then, the average travel speed of the diesel vehicles on each road was estimated to calculate the carbon dioxide emission factor through the field survey. The results revealed that the amount of carbon dioxide emissions related to truck haulage operations could be calculated by considering the carbon dioxide emission factor and travel distance of diesel vehicles determined by the GIS-based network analysis. However, this study did not consider that the carbon dioxide emission factor of each road could change according to the road gradient, load capacity of vehicles, and emission standards. Oh et al. [33] followed the new carbon dioxide emission factor suggested by the European Environment Agency. This study constructed a vector network of underground haulage roads, which includes data detailing the speed of the diesel vehicle, road gradient, and load capacity.

Baek et al. [34] developed the Bluetooth Beacon-Based Underground Navigation System to display optimal haulage route and truck location data on mobile devices in real time. A GIS-based network analysis was adopted for optimal haulage routing. After determining the optimal haulage route, the route information is updated to the web server. Moreover, GIS-based spatial visualization was used to display the installation locations of wireless sensors, which identify the dump truck locations.

4.2. Wireless Communication System Design and Management

In open-pit mines and underground mines, wireless communication systems are installed to collect data pertaining to equipment operation, work safety, and production processes on a web server in real time. Studies have been conducted to design and manage wireless communication systems using web-based GIS and spatial analysis.

For example, Li and Zhong [35] proposed a digital mine model, which is composed of a data warehouse platform and an application/analysis platform based on web-based GIS technology. When new drill hole data were acquired through the control network system, spatial feature layers could be created in a shape file format based on the coordinate information, and a new spatial data set could be converted into a geodatabase file. This model permitted remote users to search, query, visualize, and analyze mine deposit information without the need for access to the mining site.

Chai and Li [36] designed a personnel orientation system to monitor the location and status of a miner's location in an underground mine. Web-based GIS technology was adopted to store the attributes and spatial data of individual miners, which are related to the radio frequency identification system, and show a playback of traces of miners in real time through Ethernet communication. The suggested system could enhance worker safety management in underground mines.

Moridi et al. [37] modeled an underground mine monitoring and communication system, which was implemented with a wireless sensor network and a GIS server. ZigBee nodes were installed underground to sense environmental attributes, such as temperature, humidity, and gas concentrations. Then, the spatiotemporal data were joined and related to the attribute tables of the ZigBee node geographical positions in the GIS server. This study also developed a geoprocessing model that determines the emergency status according to the environmental standard, which is used to control ventilation fans and send ZigBee emergency messages using the Python programming language (ArcPy).

Baek and Choi [38] suggested a new GIS-based algorithm to design an initial wireless communications system in open-pit mines. The geoprocessing analysis was used for multiple candidate transmitter (wireless access point) selection by considering the probability of further development and environmental factors in the mine. Subsequently, a communication viewshed analysis based on the three-dimensional (3D) partial Fresnel zone was applied to quantitatively estimate the possible area of communication and analyze favorable transmitter combinations. The results revealed that the 3D Fresnel indices were calculated for 15 candidate transmitter locations, and an initial transmitter layout was designed by considering two, three, and four favorable transmitter combinations (Figure 12).

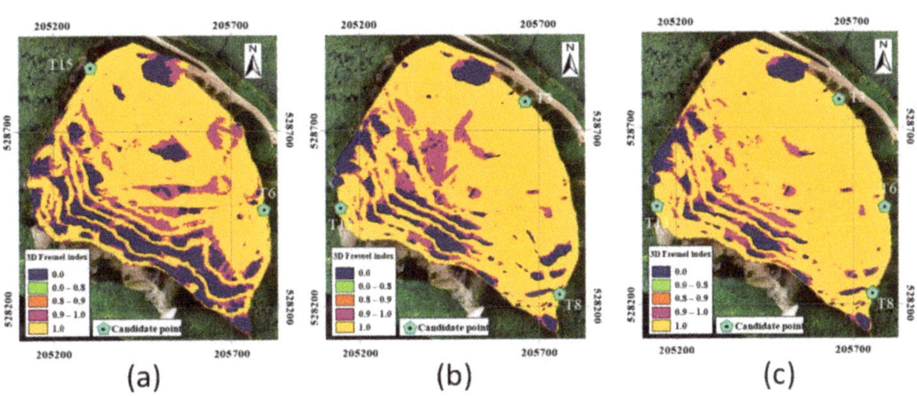

Figure 12. Three-dimensional Fresnel index overlay maps for favorable transmitter combination analysis: (**a**) combination of transmitters 6 and 15; (**b**) combination of transmitters 3, 8, and 13; and (**c**) combination of transmitters 3, 6, 8, and 13 [38].

4.3. Decision-Support Systems

GIS functions based on a spatial database and 3D visualization can also be used to manage mine data and make a high-quality mining decision for effective mining operations.

Li et al. [39] designed a Longruan GIS platform to control and monitor mine disasters induced by high crustal stress, gas thickness, and seepage water pressure in an underground coal mine. Roadway, 3D geological, and working face modeling were performed based on an irregular triangulated network using a geomechanical parameter database, which was related to hydrogeological, gas distribution, crustal stress, rock mechanical, and rock classification data. Then, rock stability and dynamic disasters were evaluated through the computation program of the application to provide support design and rescue measurements to users.

Şalap et al. [40] proposed a GIS-based monitoring and management system for underground coal mining safety. In this system, a database and metadata management system was designed to record details on mine accidents, as well as the working and operating conditions of underground workers and equipment. Various functions, including air velocity distribution analysis, thematic representations, and emergency route analysis, have been provided to users to support decision-making within a web-based GIS environment.

Gu et al. [41] suggested a dynamic ore-blending management system for effective mine production monitoring and shovel dispatching. In this system, a spatial database is constructed, which includes the coordinate data of the holes, boundaries of muck-piles, ore grade of holes, and shovel data, and the isolines of the ore grade and location of each shovel are displayed on the open-pit mine map. Through the ore weighting and truck dispatching system based on the GPS and GPRS, dispatchers could analyze the amount and average grade of ore produced in real time and control ore productivity by dispatching shovels and trucks according to the daily ore-blending plan.

Banerjee et al. [42] developed an enterprise GIS system, which was implemented with distributed desktop applications throughout an entire mining company for active mine data analysis and management. There are three types of use levels in the system: (1) operators can generate and retrieve georeferenced information, mine and civil maps, and employee and equipment databases; (2) users can convert the basic maps into digital formats and generate map features and thematic maps; and (3) company executives can analyze mining activities through map queries and make decisions concerning mine operation optimization. This study also elaborated on the implementation strategy of this system at each mine development phase.

Duncan and Rahman [43] presented a 3D mine prototype to establish a 3D spatial database and a 3D model of subsurface objects, such as mine structures and ore bodies in Ghana. A GIS-based 3D tetrahedron model was introduced to represent the 3D underground mine model. This study delineated the type of spatial data produced during mine development and operation, and suggested integrating spatial data with a 3D model for a comprehensive understanding of mine operation.

4.4. Mine Ventilation/Safety/Drainage System Design and Management

GIS can contribute to the design of an underground ventilation network model and management of drainage systems to promote safety and prevent mine accidents during mine operations. Spatial database construction and GIS-based spatial analysis have often been applied for this purpose.

Liu and Yang [44] constructed a GIS-based mine ventilation network model for the ventilation route and quantity planning using network analysis. This study generated all types of ventilation network components, such as underground tunnels, points of intersection (nodes), and ventilation facilities (air door, wall, and window), and established a spatial database related to the ventilation network and excavation. GIS-based network analysis was conducted to analyze the optimal ventilation route and the best rescue route in fire accidents by assigning resistance values to the specific ventilation network.

Massanés et al. [45] proposed a ventilation management system to analyze and visualize environmental data obtained from an underground ventilation network within the GIS environment. The investigation has focused on a Spanish mine, and the ventilation network was constructed using two perspectives: principal and auxiliary circuits. The network database parameters consisted of coordinates, date of the measure, air velocity, dry and wet temperature, and noxious gas concentrations for principal and auxiliary ventilation. The results showed that variations in air velocity and gas concentration along the principal circuit were obtained using 753 historical data records. Moreover, the most adverse working faces were analyzed by considering effective temperatures and gas concentrations. Using the same GIS-based database of the ventilation network, Bascompta et al. [46] estimated the air recirculation over time and investigated the effect of air recirculation on the underground environmental conditions in the Spanish mine. It was found that changes in the environmental conditions due to air recirculation, especially in the CO and NO_x values, depend on the ventilation layout.

Prakash and Vekerdy [47] developed a prototype GIS, called CoalMan, to manage surface and underground fires in coal mining areas. This system monitors the fire region based on remote sensing data processing functions, such as pre-processing, enhancement, and classification, provided in the GIS software package ILWIS. In addition, using spatial analysis, coal fire risk maps were generated by considering the presence or absence of mining activity, access to air, and propensity of the coal.

Choi et al. [48] examined quantitative visible areas to select optimal viewpoints near an open pit mine and analyzed hazardous sections for dump truck drivers with limited sight distance. Viewshed analysis was adopted to calculate the visibility index for all points in a DEM by considering the inter-visibility between the viewpoint and target points. The application at Pasir coal mine, Indonesia, showed that the proposed methodology could be used to select the optimal viewpoints by analyzing the ratio of the pit to the viewshed area. In addition, the potential truck collision area could be identified by conducting viewshed analysis through the centerline of a haulage route. This finding could assist in improving the safety of mining operations by considering visibility effects.

Zhang et al. [49] developed an emergency rescue command information system to design and simulate an emergency escape plan within the web-GIS environment. The architecture of the system consisted of (1) emergent incident management, (2) incident display, (3) historical accident management, and (4) emergency escape simulation to show the shortest escape path for each type of accident in a coal mine.

Choi and Park [50] suggested a new GIS-based modeling technique to design in-pit stormwater ponds in large-scale open-pit mines. The analysis model identified hydrological parameters, including the flow accumulation of rainfall in sinks and catchments, and designed in-pit stormwater ponds with pumping facilities by considering an optimal pump capacity and pond storage. Finally, a flooding simulation was performed to determine the feasibility of pond design under concentrated rainfall conditions. This model could be utilized as a powerful tool for drainage system design.

5. Environmental Management Using GISs

GIS-based research on the environmental management of mines can be classified into three topics: mine operation environment management, mine-induced hazard management, and mine rehabilitation design.

5.1. Mine Operation Environment Management

Regarding mine operation environment management, research has been conducted on the prediction of blasting vibration and dust impact, evaluation of the slope stability in an open-pit mine, and analysis of the composition of the mine area. The studies discussed here utilized GIS functions, such as spatial database construction and mapping, geoprocessing, and drainage analysis.

Bui et al. [51] examined the impact of blast-induced ground vibrations on surrounding structures (e.g., highways, water pipes, railways, and residential areas) from a quarry mine in Vietnam. They investigated a total of 25 blasting events and recorded 83 observations, which were classified using the fuzzy C-means clustering (FCM) algorithm. On this basis, quantile regression neural network (QRNN) models were developed. The combination of FCM and QRNN models created a new hybrid model with which to predict blast-induced ground vibration. The developed model was found to have a higher accuracy when compared with other models, such as that from the US Bureau of Mines, random forest, QRNN (without clustering), and ANN. The proposed model can be used to control the blast-induced ground vibration and minimize its effects.

Li et al. [52] developed an appraisal environmental system that addresses the main tasks of environmental impact assessment, such as environmental impact analysis, pollutant impact prediction, and the assessment of mitigation measures, through the use of scoping exercises (step 1) and technical assessments (step 2). Scoping assesses the qualitative impact using a knowledge-based system, while technical assessment evaluates quantitative impacts based on GIS, mathematical modeling, and evaluation. After constructing the mine model, the user can run a scoping system by importing information from both the user and mine model database to qualitatively assess the environmental impact of the operation. If there is a significant adverse impact at this stage, further quantitative investigations can be carried out using the technical assessment system. By applying the developed system, case studies were conducted for surface mining and quarrying projects, and the results are presented in several GIS images.

Francioni et al. [53] proposed a new approach using simple kinematic analysis techniques and three-dimensional finite difference method stress models for deterministic stability analysis. This approach is applied to the study of open-fit quarry slopes based on the integrated use of two remote sensing techniques, terrestrial laser scanning (TLS) and unmanned aerial vehicle (UAV). These techniques supplement the data obtained from traditional engineering geology surveys. TLS was performed using a long-range laser scanner (up to 1 km range) with three different point clouds to avoid occlusions. Digital terrestrial photogrammetry was performed using a UAV system that overcomes problems related to elevation, steepness, and complex slope geometry. In addition, a kinematic analysis was carried out using GIS techniques. The developed approach helps to overcome the common problems of complex slope geometry encountered in kinematic slope analysis.

Nelson et al. [54] used GIS software and modeling techniques to analyze slope failures in the Chuquicamata open-pit mine in Chile. They used several steps for the analysis, including initial data compilation and conversion, data validation, derivative data layer generation, and predictive modeling. In addition, the risk of slope failure was analyzed by comparing GIS-based predicted slope stability models with post-modeling slope failures. Consequently, the GIS-based modeling methods were found to be more effective at analyzing the risk of slope failure in an open-pit mine than conventional graphic-based drawing programs.

Choi et al. [55] developed and presented a GIS model that combines fuzzy theory and AHP to evaluate slope instability in open-pit coal mines. Relative collapse risk can be evaluated by considering seven influence factors simultaneously (i.e., overall slope gradient, slope height, surface flows, excavation plan, tension crack, fault, waterbody). In addition, the weight of each factor can be determined using fuzzy membership functions and the AHP. This model could be used in conjunction with the GIS database to assess the risk of slope collapse for the entire slope instability. The application of the model to the Pasir open-pit coal mine in Indonesia confirmed that it could be an effective tool for analyzing the risk of slope failure when establishing a slope management solution.

Choi et al. [56] conducted a study of flood and gully erosion problems at the Pasir open-pit coal mine in Indonesia. First, GIS was used to obtain quantitative information regarding the hydrological characteristics of the surface drainage system from a DEM. Based on the analysis results of the drainage system, raster-based spatial analysis was performed to distinguish the bench slopes with concentrated surface drainage flows and calculate the ratio of the amount of incoming water to the size of the temporary storage pond [56]. Finally, the bench slopes that are vulnerable to gully erosion were identified and an optimal design for pump placement to reduce the risk of flooding during heavy rain was proposed. Figure 13 shows the rated value map derived from the surface flow accumulation in the study area.

Yucel et al. [57] researched the detection and visualization of changes in acid mine lakes using time-series satellite image data from GIS that provided a variety of details on a year-to-year basis regarding the regional scale of the land and changes in water bodies. The data received from the remote-sensing satellites were used to verify the developed monitoring system, and various acid mine areas of 9 km^2 were selected for the case study. GIS software was used to analyze satellite images as a time series and convert the boundaries of the acid mine lakes into a vector data format. Then, the changes in the area and perimeter of the acid mine lakes were calculated and presented through tables and graphics. In addition, thematic maps of the changes in the acid mine lakes were created. Case studies indicated that the number of mine lakes increased, creating environmental risks due to their increased hydrochemical properties and area.

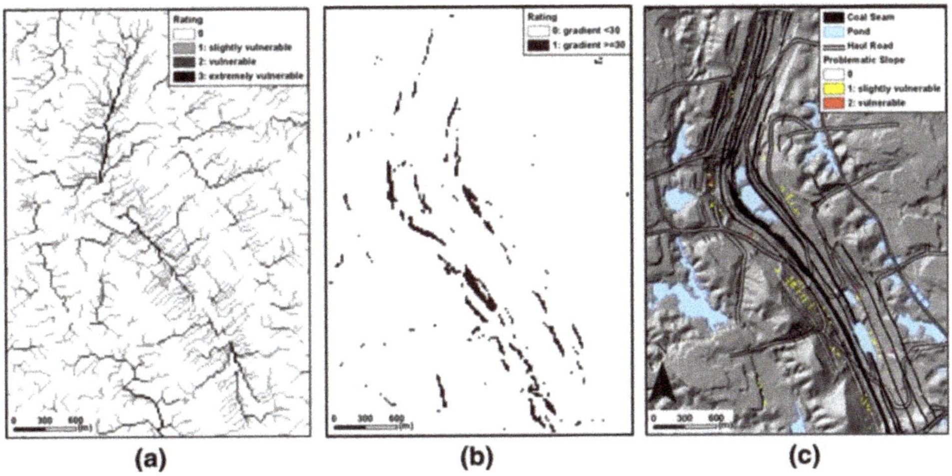

Figure 13. Rated value maps. (**a**) Rating the vulnerability to erosion from the flow accumulation of surface drainage. (**b**) Rating the slope gradient to extract steep bench slopes (30° C). (**c**) Locations of bench slopes vulnerable to gully erosion during rainfall [56].

Werner et al. [58] created detailed maps of representative samples of current larger-scale mines around the world and determined whether a generalization can be made regarding mine areas based on factors that influence their extent. In this study, commodities primarily produced by the mines were selected for comparative analysis and site mapping. At this time, the main focus was on open pits, infrastructure areas of mining sites, waste rock dumps, water storage ponds, and tailings storage facilities. Then, additional site characteristics were compiled. Finally, multiple regression analysis was used to assess the possible relationships between the extent of mining features and potential explanatory variables. As a result, the analyzed mines showed considerable diversity in their extent and configurations, and the effects of the distribution of the ores and surrounding topography on the arrangement of the features around the mines were determined. In addition, the most conservative regression models accounted for $\approx 40\% - 75\%$ of the variability in the extent of the various mining features, except for water storage ponds. The results of this study could be used to determine the demand for land in current and future mining developments.

5.2. Mine-Induced Hazard Management

GIS functions, such as watershed analysis, spatial mapping, spatial database, and spatial modeling, could be used for mine-induced hazard management. In this paper, we present examinations that have addressed the movement of soil, water inrush, and sediment; mine subsidence; and the storage and visualization of mine reclamation data.

Moomen and Dewan [59] developed early warning indicators of mining-induced land degradation to enable coordination between stakeholders in new regions that have undergone exploration and mining activities. To this end, they developed a model to quantify the relationships between mining leases and biophysical conditions that enhance land degradation. The Upper West Region of Ghana was considered as the study area, and the spatial impact relationship between mine development and land degradation was evaluated by considering vegetation displacement, the land degradation impact index, and the rainfall erosivity index in satellite images during two time periods. The model developed in this study can help to find efficient strategies for sustainable mining and the mutual benefit of rural land resources between policymakers, communities, and interested development partners. It can also enhance the understanding of companies and governments regarding the links between mining-induced land degradation and social conflict.

Duncan et al. [60] assessed the Golden Star Resources Bogoso Prestea Limited concession, where mining caused land use changes based on GIS-based spatial analysis. Area estimation and analysis of land use flow methods were performed to analyze a 20-year period (1986–2006), making it possible to identify how changes in land use in mining and agriculture over those two decades have emerged. This analysis revealed that land use from mining has stabilized owing to good reclamation practices and reduced mine development.

Wu et al. [61] proposed a GIS- and ANN-based coupling technique to evaluate the vulnerability of the water inrush controlled by geomorphology and the regional geologic structure of coal mines. To this end, a coal mine located in northern China was used to assesses the vulnerability of the water inrush through GIS, and ANN was utilized to determine the weight coefficient for each factor that affects the water inrush. The technology developed in this study could more accurately predict karst water inrushes while solving the following difficulties in assessing the vulnerability of the water inrush: (1) water inrush is controlled by several factors, (2) inaccurate weight for each factor, (3) lack of an established mathematical model to explain the process of water inrush, and (4) water inrush closely related to human activity.

Boggs et al. [62] presented the results of hydrology and landform evolution modeling techniques integrated with GIS to assess the possible impacts of mining on the Swift Creek catchment in the Northern Territory of Australia. First, a spatial and attribute database related to the study area was constructed using GIS, and hydrology and sediment transport parameters were derived from field data collected within the Swift Creek catchment. The Distributed parameter Field-Williams (DISTFW) hydrology model was then used to determine long-term hydrology parameters, and the SIBERIA landform evolution model was employed to derive annual hydrographs. Attempts to link the hydrology and landform evolution models with GIS showed that the process of parameter derivation and modeling can be simplified. In addition, linking these models to GIS provides significant benefits, as GIS can assist in the derivation, storage, manipulation, processing, and visualization of georeferenced data at the catchment-wide scale.

Miao et al. [63] investigated the feasibility of differential radar interferometry (DInSAR) using both ERS and Envisat images to monitor mine subsidence in Tang Shan, Hebei Province, China. They used GIS tools to analyze and validate the DInSAR results. They also discussed the disadvantages of using interferometric measurements for monitoring mine subsidence. Consequently, the results from Tang Shan showed that the subsidence was serious in the 1990s, and the resulting displacement was measured.

Maryati et al. [64] conducted research to design a GIS database template for environmental management in mining operations in Indonesia. Using ArcCatalog ArcGIS 9.3 software, the GIS database was designed, followed by steps such as the inventory and evaluation of government regulations, inventory and evaluation of environmental quality standards, sorting and grouping parameters, classification of design features and attributes, and creation of a GIS database and dictionary. The newly designed GIS database template has many benefits for environmental management, including integration into a single database, prevention of duplicate data, volume data reduction, data uniformity, ease in finding and tracking data, and integration of spatial and attribute data. Furthermore, the template can be used as an input for GIS analysis for decision-making and development strategies.

Álvarez et al. [65] offered a different perspective on the rationale for research, preservation, and development of a mining heritage for educational purposes. To this end, the Lullumres iron mine in Spain was selected, and documents were analyzed and classified to investigate and reconstruct the 3D geometric structure of the mine. They also proposed ways to integrate this information into GIS to improve the management of the mine data. Consequently, the closed mine was found to be a live mine with factors, traditionally regarded as negative, that need to be included as patrimonial elements with associated implications.

Radulescu and Radulescu [66] presented the main directions of implementation of GIS technology in mining through a management information system. They explained the role of GIS in Romania's mining sector and how it is used at each stage of mine development. The research shows that GIS can have a significant impact on the exploration and development stages of a mine and the mining process.

5.3. Mine Rehabilitation Design

Regarding mine rehabilitation design, studies performed using the GIS main functions of spatial modeling and analysis and decision support were reviewed.

Gorokhovich et al. [67] presented a methodology for combining spatial data on resources for coal mine reclamation and used the extrapolation of GIS analysis to prioritize a list of potential mine reclamation sites within the contiguous United States. They used four major spatial elements, namely dredged materials, abandoned mines, fly ash sources, and railroad data, such as that from the mine reclamation project in Bark Camp, Pennsylvania, USA, and also used GIS to prioritize areas where reclamation projects similar to Bark Camp are feasible. For each 1 km of the United States territory, GIS analysis identified unique occurrences of all four spatial elements within 20, 40, 60, 80, and 100 km radii from abandoned mines [67]. Consequently, the number of mines abandoned and their location in each state in the United States were identified and visualized. The methodology presented in this study could be helpful for government officials to plan mine reclamation.

Fadda et al. [68] presented a plan for the re-utilization of excavation based on the naturalistic and geographical conditions of the mining areas and their surroundings. As preparatory work, they characterized the research area with a topographic survey; established a geodetic control network with a global positioning system to describe the morphological features of the landscape; and performed three-dimensional analysis and geoprocessing using a commercial geographical system, such as GIS, and computer-aided design. The graphic data layers were then integrated with geological and morphological maps. This process helps to create and render virtual 3D environments for current, past, and future landscape development.

Kim et al. [69] developed ArcMine, a new GIS extension that can support reclamation projects in abandoned mining areas. ArcMine offers four main functions: the ability to assess the hazards of mine subsidence, estimate the erosion of mine waste, analyze the flow path of mine water on the surface, and identify tree species suitable for mine reforestation. They integrated topographical maps, geological maps, mine drift maps, and borehole data to set up a spatial database and examined the distributed mine hazards that were likely to damage the surrounding environment. It was demonstrated that the new GIS extensions developed in this study could provide useful information on mine hazards to support reclamation planning.

Pavloudakis et al. [70] proposed a spatial decision support system (SDSS) that minimizes problems such as the lack of data integration and time-consuming analysis caused by conventional methods in the reclamation plan of the mine. Based on binary integer linear programming models, the SDSS could solve these problems by integrating GIS with multiple-criteria decision-making methods. The SDSS also allows users to select appropriate land use in different parts of the post-mining area, taking into account social, technical, economic, environmental, and safety criteria. The proposed SDSS was used to carry out a case study regarding the selection of the optimal landscape reclamation strategy for the Amynteon lignite surface mine in Greece. The system showed that mining companies can efficiently determine the specific land use (agricultural land, forest, recreational area, and industrial zone) that is most suitable for each part of the research area.

Menegaki and Kaliampakos [71] proposed a new method for quantitatively evaluating the impacts on the landscape during mine operation and the improvement achieved after rehabilitation of the mine site. They named the method Landscape Evaluation Tool for Open Pit Mine Design, which focuses on measuring two main parameters: the alteration of the topographic relief and the sensitivity of observation conditions, with both making use of GIS tools. The quantitative values generated for each of the above parameters were shown to facilitate the discrimination of seemingly similar alternative designs and rehabilitation schemes. In addition, case studies showed that the partial backfilling of the quarry site could effectively reduce its visual impacts, producing low topographic relief alteration and viewing sensitivity indices.

6. Discussion

6.1. Advantages of Using GIS in Mining

To support mine planning, optimize mine operations, and analyze the environmental impact on the mining site, GIS can be effectively used for the following reasons.

(1) Effective mine data management in georeferenced, spatial, and attribute data formats: Mine data mainly include drill holes, ore bodies, mineral deposits, pit boundaries, and underground haulage roads, which have spatial characteristics. Therefore, these data can be effectively stored, manipulated, and processed in georeferenced (geographic coordinate-related characteristics), spatial (locational information), and attribute (spatial properties) data formats in the GIS environment. For this purpose, various GIS database frameworks have been established for drill hole data [35,41], block models of mineral deposits [17], underground haulage road networks [29,31–34], underground environmental attributes using ZigBee nodes [37], and underground ventilation networks [44–46].

(2) Powerful decision-making support for mine managers: GIS can be used to support decision-making during the various stages of mining development. Examples include ore reserve detection and estimation using spatial 3D query tools [13], visualization of uncertainty in open-pit boundaries [17], and land-use selection after mine closure using multi-criteria decision analysis [70]. Furthermore, mining-induced potential conflicts [24,25], disasters [47,53–55], and hazards [59–63] can be identified in advance using spatial analysis tools. Finally, all of the data and information are shared with multiple users to discuss optimal decisions through enterprise GIS technology [42].

(3) Flexibility of multi-parameter considerations: GIS offers numerous data integration tools, map overlapping tools, multi-criteria analysis tools, and mapping algebra functions for multi-parameter consideration. For example, various spatial analysis models have been developed for optimal truck haulage route determination by considering truck movement impact factors [27], multiple-transmitter candidate location selection by considering the possibility of additional mine development and environmental factors [38], and in-pit stormwater pond design by considering the hydrological characteristics of surface drainage systems [50].

(4) Fast and objective mine site investigation without on-site access: GIS can reduce the time and effort required for detailed mine site investigation by using remote-sensing data and various types of maps (e.g., topographic contour maps, geological maps, hydrological maps, and infrastructure maps). For mine reclamation planning, these methodologies are applied to determine mine-land reclamation activities [70], access mine subsidence and waste erosion [69], and evaluate the visibility impact of mining sites [71]. Consequently, mine investigators do not need to access undeveloped, dangerous, and hazardous areas for a site investigation.

6.2. Future Direction of GIS in Mining

In recent years, new technologies, such as sensor networks, mobile devices, wireless communication, and UAVs, have been introduced into mining sites in accordance with the fourth industrial revolution. In particular, information-communication-technology-based mine safety management systems have been actively utilized in open-pit and underground mines to track equipment and worker locations and monitor the mine operation environment. To transmit location data to the web server, a wireless communication network is installed in the open-pit and underground mines. Several studies have been conducted to simulate the truck haulage system using data derived from the mine safety management system. For instance, Baek and Choi [72] presented a knowledge-based simulation methodology for truck haulage systems in underground mines by considering the truck travel time, which was extracted from big data of a mine safety management system. Moreover, Baek and Choi [73] developed a deep neural network (DNN) model, which was trained using a large set of truck haulage system operation conditions and truck cycle times to predict the ore production and crusher utilization of a truck haulage system in an underground mine.

However, the aforementioned studies manually generated input data, which were required for the truck haulage system simulation and DNN model training of big data from a mine safety management system without GIS-based spatial and attribute database generation for the system. Therefore, in the future, current research needs to be expanded as follows: (a) the introduction of a GIS-based spatial and attribute database for a mine safety management system with consideration of the packet data frame (e.g., equipment detection time, ID of detected equipment); (b) the development of algorithms to automatically measure input parameters (e.g., haulage operation time, number of dispatched trucks, truck travel time) of a truck haulage system simulation and DNN model training from the spatial and attribute database and generate input data in a matrix format; and (c) the development of database query functions to extract specific input data according to the haulage operation time, date, and type of equipment.

7. Conclusions

In this review, numerous published studies on GIS-based methods and applications for planning, optimizing, and managing mine development processes were presented. A total of 58 recently published articles were found, and the detailed contents, primarily regarding GIS functions, were analyzed for this literature review. For systematic reviews, three topics were used for classification, namely mine planning, mine operation, and environmental management, by taking into account various mine development phases and a range of environments to be managed. It was observed that GIS-based methods, including spatial database, spatial analysis, spatial mapping, and visualization, can be effectively used at all mine development stages at the global, regional, and mine scales. The conclusions drawn from the detailed review of each topic are as follows.

(1) In the mine planning phase, GIS-based methods were adopted for ore reserve estimation, open-pit boundary optimization, mine infrastructure design, and potential conflict analysis. There were three main characteristics of GIS considered for this topic. First, GIS capabilities, such as a spatial database and the associated analysis, were integrated with geostatistics interpolation functions and various weighting methods (AHP, Delphi, logistic regression, and ANN) to quantitatively estimate ore deposits in the feasibility study stage. Second, it was confirmed that 3D spatial query functions would be effective searching tools for mineral target visualization above a cut-off grade. Finally, spatial analysis, including LCPA, corridor analysis, and hydrologic modeling, could be widely applied at the regional scale to design the mine-port haulage route and prevent potential social and environmental conflict.

(2) GIS-based methods can be widely used for mine operation system optimization and management. Various mine operation systems have been implemented in mining sites for ore haulage operations, wireless communication, ore management, safety monitoring, underground ventilation, and drainage systems. There are three steps to designing and managing mine operation systems using GIS. First, the spatial database and metadata of workers and equipment, haulage roads and ventilation networks, production rates, and environmental factors are built in the GIS environment. Then, all mine members are allowed to share mine operation data in real time by uploading and displaying it on the GIS web server. Finally, mine operation systems can be optimized by determining the ore haulage route, ventilation direction and quantity, and rescue planning in an emergency through GIS-based spatial analysis.

(3) Numerous GIS-based applications for mine environmental management were observed in the mine operation and reclamation phase. The impact of mine-induced disasters and hazards on mining environments and human health has been predicted and managed using spatial databases, spatial analysis, and visualization techniques. Remote sensing images were principally used to analyze the variation in the mine environment (mine subsidence, slope failure, and mine lake) as a time series. Moreover, various GIS extension systems were developed to support decision-making in mine reclamation planning and re-utilization designs.

GIS has powerful capabilities for designing and optimizing mine development stages in terms of spatial data management, decision-making support, and multi-parameter consideration. Nonetheless, the practical use of GIS-based methods for mine planning, operation, and environmental management remains difficult. Consequently, to maximize the practicality and applicability of GIS in the mining site, a general understanding and training of the GIS platform is required through continuous communication and feedback between engineers and researchers.

Author Contributions: Y.C. gathered the published literature; all the authors reviewed and analyzed the published literature; all the authors wrote the paper. Y.C. implemented the final work for the paper submission. All authors have read and agreed to the published version of the manuscript.

Funding: This work was supported by the KETEP grant funded by the Korea Government's Ministry of Trade, Industry and Energy (project no. 20182510102370).

Conflicts of Interest: The authors declare no conflict of interest.

References

1. Carvalho, F.P. Mining industry and sustainable development: time for change. *Food Energy Secur.* **2017**, *6*, 61–77. [CrossRef]
2. Chang, K.-T. *Introduction to Geographic Information Systems*, 5th ed.; McGraw-Hill higher education: New York City, NY, USA, 2010; pp. 1–448.
3. Lewis, P. A Review of GIS Techniques for Handling Geoscience Data within Australian Geological Surveys. In *Geophysics and geochemistry at the millenium. Proceedings of Exploration 97: Fourth Decennial International Conference on Mineral Exploration, Toronto, Canada, 14-18 September 1997*; Gubins, A.G., Ed.; Prospectors and Developers Associaton: Toronto, Canada, 1997; pp. 81–86.
4. Lukasheh, A.F.; Droste, R.L.; Warith, M.A. Review of Expert System (ES), Geographic Information System (GIS), Decision Support System (DSS), and their applications in landfill design and management. *Waste Manage Res.* **2001**, *19*, 177–185. [CrossRef] [PubMed]
5. Wang, H.; Liu, G.; Xu, W.; Wang, G. GIS-based landslide hazard assessment: an overview. *Prog. Phys. Geogr.* **2005**, *29*, 548–567. [CrossRef]
6. Gajos, M.; Sierka, E. GIS Technology in Environmental Protection: Research Directions Based on Literature Review. *Pol. J. Environ. Stud.* **2012**, *21*, 241–248.
7. Hou, d.; O'Connor, D.; Nathanail, P.; Tian, L.; Ma, Y. Integrated GIS and multivariate statistical analysis for regional scale assessment of heavy metal soil contamination: A critical review. *ENVIRON. POLLUT.* **2017**, *231*, 1188–1200. [CrossRef]
8. Wang, X.; Xie, H. A Review on Applications of Remote Sensing and Geographic Information Systems (GIS) in Water Resources and Flood Risk Management. *Water* **2018**, *10*, 608. [CrossRef]
9. Hustrulid, W.A.; Kuchta, M.; Martin, R.K. *Open Pit Mine Planning and Design*, 3rd ed.; CRC Press: Leiden, The Netherlands, 2013; pp. 1–1288.
10. Darling, P. *SME Mining Engineering Handbook*, 3rd ed.; Society for Mining, Metallurgy and Exploration (SME): Littleton, CO, USA, 2011; pp. 1–1837.
11. Environmental Code of Practice for Metal Mines. Available online: https://www.ec.gc.ca/lcpe-cepa/documents/codes/mm/mm-eng.pdf (accessed on 19 February 2020).
12. Suh, J.; Kim, S.-M.; Yi, H.; Choi, Y. An Overview of GIS-Based Modeling and Assessment of Mining-Induced Hazards: Soil, Water, and Forest. *Int. J. Environ. Res. Public Health* **2017**, *14*, 1463. [CrossRef]
13. Sprague, K.; de Kemp, E.; Wong, W.; McGaughey, J.; Perron, G.; Barrie, T. Spatial targeting using queries in a 3-D GIS environment with application to mineral exploration. *Comput. Geosci.* **2006**, *32*, 396–418. [CrossRef]
14. Uygucgil, H.; Konuk, A. Reserve estimation in multivariate mineral deposits using geostatistics and GIS. *J. Min. Sci.* **2015**, *51*, 993–1000. [CrossRef]
15. Hosseinali, F.; Alesheikh, A.A. Weighting Spatial Information in GIS for Copper Mining Exploration. *Am. J. Appl. Sci.* **2008**, *5*, 1187–1198. [CrossRef]
16. Kim, S.M.; Choi, Y.; Park, H.D. New Outlier Top-Cut Method for Mineral Resource Estimation via 3D Hot Spot Analysis of Borehole Data. *Minerals* **2018**, *8*, 348. [CrossRef]

17. Baek, J.; Choi, Y.; Park, H.-S. Uncertainty Representation Method for Open Pit Optimization Results Due to Variation in Mineral Prices. *Minerals* **2016**, *6*, 17. [CrossRef]
18. Sinha, N.; Deb, D.; Pathak, K. Development of a mining landscape and assessment of its soil erosion potential using GIS. *Eng. Geol.* **2017**, *216*, 1–12. [CrossRef]
19. Grenon, M.; Hadjigeorgiou, J. Integrated structural stability analysis for preliminary open pit design. *Int. J. Rock Mech. Min.* **2010**, *47*, 450–460. [CrossRef]
20. Grenon, M.; Laflamme, A.-J. Slope orientation assessment for open-pit mines, using GIS-based algorithms. *Comput. Geosci.* **2011**, *37*, 1413–1424. [CrossRef]
21. Lechner, A.M.; Devi, B.; Schleger, A.; Brown, G.; McKenna, P.; Ali, S.H.; Rachmat, S.; Syukril, M.; Rogers, P. A Socio-Ecological Approach to GIS Least-Cost Modelling for Regional Mining Infrastructure Planning: A Case Study from South-East Sulawesi, Indonesia. *Resources* **2017**, *6*, 7. [CrossRef]
22. Blachowski, J. Spatial analysis of the mining and transport of rock minerals (aggregates) in the context of regional development. *Environ. Earth Sci.* **2014**, *71*, 1327–1338. [CrossRef]
23. Baek, J.; Choi, Y. A New Method for Haul Road Design in Open-Pit Mines to Support Efficient Truck Haulage Operations. *Appl. Sci.* **2017**, *7*, 747. [CrossRef]
24. Preciado Jeronimo, R.; Rap, E.; Vos, J. The politics of Land Use Planning: Gold mining in Cajamarca, Peru. *Land Use Policy* **2015**, *49*, 104–117. [CrossRef]
25. Craynon, J.R.; Sarver, E.A.; Ripepi, N.S.; Karmis, M.E. A GIS-based methodology for identifying sustainability conflict areas in mine design – a case study from a surface coal mine in the USA. *Int. J. Min., Reclam. Environ.* **2016**, *30*, 197–208. [CrossRef]
26. Gu, Q.; Lu, C.; Li, F.; Wan, C. Monitoring dispatch information system of trucks and shovels in an open pit based on GIS/GPS/GPRS. *J. China Univ. Min. Technol.* **2008**, *18*, 288–292. [CrossRef]
27. Choi, Y.; Park, H.D.; Sunwoo, C.; Clarke, K.C. Multi-criteria evaluation and least-cost path analysis for optimal haulage routing of dump trucks in large scale open-pit mines. *Int. J. Geogr. Inf. Sci.* **2009**, *23*, 1541–1567. [CrossRef]
28. Choi, Y.; Nieto, A. Optimal haulage routing of off-road dump trucks in construction and mining sites using Google Earth and a modified least-cost path algorithm. *Autom. Constr.* **2011**, *20*, 982–997. [CrossRef]
29. Park, B.; Choi, Y.; Park, H.S. Creation of Vector Network Data with Considering Terrain Gradient for Analyzing Optimal Haulage Routes of Dump Trucks in Open Pit Mines. *Tunn. Undergr. Space* **2013**, *23*, 353–361. [CrossRef]
30. Dijkstra, E.W. A Note on Two Problems in Connexion with Graphs. *Numer. Math.* **1959**, *1*, 269–271. [CrossRef]
31. Park, B.; Choi, Y.; Park, H.S. Optimal Routes Analysis of Vehicles for Auxiliary Operations in Open-pit Mines using a Heuristic Algorithm for the Traveling Salesman Problem. *Tunn. Undergr. Space* **2014**, *24*, 11–20. [CrossRef]
32. Park, B.; Park, S.; Choi, Y.; Park, H.S. Calculation of a Diesel Vehicle's Carbon Dioxide Emissions during Haulage Operations in an Underground Mine using GIS. *Tunn. Undergr. Space* **2015**, *25*, 373–382. [CrossRef]
33. Oh, S.; Choi, Y.; Park, H.D. Calculation of a Diesel Truck's Carbon Dioxide Emissions in an Underground Mine Considering Road Gradient, Load Capacity and Emission Standards. *J. Korean Soc. Miner. Energy Resour. Eng.* **2017**, *54*, 100–109. [CrossRef]
34. Baek, J.; Choi, Y.; Lee, C.; Suh, J.; Lee, S. BBUNS: Bluetooth beacon-based underground navigation system to support mine haulage operations. *Minerals* **2017**, *7*, 228. [CrossRef]
35. Li, X.; Zhong, S. *Digital mine design based on Data Warehouse and GIS. Proceedings of the 2nd International Conference on Information Science and Engineering, Hangzhou, China, 4-6 December 2010*; IEEE: New York, NY, USA, 2010; pp. 3653–3656.
36. Chai, Y.; Li, D.; Chai, Y. *Research and Design of Personnel Orientation System in Mine Based on GIS/MIS. Proceedings of the 2010 Fourth International Conference on Genetic and Evolutionary Computing, Shenzhen, China, 13-15 December 2010*; IEEE: New York, NY, USA, 2010; pp. 268–271.
37. Moridi, M.A.; Kawamura, Y.; Sharifzadeh, M.; Chanda, E.K.; Wagner, M.; Jang, H.; Okawa, H. Development of underground mine monitoring and communication system integrated ZigBee and GIS. *Int. J. Min. Sci. Technol.* **2015**, *25*, 811–818. [CrossRef]
38. Baek, J.; Choi, Y. A New GIS-Based Algorithm to Support Initial Transmitter Layout Design in Open-Pit Mines. *Energies* **2018**, *11*, 63. [CrossRef]

39. Li, M.; Zhang, X.; Mao, S.; Liu, Q. Study of deep mining safety control decision making system. *Procedia Earth Planet. Sci.* **2009**, *1*, 377–383. [CrossRef]
40. Şalap, S.; Karslioğlu, M.O.; Demirel, N. Development of a GIS-based monitoring and management system for underground coal mining safety. *Int. J. Coal Geol.* **2009**, *80*, 105–112. [CrossRef]
41. Gu, Q.; Lu, C.; Guo, J.; Jing, S. Dynamic management system of ore blending in an open pit mine based on GIS/GPS/GPRS. *Min. Sci. Technol.* **2010**, *20*, 132–137. [CrossRef]
42. Banerjee, T.K.; Roy, S.; Dey, S. A GIS Solution for an Integrated Underground Coal Mine Management: A Conceptual Framework. *J. Manag. Policies Pract.* **2014**, *2*, 129–143.
43. Duncan, E.E.; Rahman, A.A. 3D GIS for mine development – integrated concepts. *Int. J. Min., Reclam. Environ.* **2015**, *29*, 3–18. [CrossRef]
44. Liu, H.; Yang, D. *GIS-based mine ventilation network and safety analysis. Proceedings of 2004 IEEE International Geoscience and Remote Sensing Symposium, Anchorage, AK, USA, 20-24 September 2004*; IEEE: New York, NY, USA, 2004; pp. 2945–2948.
45. Massanés, M.B.; Pera, L.S.; Moncunill, J.O. Ventilation management system for underground environments. *Tunn. Undergr. Sp. Technol.* **2015**, *50*, 516–522. [CrossRef]
46. Bascompta, M.; Castañón, A.M.; Sanmiquel, L.; Oliva, J. A GIS-based approach: Influence of the ventilation layout to the environmental conditions in an underground mine. *J. Environ. Manage.* **2016**, *182*, 525–530. [CrossRef]
47. Prakash, A.; Vekerdy, Z. Design and implementation of a dedicated prototype GIS for coal fire investigations in North China. *Int. J. Coal Geol.* **2004**, *59*, 107–119. [CrossRef]
48. Choi, Y.; Suh, J.; Park, H.D.; Sunwoo, C. Application of GIS-based Viewshed Analysis for Selecting Optimal Viewpoints and for Evaluating the Sight Distance on a Haul Road in Open-pit mines. *J. Korean Soc. Geosystem Eng.* **2008**, *45*, 265–275.
49. Zhang, M.-y.; Shan, L.-y.; Zhao, Z.-p. *The design of coal mine emergency rescue command information system based on web GIS. Proceedings of the 2011 2nd International Conference on Artificial Intelligence, Management Science and Electronic Commerce (AIMSEC), Dengleng, China, 8-10 August 2011*; IEEE: New York, NY, USA, 2011; pp. 4455–4458.
50. Choi, Y.; Park, H.D. GIS Modeling for Design of In-pit Stormwater Ponds in Large Scale Open-pit Mines. *J. Korean Soc. Geosystem Eng.* **2011**, *48*, 165–177.
51. Bui, X.-N.; Choi, Y.; Atrushkevich, V.; Nguyen, H.; Tran, Q.-H.; Long, N.Q.; Hoang, H.-T. Prediction of blast-induced ground vibration intensity in open-pit mines using unmanned aerial vehicle and a novel intelligence system. *Nat. Resour. Res.* **2019**, *28*, 1–20. [CrossRef]
52. Li, S.; Dowd, P.A.; Birch, W.J. Application of a knowledge- and geographical information-based system to the environmental impact assessment of an opencast coal mining project. *Int. J. Surf. Min., Reclam. Environ.* **2000**, *14*, 277–294. [CrossRef]
53. Francioni, M.; Salvini, R.; Stead, D.; Giovannini, R.; Riccucci, S.; Vanneschi, C.; Gullì, D. An integrated remote sensing-GIS approach for the analysis of an open pit in the Carrara marble district, Italy: Slope stability assessment through kinematic and numerical methods. *Comput. Geotech.* **2015**, *67*, 46–63. [CrossRef]
54. Nelson, E.P.; Connors, K.A.; Suárez, S.C. GIS-Based Slope Stability Analysis, Chuquicamata Open Pit Copper Mine, Chile. *Nat. Resour. Res.* **2007**, *16*, 171–190. [CrossRef]
55. Choi, Y.; Park, H.D.; Sunwoo, C.; Jung, Y.B. Application of Fuzzy Theory and AHP to Evaluate the Slope Instability at Pasir Open Pit Coal Mine, Indonesia. *J. Korean Soc. Miner. Energy Resour. Eng.* **2009**, *46*, 45–60.
56. Choi, Y.; Park, H.D.; Sunwoo, C. Flood and gully erosion problems at the Pasir open pit coal mine, Indonesia: A case study of the hydrology using GIS. *Bull. Eng. Geol. Environ.* **2008**, *67*, 251–258. [CrossRef]
57. Yucel, D.S.; Yucel, M.A.; Baba, A. Change detection and visualization of acid mine lakes using time series satellite image data in geographic information systems (GIS): Can (Canakkale) County, NW Turkey. *Environ. Earth Sci.* **2014**, *72*, 4311–4323. [CrossRef]
58. Werner, T.T.; Mudd, G.M.; Schipper, A.M.; Huijbregts, M.A.J.; Taneja, L.; Northey, S.A. Global-scale remote sensing of mine areas and analysis of factors explaining their extent. *Glob. Environ. Change* **2020**, *60*, 1–10. [CrossRef]
59. Moomen, A.-W.; Dewan, A. Assessing the spatial relationships between mining and land degradation: evidence from Ghana. *Int. J. Min., Reclam. Environ.* **2017**, *31*, 505–518. [CrossRef]

60. Duncan, E.E.; Kuma, J.S.; Primpong, S. Open Pit Mining and Land Use Changes: An Example from Bogosu-Prestea Area, South West Ghana. *J. Inf. Syst. Dev. Ctries.* **2009**, *36*, 1–10. [CrossRef]
61. Wu, Q.; Xu, H.; Pang, W. GIS and ANN coupling model: An innovative approach to evaluate vulnerability of karst water inrush in coalmines of north China. *Environ. Geol.* **2008**, *54*, 937–943. [CrossRef]
62. Boggs, G.S.; Evans, K.G.; Devonport, C.C.; Moliere, D.R.; Saynor, M.J. Assessing catchment-wide mining-related impacts on sediment movement in the Swift Creek catchment, Northern Territory, Australia, using GIS and landform-evolution modelling techniques. *J. Environ. Manage.* **2000**, *59*, 321–334. [CrossRef]
63. Miao, F.; Yan, M.; Qi, X.; Ye, C.; Wang, B.; Liu, R.; Jianhua, C. Application of DInSAR and GIS for underground mine subsidence monitoring. *Int. Arch. Photogramm. Remote Sens. Spat. Inf. Sci.* **2008**, *37*, 251–255.
64. Maryati, S.; Shimada, H.; Sasaoka, T.; Hamanaka, A.; Matsui, K.; Nagawa, H. GIS Database Template for Environmental Management of Mining in Indonesia. *J. Geogr. Inf. Syst.* **2012**, *4*, 62–70. [CrossRef]
65. Álvarez, J.P.F.; Pieiga, A.G.-L.; Suárez-Lázare, C.J. New concepts in reassessing mining heritage: A study and its implications from the Ancient Iron Mine of Llumeres (North Spain). *J. Cult. Herit.* **2010**, *11*, 172–179. [CrossRef]
66. Radulescu, C.; Radulescu, V.M.G. Approaches of the management informational systems regarding the implementation of the geographic information systems (GIS) in the mining basins of Romania. In Proceedings of the 11th International Multidisciplinary Scientific Geoconference and EXPO, Albena Seaside & Spa Resort, Bulgaria, 19–25 June 2011; pp. 215–222.
67. Gorokhovich, Y.; Voros, A.; Reid, M.; Mignone, E. Prioritizing Abandoned Coal Mine Reclamation Projects Within the Contiguous United States Using Geographic Information System Extrapolation. *Environ. Manage.* **2003**, *32*, 527–534. [CrossRef]
68. Fadda, S.; Fiori, M.; Matzuzzi, C. Developing rehabilitation design for the abandoned mine excavations in central Sardinia, Italy. *Int. J. Min., Reclam. Environ.* **2010**, *24*, 286–306. [CrossRef]
69. Kim, S.M.; Choi, Y.; Suh, J.; Oh, S.; Park, H.D.; Yoon, S.H.; Go, W.R. ArcMine: A GIS extension to support mine reclamation planning. *Comput. Geosci.* **2012**, *46*, 84–95. [CrossRef]
70. Pavloudakis, F.; Galetakis, M.; Roumpos, C. A spatial decision support system for the optimal environmental reclamation of open-pit coal mines in Greece. *Int. J. Min., Reclam. Environ.* **2009**, *23*, 291–303. [CrossRef]
71. Menegaki, M.E.; Kaliampakos, D.C. Evaluating mining landscape: A step forward. *Ecol. Eng.* **2012**, *43*, 26–33. [CrossRef]
72. Baek, J.; Choi, Y. Simulation of Truck Haulage Operations in an Underground Mine Using Big Data from an ICT-Based Mine Safety Management System. *Appl. Sci.* **2019**, *9*, 639. [CrossRef]
73. Baek, J.; Choi, Y. Deep Neural Network for Ore Production and Crusher Utilization Prediction of Truck Haulage System in Underground Mine. *Appl. Sci.* **2019**, *9*, 4180. [CrossRef]

© 2020 by the authors. Licensee MDPI, Basel, Switzerland. This article is an open access article distributed under the terms and conditions of the Creative Commons Attribution (CC BY) license (http://creativecommons.org/licenses/by/4.0/).

applied sciences

MDPI

Article

A Tightly Coupled GIS and Spatiotemporal Modeling for Methane Emission Simulation in the Underground Coal Mine System

Hui Liu [1], Shanjun Mao [1,*], Mei Li [1] and Shuangyong Wang [2]

1 Institute of Remote Sensing and Geographic Information System, Peking University, Beijing 100871, China; huil@pku.edu.cn (H.L.); mli@pku.edu.cn (M.L.)
2 Beijing LongRuan Technologies Inc., Beijing 100190, China; wangshuangyong@longruan.com
* Correspondence: sjmao_@pku.edu.cn; Tel.: +86-10-62755420

Received: 26 March 2019; Accepted: 7 May 2019; Published: 10 May 2019

Abstract: Mine safety is of primary concern in the underground coal mining system. At present, there is a lack of an efficient platform to manage the numerical simulation procedure and inherent spatiotemporal data for coal mine disasters. This necessitates the coupling of spatiotemporal model with geographic information system (GIS) in practical application. Here, a novel spatiotemporal model tightly coupled with GIS is presented to improve the model-data integration. Such tight coupling is achieved by developing a lattice Boltzmann method (LBM) based turbulent model with an underlying shared FluentEntity model within the LongRuanGIS platform. The case study and comparison with the traditional computational fluid dynamics (CFD) method demonstrated that the platform is capable and effective in providing functionalities for lattice domain decomposition, simulation, visualization and analyses, as well as improving the computational efficiency. The proposed approach and platform, promising for the disaster prevention, offer a template for future GIS-Model integration and also applicable for other underground coal mine disasters.

Keywords: mine safety; GIS-coupled; spatiotemporal model; LBM; methane gas emission

1. Introduction

Mine safety issue has long been a paramount concern in the underground coal mining business, since mine accidents can lead to serious injuries for personnel, substantial economic losses, and delayed production. Numerical modelling offers a useful tool for prediction and control of various disasters in mine workings. The principles of computational fluid dynamics are widely applied to the underground coal mine systems and numerous computational fluid dynamics (CFD) methods are utilized to simulate various ventilation-related safety and health issues [1–4]. It is remarkably noted that the procedure of CFD simulation interrelated with geometric modeling and numerical calculation, is inherently geospatial. The spatiotemporally simulated data requires effective management and analysis, which is essential to reduce the risk associated with mine accidents. However, the current simulations of coal mine disasters mostly depend on the independent third-party numerical simulation software, and there is no professional simulation platform developed for specific problems. For instance, numerical platforms such as Fluent, COSFLOW, FLAC3D, and AutoReaGas are utilized by many researchers in simulating gas emission in mining face, the hole wall, and other ventilation related activities [5–8]. There is a lack of an integrative and effective platform to store and manage the massive amount of spatiotemporal simulated data, as well as the data visualization and analysis, which makes the whole analysis procedure difficult to direct the on-site production. According to researches proposed by Goodchild, one of the key challenges in the application of physics based models is the lack of a platform for efficient prototyping of model simulations, evaluation of a-priori parameters, and for simulation, analysis, and visualization [9,10]. Fortunately, geographic information system (GIS) technology is able to acquire, develop, and interpret the complex spatial datasets used for mining and the earth sciences. [11], which is suitable to provide an ideal platform to organize the numerical simulation procedure as well as data management. Therefore, coupling GIS with physics based numerical models is of paramount significance.

Application of GIS in mining is a relatively new and fast evolving concept [12]. GIS technologies create efficiency and productivity opportunities in all aspects of mineral exploration and mining, which enables a mine operator to mine intelligently, efficiently, competitively, safely, and environmentally. In this way, GIS integrates exploration, operation, and environmental issues with mine management [13]. To date, many GIS-based systems for underground coal mining safety were developed. For example, a GIS-based monitoring and management system was developed for underground coal mining safety by Seda Şalap [14], and many researchers have conducted the modeling and risk assessment of mining-induced hazards based on GIS [15,16]. However, the system development is mostly based on the open source GIS framework and focused on information management and monitoring, which are limited to the display and query of the coal mine map [17], incapable of displaying the models of various disasters from the perspective of underlying physical mechanism [18,19]. Nowadays, the coupling of numerical modeling with GIS system has already been investigated in other environmental issues, and how to effectively integrate them is regarded as one of the most increasingly focused issues. It is noted that a great number of related research were carried out in the fields of geomechanics, environmental monitoring, and hydrology [20–23]. For instance, the integration of finite element simulation with GIS was investigated and applied in the tunnel engineering by Zhenping Liu [24]. Maohui Zhen studied the block gas diffusion process within GIS based on the open source CFD software OpenFOAM [25]. Junting Ma proposed finite element numerical simulation of the groundwater in the GIS technical platform [26]. Mar Alcaraz studied the configuration of geometry and parameterization for groundwater numerical models, and proposed a loosely coupled GIS and hydrogeological modeling platform [27]. Gopal Bhatt presented an open-source, platform independent, tightly coupled GIS and distributed hydrologic modeling platform (PIHMgis), which was aimed to improve the model-data integration [28]. However, the existing coupling modes mainly depend on the third-party professional software to implement the numerical simulation procedure, which requires data transformation in different platforms due to the independence of the traditional CFD methods with respect to the mesh generation and simulation procedure [29]. And this transition needs specific methodologies to adapt the geometries and alphanumerical data from the conceptual model to the numerical model for obtaining optimal numerical results. To overcome this gap and to make the integrated system more flexible, a more efficient way to integrate the numerical CFD model with the GIS is mandatory [30–32]. By comparing the existing CFD methods, it was demonstrated that the property of the unnecessary explicit mesh generation and the inherent simplicity of the algorithm make LBM significantly popular in CFD community [33–37], which is expected to be promising and more efficient to integrate with the GIS platform.

Therefore, this study aims to apply the coupling concept of GIS and numerical simulation in the underground mining for the first time by proposing an applicable numerical model based on the non-traditional lattice Boltzmann method (LBM). By considering the spatiotemporal characteristics of coal mining activities, major accidents, and its interrelated spatiotemporal data analysis, a robust and tight GIS-coupled spatiotemporal modeling system is a powerful and thorough solution to simulate the major accidents by providing more intuitive visualization, query, and analysis tools. Specifically, the system should: (i) perform numerical model, (ii) display the simulated field data with inherent geospatial data such as coal mine map, seamlessly, (iii) provide data storage, management and access of the surveillance data based on database management system, which allows the data verification and validation of the numerical model, (iv) provide managers and miners the access to large amount of location-based information by spatiotemporal query and analysis to guide the coal mine operation.

Longruan GIS (LrGIS) developed by our team, is professionally designed and ideally suited to assist mining professionals in meeting the complex challenges of running the mine operation. With tools to compile, process, display, analyze, and archive massive volumes of data, the LrGIS platform is increasingly applied in the business of mining industry in China, which allows us to develop the mathematical model and manage volumes of data through its secondary development mode [38]. Thus, this study chooses LrGIS as the supporting GIS software package, and it takes the methane gas emission as the case study, which is one of the most crucially important mine safety issues. It is worth noted that the numerical model can be extended and improved to simulate other mining activities, such as ventilation solution, mine fire, and methane explosion.

We performed the model coupling and numerical simulation from the following ways. First, the U-type ventilation system-15116 in Sijiazhuang coal mine is specified as the computational domain, and a novel turbulent LBM based velocity-concentration coupling model is proposed to investigate the spatiotemporal characteristics of the airflow behavior and the methane gas emission in a coal mine working face. Second, the numerical model is integrated with LrGIS platform via FluentEntity model, and the graphical representations of simulated field data as well as various map layers are demonstrated. From this, the spatial characteristics of methane distribution can be intuitively visualized, and detailed location-based information for the practical mining production can be provided by spatiotemporal query and analysis. Third, the numerical simulation results are verified by field measurements in Sijiazhuang coal mine and compared with traditional CFD methods, demonstrating the accuracy and reliability of the proposed model. Thus, the developed system can assist decision makers to simulate, understand, and forecast ventilation-related activities and its dynamic change inside the laneway, which is expected to be an efficient tool for improving and maintaining the health standards in the underground coal mining industry.

2. Case Study

A fully mechanized U-type working face (15116 laneway) in Sijiazhuang coal mine was selected as the physical prototype to implement the GIS-coupled spatiotemporal modeling, which is located in Shanxi province, China. This mine working face is selected for access to precisely and continuously recorded mine data. The layout of the working face laneway is shown in Figure 1. It has a working face zone, wind intake laneway, and wind outtake laneway with the rectangular cross-section of 4.2 m wide and 4 m high. The working face is 220 m long, while the wind intake and outtake laneways are all 600 m long.

According to the safety operation regulations in Sijiazhuang coal mine, the airflow quantity in the 15116 working face laneway is 53.3 m^3/s, which is calculated based on the requirement of the mine gas and carbon dioxide emission quantity, the temperature and the number of maximum miners in the working face area, and it has been checked by the maximum and minimum wind speed. The estimated methane emission flow rate is about 0.2 m^3/s. During the procedure of coal mining in the U-type laneway system, the working face is the main source of the methane gas emission. A large amount of the methane gas generated from the working face disperses with the pressing air from the wind inflow side, finally, part of it is removed through the wind inflow laneway, while the rest stays in the laneway space. During the delivering process, movement of methane coupled with the airflow exhibits a species transportation flow. Here, we have three model assumptions consistent with the published research [39]: (i) the airflow in the working face zone belongs to the turbulent flow; (ii) the airflow and methane gas are continuous medium gas; (iii) heat and mass transfer are ignored during the process of air flow.

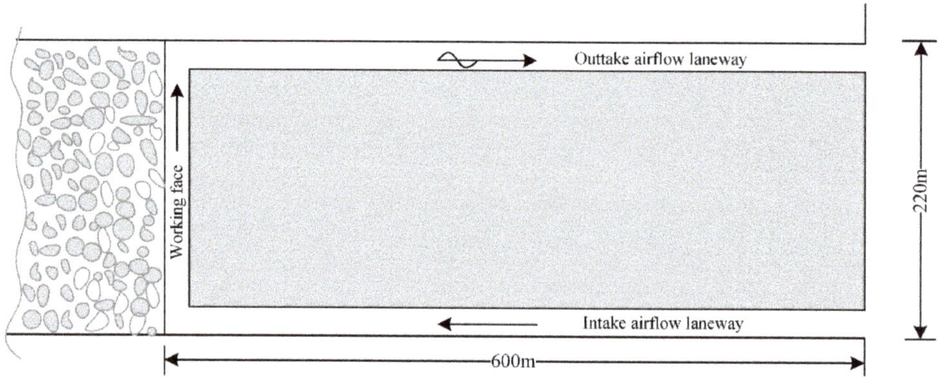

Figure 1. Layout of 15116 workface in Sijiazhuang mine.

3. Model-GIS Coupling Strategy

In this study, the numerical model and LrGIS platform are linked to achieve the model-GIS integration. Specifically, the proposed numerical model is the velocity-concentration coupled 2D LBM, which is integrated with LrGIS platform by developing the FluntEntity model, so that both GIS platform and model functions can access the geospatial simulation results. The GIS platform, the numerical model, and the integration strategies are described in detail in the following sections.

3.1. LrGIS Platform

LrGIS (Longruan GIS 3.2) is a professional coal mine GIS platform (Beijing LongRuan Technologies Inc.) with its own intellectual property right, which is widely applied in the business of mining industry in China and well supported by professional developers. It is well designed for coal mine system with respect to the geospatial features of coal mine laneway, ventilation network, and etc., and professionally focuses on large amount of mine spatial data, such as surveying, geology, hydrology reserves, transportation, ventilation network, mine design schedule. Besides, it fulfills the integrated management of massive spatial data and information sharing among different platforms, and provides flexible secondary development mode, such as API, DLL, ActiveX, and application template [40]. Therefore, LrGIS is suitable to perform GIS-numerical model integration in this study.

Despite all these advantages, these functions rely on inherent and advanced GIS capabilities, and there is a lack of defined conceptual model associated with spatiotemporal modeling to simulate various kinds of disasters to satisfy some actual needs, such as the methane gas emission simulation, the gas explosion prediction in emergency response, ventilation network design based on the pre-simulation, and mine fire simulation to guide miners to escape from the disaster and etc. The integration of the spatiotemporal modeling with GIS can provide great support for practical production needs. These inbuilt GIS functions are insufficient in terms of spatiotemporal modeling, because the data require specific treatments to be used in spatiotemporal modeling. In this study, explicit mesh creation is not needed, which is usually troublesome in traditional CFD method and inconsistencies of input geometries always lead to labor intensive and time-consuming process for mesh generation [41,42]. The integration mode proposed in this study avoids this problem and improves the computational efficiency.

3.2. Mathematical Models

The essential regularity of methane dispersion and its coupling with air flow is a key issue for the methane gas emission problem, and its flowing regulation belongs to viscous Newtonian fluid, which is governed by the Navier-Stokes (N-S) equations. There are many numerical approaches to solve the N-S equations, such as the traditional CFD methods and the LBM. Here the LBM based numerical method is adopted. In phase space, the Boltzmann equation is discretized in threefold, including space, time, and velocities. The movement and distributions of the fluid are described as particle distribution functions residing at the sites of a regular lattice of points, which encompasses the entire computational field. The particle distribution functions stand for the probability of particle presence with a specific velocity at each lattice or grid site. The macroscopic variables of the fluid, such as velocity or density, can be derived from these distribution functions.

The LBM consists of two fundamental steps, namely, the streaming-step and the collision-step. For every time step, distribution functions are distributed and streamed from each site to the neighboring sites in the streaming-step, and then the collision-step is conducted with the distribution functions relaxed towards a local equilibrium on the basis of the new macroscopic variables [32]. Here, the coupling flow of the methane gas and airflow belongs to the species transportation. The coupling model of the velocity and concentration based on 2D LBM was deduced, and the simulation procedure for velocity field and concentration field is presented. The Bossinesq approximation method is utilized to couple these two fields.

3.2.1. Lattice Boltzmann Method for Velocity Field

The LBM is commonly labelled as DdQq, where d stands for the space dimension and q is the number of microscopic lattice velocity directions. The possible nodes for 2D lattices are D2Q5, D2Q9. In this study, the D2Q9 model is employed to carry out the airflow velocity field, because it maintains good isotropy of the lattice. The airflow particles distribution on a lattice is shown in Figure 2a.

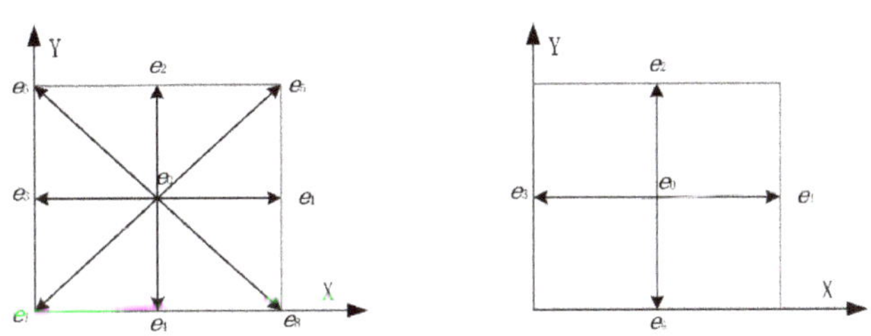

(**a**) Velocity field lattice (D2Q19) (**b**) Concentration field lattice (D2Q5)

Figure 2. Schematic views of particles distribution in 2d lattice.

The fictitious fluid particles are subject to the lattice Boltzmann equation

$$f_i(x + ce_i\Delta t, t + \Delta t) - f_i(x, t) = \Omega_i(f), \qquad i = 0, 1, \ldots, 8, \tag{1}$$

which describes the evolution of distribution function f_i on Cartesian grids with a selective velocity e_i, c is the movement velocity, $\Omega_i(f)$ denotes the discrete collision operator, and Δt is the time step. The complex collision operator is approximated by using the standard Bhatnagar-Gross-Krook (BGK) scheme [43], which states that the distribution functions is close to a local equilibrium distribution function f_i^{eq} and relaxes toward this equilibrium with a characteristic time τ. The evolution of the distribution functions using the BGK collision is described by the following equation:

$$\Omega_i(f) = -\frac{1}{\tau}[f_i(x, t) - f_i^{eq}(x, t)] \tag{2}$$

where τ is the relaxation factor (viscosity of fluid is related to the relaxation time), f_i^{eq} depends on the density ρ and velocity u of the gas, and is typically defined as

$$f_i^{eq}(\boldsymbol{x},t) = \omega_i \rho (1 + \frac{\boldsymbol{e}_i \boldsymbol{u}}{c_s^2} + \frac{(\boldsymbol{e}_i \boldsymbol{u})^2}{2c_s^4} - \frac{u^2}{2c_s^2}), \qquad i = 0,1,\ldots,8, \tag{3}$$

where ω_i is the model-dependent weight coefficient $c_s = \sqrt{RT}$ (R is the gas constant) is the lattice sound speed. For isothermal flows, c_s is set to be $c/\sqrt{3}$ with $c = \Delta x/\Delta t$, where Δx is the lattice spacing (c = 1 in this paper). Through the Chapman-Enskog expansion, the macroscopic fluid density ρ, and velocity u, can be derived as the zeroth and first order moments of f_i respectively,

$$\rho = \sum_{i=0}^{b-1} f_i, \qquad \rho \boldsymbol{u} = \sum_{i=0}^{b-1} \boldsymbol{e}_i f_i \tag{4}$$

The fluid pressure is defined directly as $p = c_s^2 \rho$, the velocity set and the corresponding weight coefficients are defined as

$$\boldsymbol{e}_i = \begin{bmatrix} 0 & 1 & 0 & -1 & 0 & 1 & -1 & -1 & 1 \\ 0 & 0 & 1 & 0 & -1 & 1 & 1 & -1 & -1 \end{bmatrix} \tag{5}$$

$$\omega_i = \begin{cases} 4/9, & i = 0 \\ 1/9, & i = 1,2,3,4 \\ 1/36, & i = 5,6,7,8 \end{cases} \tag{6}$$

The above model can deduce the Navier-Stokes equation, which describes airflow regularity in the laneway.

$$\frac{\partial \rho}{\partial t} = -\nabla \cdot (\rho \boldsymbol{u}) \tag{7}$$

$$\frac{\partial \rho}{\partial t} + \nabla p + \nabla \cdot (\rho \boldsymbol{u}) = \nabla [\rho v (\nabla \boldsymbol{u} + (\nabla \boldsymbol{u})^T)] \tag{8}$$

3.2.2. Concentration Field through a Coupled Model

A coupled model is used to simulate the methane concentration field, where a D2Q9 lattice with a BGK collision operator is also used to solve the velocity and density, while the methane gas concentration is solved on a smaller and separate D2Q5 lattice. The distribution of concentration particles on a lattice is shown in Figure 2b.

The evolution equation for the methane concentration field is described by the evolution of distribution function C_i on the Cartesian grids with selective velocity $\boldsymbol{e}_i$,

$$C_i(x + ce_i \Delta t, t + \Delta t) - C_i(x,t) = \Psi_i(C) \qquad i = 0,1,\ldots,4 \tag{9}$$

where c denotes the velocity, Δt is the time step, and $\Psi_i(C)$ is the discrete collision operator, which is given by,

$$\Psi_i(C) = -\frac{1}{\tau_c}[C_i(\boldsymbol{x},t) - C_i^{eq}(\boldsymbol{x},t)] + \Delta t \frac{R}{4} \tag{10}$$

$$\tau_c = 2\Delta t D/\Delta x^2 + 0.5 \tag{11}$$

in which R is the source item, C_i^{eq} is the equilibrium distribution function of the methane concentration, and depends on the velocity u, temperature T of the gas, which is typically defined as

$$C_i^{eq}(x,t) = \frac{C(x,t)}{4}[1 + 2\frac{\boldsymbol{e}_j \boldsymbol{u}}{c_s^2}], \qquad i = 0,1,\ldots,4, \tag{12}$$

τ_c is the relaxation factor, D is the methane gas diffusion coefficient, c_s and Δx is the same as mentioned above, the direction set is defined as

$$\boldsymbol{e}_i = \begin{bmatrix} 1 & 0 & -1 & 0 \\ 0 & 1 & 0 & -1 \end{bmatrix} \tag{13}$$

$C(x,t)$ is the methane concentration in lattice x at time t, which is calculated by

$$C = \sum_{i=1}^{4} C_i(x,t) \tag{14}$$

This model can also deduce the macroscopic equations for gas concentration diffusion.

$$\frac{\partial C}{\partial t} = -\nabla \cdot (\boldsymbol{u}C) + D\nabla^2 C + R \tag{15}$$

The Boussinesq approximation method is employed to implement the coupling mode of the airflow velocity and the methane concentration field; the evolution equation after coupling is

$$f_i(x + ce_i\Delta t, t + \Delta t) - f_i(x,t) = -\frac{1}{\tau}[f_i(x,t) - f_i^{eq}(x,t)] + g_i \tag{16}$$

$$g_i = \frac{k_i}{2} e_i g \beta_s (C - C_0) \tag{17}$$

where g_i, k_i and C_0 are the gravitational acceleration, the volume expansion coefficient of concentration and reference concentration constant respectively.

3.2.3. The Turbulence Model

Airflow behavior in underground coal mine laneway is generally turbulent, hence our flow solver based on LBM should take the effect of turbulence into account. Therefore, a wide range of scales of fluid motion existed in the coupling flow of the air and methane concentration is necessary to resolve. The simulation in all scales in a turbulence flow requires a very fine lattice and long computation time. Instead, a Smagorinsky sub-lattice model is utilized to simulate the effects of the unresolved sub-lattice motion on the resolved motion, which is similar to the large eddy simulation (LES) in traditional CFD methods [44]. As for the LBM evolution equation, the effect of the sub-lattice can be addressed by local relaxation time [45–47]. Then the modified relaxation time is adopted in the relaxation process, and every node of lattice would relax at different rates.

3.3. Implementation of the LBM Integrated with LrGIS Platform

Numerical model and GIS integration can be achieved using one of the three potential coupling strategies outlined in Table 1 [28]. Here, a tight coupling approach is used to integrate LBM in LrGIS platform, which requires a straightforward connection from the geometric laneway data to LBM based numerical model. An applicable numerical coupling model was developed, which was specifically designed to adapt geometric laneway data to be integrated with LrGIS platform. The implementation procedure focuses on improving the geometry of the input features that configure the model geometry, the boundary conditions, and other aspects of spatiotemporal modeling, which avoids explicit mesh generation procedure that is necessary in traditional CFD method. It will help modelers in the development of future projects.

Table 1. Characteristics of different strategies for GIS and model integration.

Characteristics	Coupling Type		
	Loose	Tight	Embedded
Shared data and method base		√	√
Intra-simulation model modification			√
Intra-simulation query and control			√
Advantages & disadvantages	Different GIS and Modeling packages have independent interfaces. Information sharing is based on exchange of files, which can be error prone and inefficient. Different tools and libraries facilitate independent development.	The coupled platform merges different modules in a single powerful system, which avoids inconsistency and data loss originating from redundancy and heterogeneity of method base. Data exchange is automatic between GIS platform and model.	Programming and data management is significantly complex, and the code base is not easy to be changed due to embedded large source code structure. Steerable simulation according to the changes of parameters or processes.
Examples	[48]	[49]	[50]

Figure 3 shows the entire architecture of numerical model integrated with LrGIS platform. The platform consists of three layers: Data source, GIS kernel, and desktop application. Different kinds of data source can be quickly stored, retrieved, indexed, and searched by SDE interface layer. GeoLattice can be constructed on LrGIS platform to prepare for the numerical simulation, like other geometric models such as GeoPoint, GeoLine. In terms of the data management, LrGIS contains many entity models, such as GeoEntity, RasterEntity. Similarly, FluentEntity is specifically designed to manage the field data obtained by fluid simulation. The physical attributes consist of velocity, pressure, temperature, gas concentration, and display attributes comprise geometrical shape, color rendering, profile. The simulated result can be visualized on LrGIS GUI and execute the spatial query function. Specifically, two main parts have been solved: (i) GeoLattice and FluentEntity setup, which constructs the laneway model and configures the geometry and attributes of each lattice to make them valid for numerical computation; (ii) Database organization, which controls and traces the output data that are obtained from the numerical simulation. The data is saved as time series grid table files, which can be called by FluentEntity.

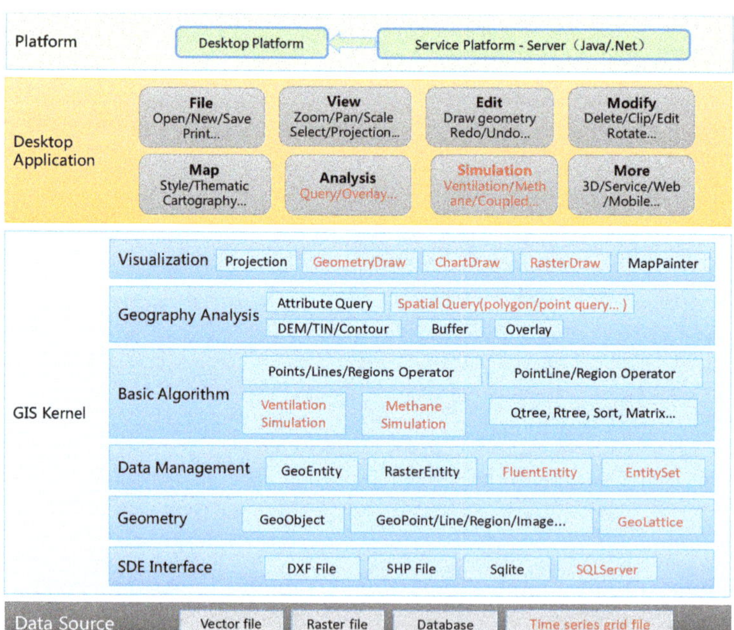

Figure 3. The entire architecture of numerical model on LrGIS platform.

The lattice attributes and the corresponding boundary conditions are described in details: different kinds of lattice nodes on the geometry are associated with different material numbers in their own coordinates. Figure 4 presents a 2D top view of the laneway, with material number one, two, three, four and five corresponding to fluid region, inflow, outflow, methane mass flow, and bounce back boundary, respectively. Then different material numbers are defined with the corresponding dynamic conditions, which are described, as follows:

(i) fluid region: all species can move in this area, namely, the mesoscopic streaming and collision step of particles happened;
(ii) inflow: the velocity inflow boundary condition is adopted, which means that the airflow is through this boundary with specified velocity;
(iii) outflow: the outlet of the laneway is set to be under the constant pressure boundary condition;
(iv) methane mass flow: the methane gas is released evenly from working face area with a total flow rate of 0.2 m^3/s;
(v) bounced back boundary: the no-slip boundary is prescribed at all laneway walls.

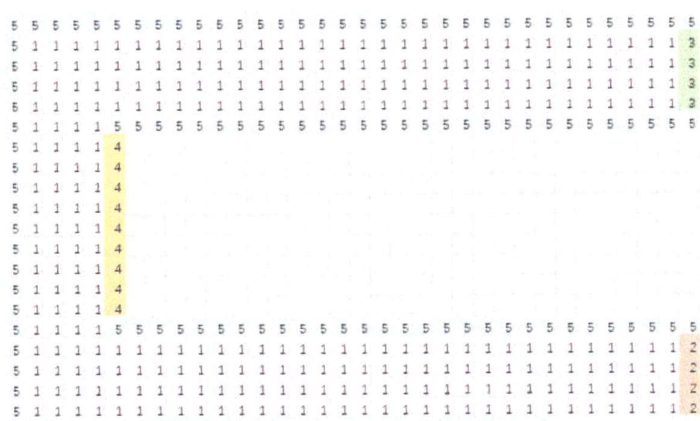

Figure 4. 2D top view of the laneway with associated material numbers.

The detailed algorithm of LBM based numerical model integrated with LongRuanGIS platform is implemented as follows (Table 2), where the computational complexity of this numerical model is the product of lattice amount and number of iterations.

Table 2. Algorithm steps of the numerical model.

Algorithm. The LBM Based Turbulent Velocity-Concentration Coupling Model Integrated with LrGIS Platform

Step 1: Geometric model construction with specific geospatial coordinates.
Step 2: Lattice attributes configuration for two different lattice (D2Q9 and D2Q5)
Step 3: Main loop starts
    Parameters initialization: $f_i, f_i^{eq}, C_i, C_i^{eq}, \Delta x, \Delta t$, maxIter
    For (it = 0; it < maxIter; ++it)
        If (it = 0)
            Set static boundary conditions (fluid region, wall).
        Else if (it != 0)
Set dynamic boundary conditions(inflow, outflow and methane emission rate).
Do collision step
            $\widetilde{f_i}(x,t) = f_i(x,t) - 1/\tau[f_i(x,t) - f_i^{eq}(x,t)]$
            $\widetilde{\Psi}_i(C) = -1/\tau_c[C_i(\boldsymbol{x},t) - C_i^{eq}(\boldsymbol{x},t)] + \Delta t\omega/4$
        Do streaming step
            $f_i(x + ce_i\Delta t, t + \Delta t) = \widetilde{f_i}(x,t)$
            $C_i(x + ce_i\Delta t, t + \Delta t) - C_i(\boldsymbol{x},t) = \widetilde{\Psi}_i(C)$
        Do coupling step
            $f_i(x + ce_i\Delta t, t + \Delta t) - \widetilde{f_i}(x,t) = -1/\tau[\widetilde{f_i}(x,t) - f_i^{eq}(x,t)] + k_i e_i g\beta_s(C - C_0)/2$
Step 4: Save the result data as time series grid files
Step 5: FluentEntity call time series grid files and display field data on LrGIS GUI integrated with various coal mine map objects

4. Simulation and Integration Results

4.1. Spatiotemporal Characteristics of Airflow and Methane Distribution

The established model was numerically simulated and all terms of the evolution equations, including streaming step and collision step were solved. The coupling mode is solved using Boussinesq approximation method. The detailed parameters are presented in Table 3. The simulation required around 1.6 h on workstations with four core processors and 8 GB RAM.

Table 3. Parameters for LBM simulation.

Parameters	Setting
Air density (kg/m^3)	1.225
Methane gas density (kg/m^3)	0.716
Turbulent viscosity(m^2/s)	1.7894×10^{-5}
Turbulent kinetic energy	1.3
Convergence criteria	10e^{-6}
Calculation steps	10000
Lattice size (m)	0.1
Time step size (s)	1
Renolds number	500
Initial pressure	1/3
Air velocity of inlet (m/s)	3.17
Methane volume flow at the working face (m^3/s)	0.2

The velocity field of airflow is one of the main factors which directly influence the methane distribution and mitigation. As presented in Figures 5 and 6, the predicted air velocity and methane concentration profiles were investigated at different time points with a fixed methane emission rate of 0.2 m^3/s. Here, we analysed the overall spatiotemporal characteristics of velocity behaviour and methane distribution inside the laneway. The pressing air is flowing from the intake side of the laneway at high velocity, and there exists an obvious jet flow, which goes along the laneway and flows into the working face. When the airflow arrives at the working face, it is obstructed by the face and the airflow changes its flow direction and flows along the working face area, as can be seen in Figure 5a–c about the airflow distribution at different times. Correspondingly, the methane gas gushes from the coal seam and disperses into the working face with the impact of airflow after about 5 min. (Figure 6a), when the airflow is distributed fully in the working face area. Then the methane gas gradually diluted and decreased to a low concentration level with the effect of airflow in the wind outtake side of the laneway and approximately becomes stable at 10 min. (Figure 6b–d).

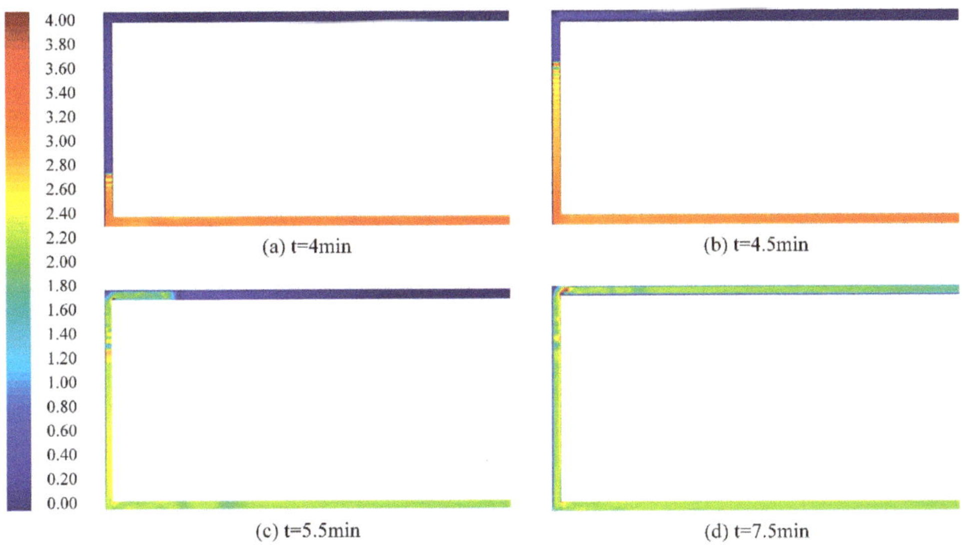

Figure 5. Airflow velocity (m/s) profiles at different time points.

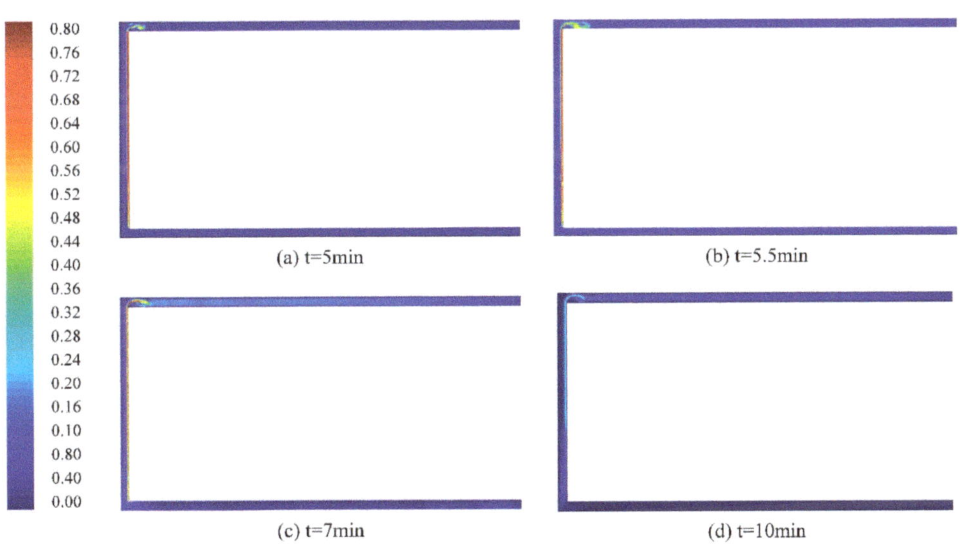

Figure 6. Methane concentration (v/v%) profiles at different time points.

4.2. The Spatiotemporal Analysis of Methane Concentration Based on LrGIS

Based on the proposed integration method, the CFD simulation function is developed on LrGIS platform, which supports the numerical simulation and can display the simulated data on the ventilation system map on the basis of coordinate matching, and also facilitate the visualization of additional relevant data. The effective user access, the location-based information, and means of viewing and analyzing for the simulated data are also the main focuses. Thus, the numerical model is integrated with LrGIS platform, of which the query and analysis functions can also be operated based on the numerical simulation results. Figure 7 shows the map of ventilation system of 5# Sijiazhuang coal mine, the study area of 15116 working face in this case is located in the bottom right corner on the map, with the latitude and longitude ranging from 105,800 to 106,020 m and from 67,125 to 67,725 m. The roam and move functions allow users to observe more detailed features on the coal mine map.

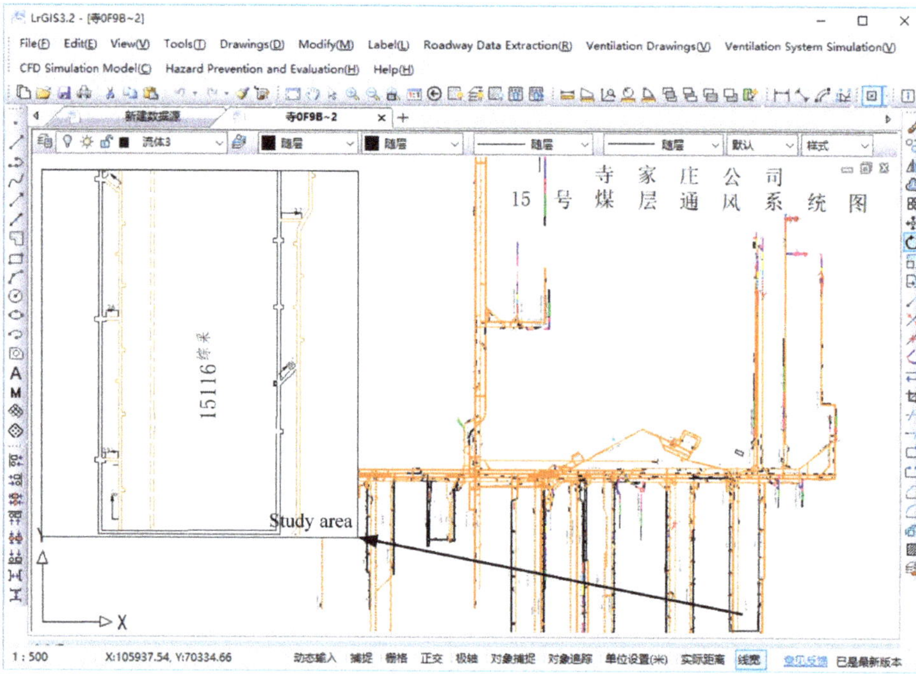

Figure 7. Study area of the ventilation system map in 15# Sijiazhuang coal mine.

To demonstrate the applicability of the CFD simulation function in LrGIS platform, the spatiotemporal analysis for our case study is performed. First, based on the specified numerical parameters, the geometry is prepared and lattice is generated, which are then used to perform the numerical simulating procedure. It is noted that the properties of lattices include the lattice ID, material numbers, and the coordinates. The output data of the simulation is saved as a series of time ordered text files, which contains the X, Y coordinates, the airflow velocity, and methane concentration of each lattice. The tool of methane emission analysis in the CFD simulation model allows for us to demonstrate the spatiotemporal distribution of the airflow behavior and methane concentration. By specifying the time point, the corresponding spatiotemporal distribution of the methane concentration can be visualized on the ventilation system map, together with various spatial coal mine layers, including geographical locations of fans, methane sensors and first aid stations, contour line, measured faultage, river and railway, and also the fictitious graticule, as presented in Figure 8. These information provide managers and miners the access to the large quantities of geographic field data, as well as its query and analysis for the location-based information integrated with simulated data.

Basically, the managers or decision makers can use the program to perform several queries according to the specifications, such as, (i) basic mine mapping operations integrated with simulated field data, (ii) finding operation, and (iii) location-based information operation. For example, the maximum methane concentration with its coordinates in an area of interest will be demonstrated immediately by the function of polygonal query, which also shows whether the predicted concentration exceeds the threshold value, as can be seen in Figure 9.

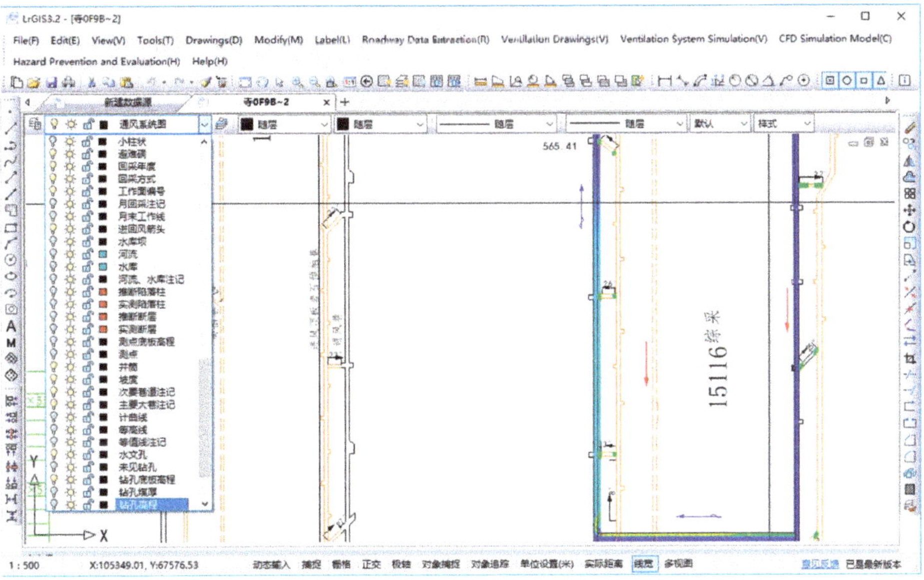

Figure 8. Methane concentration integrated with various spatial coal mine layers.

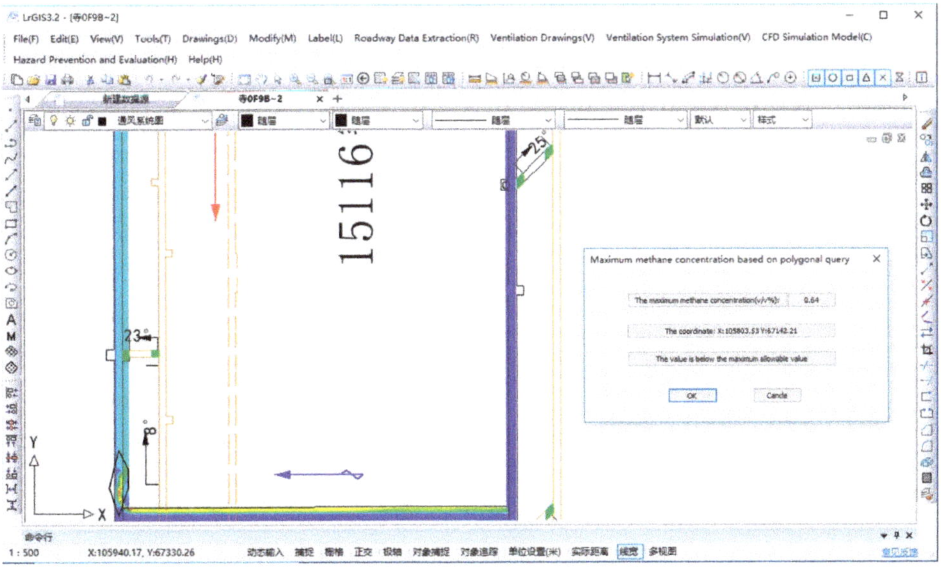

Figure 9. Methane concentration visualization and polygonal query analysis.

Besides, if a specific point is of interest in an actual field, for example, the position is where miners always work, then the point can be clicked or the coordinate can be input on the interface to get the methane concentration curve changed over time, which will be displayed on the interface and provides intuitive information for the dynamic regularity of methane concentration. The time series data can also be validated with monitored data based on the database management system. Here, one of the methane sensors located 10 m away from the working face in the outtake side of the laneway is selected to perform the data validation with the monitored data, as presented in Figure 10. The numerical simulation shows that, after the algorithm converges in 10 min, and the airflow distributes fully and stably within the laneway. The simulated methane concentration maintains at a stable level and agrees well with the monitored data after 10 min. Thus, some detailed information can be provided for decision makers, such as how long the methane distribution will be stable with the effect of the ventilation system, the peak value of methane concentration, and its specific position, which can guide miners about when to avoid the methane accident if the value is beyond the maximum allowable value. Besides, the methane distribution can also guide the production intensity, and reasonable arrangement of underground coal mine equipment.

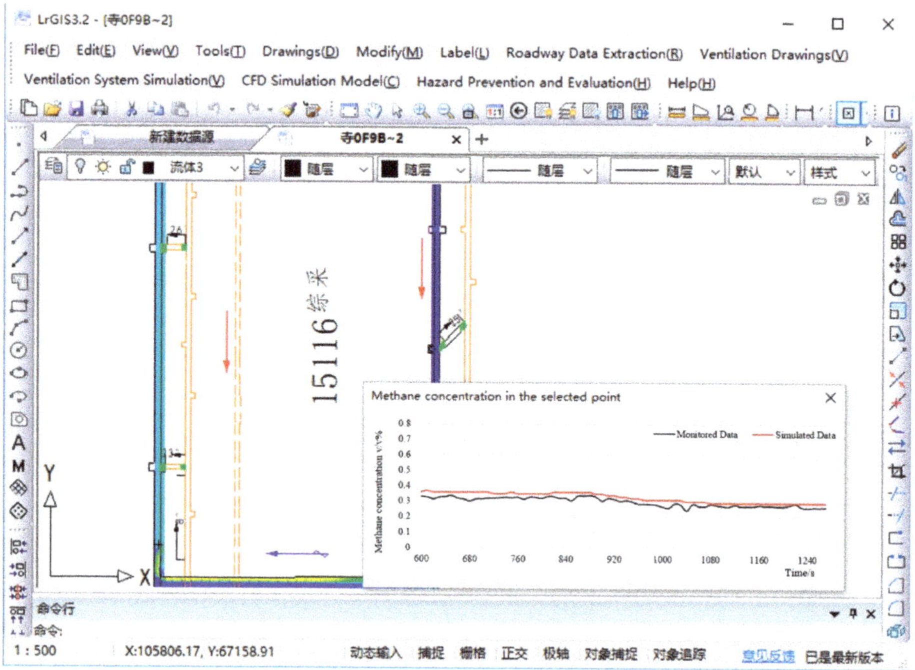

Figure 10. Point query analysis and data validation.

4.3. The Comprehensive Comparison of Traditional CFD Method and LBM

According to the kinetic theory [40], both traditional CFD method and LBM can satisfy the macroscopic Navier-Stokes equation in their continuous forms. However, Navier-Stokes equations are the target for LBM while the starting point for the traditional CFD methods, and these two methods also differ, not only in the spatial discretization forms, but also in the temporal discretization forms. Although a number of studies have been conducted to comprehensively compare their difference in essence, there is still a lack of comparison based on the actual simulation example. In this case study, comprehensive comparison and analysis between the proposed turbulent LBM based velocity-concentration coupling model and traditional CFD method are conducted from the following three aspects: accuracy, simulation efficiency, simplicity, and advantages of the integration with LrGIS platform.

Here, the traditional CFD method was implemented based on Computational Fluid Dynamics platform (Fluent) by employing the Euler-Lagrange model, all terms of the governing equations and constitutive relations were solved with the semi-implicit pressure-linked equation (SIMPLE) algorithm and second-order upwind scheme. The turbulence model and boundary conditions were solved using finite volume solver. The dynamic boundary parameters are the same with the LBM. Here, a line at 3 m high in the middle of the outtake side of laneway was chosen to investigate the airflow velocity and methane concentration at two different time points. As can be seen in Figures 11 and 12, it shows that both methods agree well with each other in the overall tendency of airflow velocity and methane concentration.

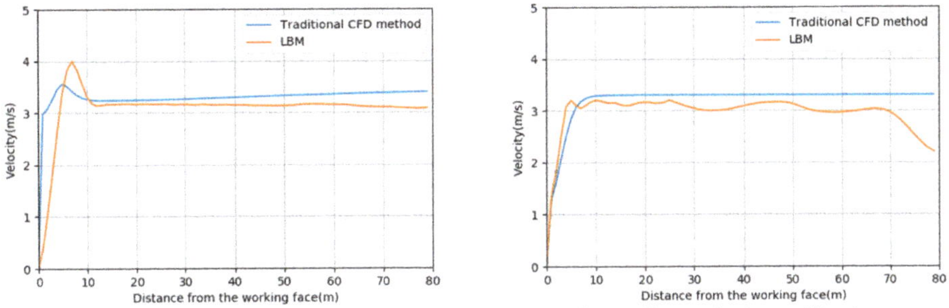

(**a**) Airflow velocity at 10 min. (**b**) Airflow velocity at 15 min.

Figure 11. Airflow velocity at the selected line.

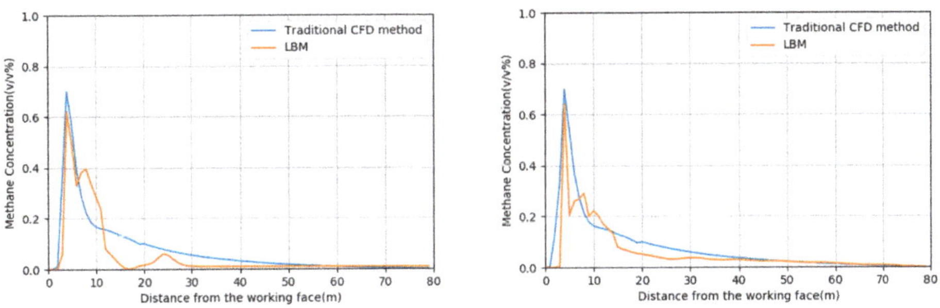

(**a**) Methane concentration at 10 min. (**b**) Methane concentration at 15 min.

Figure 12. Methane concentration at the selected line.

The traditional CFD simulation requires about 20 min on computers with four core processors and 8 GB RAM, while the model adopted in this paper uses around 15 min. It is noted that Fluent belongs to the third-party professional fluid dynamics simulation software, its independent geometric modeling and numerical simulation process makes it significantly difficult to directly guide the on-site problems, and the required data have to be preset and saved manually in the early stage of the simulation, which shows the poor practicability when applied in the on-site guidance. However, in this paper, the property of the unnecessary explicit mesh generation and the inherent simplicity of the algorithm make LBM more efficient and easy to integrate with the LrGIS platform. Specifically, the integrated system essentially implemented four objectives: (i) enabling managers to simulate case study directly on the GIS system and visualize the simulated results intuitively; (ii) storing the simulated spatiotemporal field data in the database system, which can be used for post analysis; (iii) demonstrating the simulated field data with various coal mine map objects, and providing miners the spatiotemporal analyze and query operation tools to obtain the useful location-based information; (iv) validating the results based on the coal mine monitoring system of Sijiazhuang coal mine. Therefore, by utilizing the synergies between GISs and spatiotemporal modeling, the integrated system is effective in providing decision supports for safety production, which is expected to improve the health standards in the underground coal mines.

The comparison case study demonstrates that the model and integration mode carried out in this study is more efficient in improving the computational efficiency and simplifying the integration mode of numerical simulation model with the GIS platform. It can also be concluded that the integrated system is more practical in on-site application compared with numerical simulation software.

5. Conclusions

Due to the lack of an effective platform for the practical application of numerical simulation methods in the underground coal mine systems, the concept of tight coupling of GIS and spatiotemporal modeling is introduced in this area for the first time.

Specifically, a turbulent LBM based velocity-concentration coupling model is proposed and tightly integrated with LrGIS platform by developing the FluentEntity model, which allows mine managers to simulate ventilation-related case study directly on the LrGIS system, visualize the simulated field data on the ventilation system map and other coal mine layers intuitively, and operate the spatiotemporal analyze and query to get location-based information, and therefore to guide the practical operation. Besides, the simulated result is validated with monitored data of Sijiazhuang underground coal mine, which shows the accuracy and reliability of the proposed model. In addition, by comparing with the traditional CFD method, the developed coupling scheme is demonstrated to be more efficient and easy to be integrated with GIS system.

In conclusion, this study provides a possibility to integrate the numerical simulation model with the GIS platform effectively, and shows that GIS could not only serve to carry out general constructive safety, but also generate case scenarios, quantify spatiotemporal processes, and therefore improve the safety standards for coal mines by GIS-coupled spatiotemporal modeling, analysis, and query tools. Furthermore, the system is more flexible and potentially applicable to other case study, like ventilation network solution, including both local network and the whole mine ventilation system.

Author Contributions: During this study, H.L. designed and performed the numerical simulation, and wrote the paper, S.M. and M.L. contributed materials, S.W. integrated the numerical model with the LrGIS platform.

Funding: This work is financially supported by the National Key Research and Development Program of China: [Grant Number 2016YFC0803108].

Acknowledgments: The authors would like to thank the anonymous reviewers who contributed to the quality of this letter by providing helpful suggestions.

Conflicts of Interest: No potential conflict of interest was reported by the authors.

References

1. Nakayama, S.; Kim, Y.K.; Jo, Y.D. Simulation of methane gas distribution by computational fluid dynamics. In *Mining and Science Technology*; Xie, H.P., Golosinski, T.S., Eds.; Balkema Publisher: Brookfield, VT, USA, 1999; pp. 259–262.
2. Kurnia, J.C.; Sasmito, A.P.; Wong, W.Y.; Mujumdar, A.S. Prediction and innovative control strategies for oxygen and hazardous gases from diesel emission in underground mines. *Sci. Total Environ.* **2014**, *481*, 317–334. [CrossRef]
3. Kurnia, J.C.; Xu, P.; Sasmito, A.P. A novel concept of enhanced gas recovery strategy from ventilation air methane in underground coal mines—A computational investigation. *J. Nat. Gas Sci. Eng.* **2016**, *35*, 661–672. [CrossRef]
4. Liu, H.; Wu, X.; Mao, S.; Li, M.; Yue, J. A Time Varying Ventilation and Dust Control Strategy Based on the Temporospatial Characteristics of Dust Dispersion. *Minerals* **2017**, *7*, 59. [CrossRef]
5. Xu, G.; Luxbacher, K.D.; Ragab, S.; Xu, J.; Ding, X. Computational fluid dynamics applied to mining engineering: A review. *Int. J. Min. Reclam. Environ.* **2017**, *31*, 251–275. [CrossRef]
6. Sun, Z.; Mao, S.; Wu, C. Assistant Decision System of Gas Explosion Emergency Rescue Based on Monitoring Data and CFD. *Saf. Coal Mines* **2016**, *47*, 83–86. [CrossRef]

7. Zhu, C.; Lin, B.; Ye, Q.; Zhai, C. Effect of roadway turnings on gas explosion propagation characteristics in coal mines. *Min. Sci. Technol.* **2011**, *21*, 365–369. [CrossRef]
8. Tian, Z.; Liang, Z.; Niu, Y.; Li, C.; Tu, X.; Fan, D. Simulation of the methane emission in rice fields in China during the past 40 years by DNDC model and GIS technical. In Proceedings of the 2014 Third International Conference on Agro-Geoinformatics, Beijing, China, 11–14 August 2014.
9. Goodchild, M.F.; Haining, R.; Wise, S. Integrating GIS and spatial data analysis: Problem and possibilities. *Int. J. Geogr. Inf. Syst.* **1992**, *6*, 407–423. [CrossRef]
10. Duffy, C.; Leonard, L.; Bhatt, G.; Yu, X.; Giles, L. Watershed Reanalysis: Towards a National Strategy for Model-Data Integration. In Proceedings of the 2011 IEEE Seventh International Conference on e-Science Workshops (eScienceW 2011), Stockholm, Sweden, 5–8 December 2011; pp. 61–65.
11. Juanle, W.; Kan, W.; Kuan, W. Integrated GIS solution to mining subsidence assistant decision in mining area. In Proceedings of the IGARSS 2004, 2004 IEEE International Geoscience and Remote Sensing Symposium, Anchorage, AK, USA, 20–24 September 2004; pp. 2868–2871.
12. Kolli, S.S.; Damodaran, P.S.; Evans, G.W. Geographic Information System Based Decision Support Systems for Facility Location, Routing and Scheduling. *Comput. Ind. Eng.* **1993**, *25*, 369–372. [CrossRef]
13. Banerjee, T.K. A GIS Solution for an Integrated Underground Coal Mine Management: A Conceptual Platform. *J. Manag. Policies Pract.* **2014**, *2*, 129–143.
14. Şalap, S.; Karslıoğlu, M.O.; Demirel, N.; Karslioglu, M.O. Development of a GIS-based monitoring and management system for underground coal mining safety. *Int. J. Coal Geol.* **2009**, *80*, 105–112. [CrossRef]
15. Suh, J.; Kim, S.-M.; Yi, H.; Choi, Y. An Overview of GIS-Based Modeling and Assessment of Mining-Induced Hazards: Soil, Water, and Forest. *Int. J. Environ. Res. Public Health* **2017**, *14*, 1463. [CrossRef]
16. Yenilmez, F.; Kuter, N.; Emil, M.K.; Aksoy, A. Evaluation of pollution levels at an abandoned coal mine site in Turkey with the aid of GIS. *Int. J. Coal Geol.* **2011**, *86*, 12–19. [CrossRef]
17. Ni, X.-M.; Chen, P.; Guo, L. Development of a decision-making system for coalbed methane development based on GIS. In Proceedings of the 2010 2nd Conference on Environmental Science and Information Application Technology, ESIAT 2010, Wuhan, China, 17–18 July 2010; pp. 348–351.
18. Yao, Y.; Liu, D.; Tang, D.; Huang, W.; Tang, S. Evaluation of the Coalbed Methane Potential by a GIS-Based Fuzzy AHP Model. In Proceedings of the 2009 Sixth International Conference on Fuzzy Systems and Knowledge Discovery, Tianjin, China, 14-16 August 2009; pp. 281–285.
19. Fang, K. GIS network analysis in rescue of coal mine. In Proceedings of the 21st International Society for Photogrammetry and Remote Sensing (ISPRS) Congress, Beijing, China, 3–11 July 2008.
20. Van Dijk, P.M.; Wang, H.Y.; Van Genderen, J.L. Earth observation knowledge transfer: The example of ITC's coal fire project. In Proceedings of the 20th International Society for Photogrammetry and Remote Sensing (ISPRS) Congress, Istanbul, Turkey, 12–23 July 2004.
21. Kresic, N.; Mikszewski, A. Hydrogeological Conceptual Site Models: Data Analysis and Visualization. 2012. Available online: https://books.google.com/books?id=6H383WiTiSQC&pgis=1 (accessed on 10 May 2019).
22. Ahm, Z.; Ashraf, A. Integration of Groundwater Flow Modeling and GIS. In *Water Resources Management and Modeling*; Nayak, P., Ed.; InTech: Rijeka, Croatia, 2012; pp. 239–262.
23. Li, M.; Liu, H.; Yang, C. A Real-Time GIS Platform for High Sour Gas Leakage Simulation, Evaluation and Visualization. *ISPRS Ann. Photogramm. Remote Sens. Spat. Inf. Sci.* **2015**, *2*, 225–231. [CrossRef]
24. Liu, Z.-P.; Liu, J.; He, Y.-w. Seamless coupling of 3D GIS techniques with FEM and its application to tunneling engineering. *Rock Soil Mech.* **2017**, *38*, 865–874. [CrossRef]
25. Zheng, M.; Jin, M. Modeling and Simulation of Toxic Gas Dispersion in Urban Street Supported by GIS. *Geomat. Inf. Sci. Wuhan Univ.* **2013**, *38*, 935–939.
26. Ma, J.T.; Chen, S.Z.; Zhu, X.T.; He, Z.C. Finite element numerical simulation method of groundwater flow and its application under 3D GIS. *J. Geo-inf. Sci.* **2016**, *18*, 749–757. [CrossRef]
27. Alcaraz, M.; Vázquez-Suñé, E.; Velasco, V.; Criollo, R. A loosely coupled GIS and hydrogeological modeling framework. *Environ. Earth Sci.* **2017**, *76*, 1–13. [CrossRef]
28. Bhatt, G.; Kumar, M.; Duffy, C.J. A tightly coupled GIS and distributed hydrologic modeling framework. *Environ. Model. Softw.* **2014**, *62*, 70–84. [CrossRef]
29. Gebbert, S.; Pebesma, E. A temporal GIS for field based environmental modeling. *Environ. Model. Softw.* **2014**, *53*, 1–12. [CrossRef]

30. Bai, L.; Yan, L.; Ma, Z. Interpolation and Prediction of Spatiotemporal Data Based on XML Integrated with Grey Dynamic Model. *ISPRS Int. J. Geo-Inf.* **2017**, *6*, 113. [CrossRef]
31. Papadimitriou, F. Modelling spatial landscape complexity using the Levenshtein algorithm. *Ecol. Informat.* **2009**, *4*, 48–55. [CrossRef]
32. Papadimitriou, F. The Algorithmic Complexity of Landscapes. *Landsc. Res.* **2012**, *37*, 591–611. [CrossRef]
33. He, X.; Doolen, G.D.; Clark, T. Comparison of the Lattice Boltzmann Method and the Artificial Compressibility Method for Navier–Stokes Equations. *J. Comput. Phys.* **2002**, *179*, 439–451. [CrossRef]
34. Khan, M.A.I.; Delbosc, N.; Noakes, C.J.; Summers, J. Real-time flow simulation of indoor environments using lattice Boltzmann method. *Build. Simul.* **2015**, *8*, 405–414. [CrossRef]
35. Perumal, D.A.; Dass, A.K. Simulation of incompressible flows in two-sided lid-driven square cavities. Part I—FDM. *CFD Lett.* **2010**, *2*, 13–24.
36. Chen, S.; Doolen, G.D. Lattice Boltzmann method for fluid flows. *Annu. Rev. Fluid Mech.* **1998**, *30*, 282–300. [CrossRef]
37. Elhadidi, B.; Khalifa, H.E. Comparison of coarse grid lattice Boltzmann and Navier Stokes for real time flow simulations in rooms. *Build. Simul.* **2013**, *6*, 183–194. [CrossRef]
38. Heinzer, T.J.; Williams, M.D.; Dogrul, E.C.; Kadir, T.N.; Brush, C.F.; Chung, F.I. Implementation of a feature-constraint mesh generation algorithm within a GIS. *Comput. Geosci.* **2012**, *49*, 46–52. [CrossRef]
39. Lu, Q. 3D Simulation of Concentrative Emission Gas Spreading in Ventilation Networks Based on LBM. *J. Syst. Simul.* **2011**, *23*, 2384–2390. [CrossRef]
40. Li, M.; Zhang, X.-p.; Mao, S.-j. Study on deep mining safety control decision making system. The 6th International Conference on Mining Science & Technology. *Procedia Earth Planet. Sci.* **2009**, *1*, 377–383. [CrossRef]
41. Lunarzewski, L.; Les, W. Gas emission prediction and recovery in underground coal mines. *Int. J. Coal Geol.* **1998**, *35*, 117–145. [CrossRef]
42. Bhatnagar, P.L.; Gross, E.P.; Krook, M. A Model for Collision Processes in Gases. I. Small Amplitude Processes in Charged and Neutral One-Component Systems. *Phys. Rev.* **1954**, *94*, 511–525. [CrossRef]
43. Guo, Z.; Shi, B.; Zheng, C. A coupled lattice BGK model for the Boussinesq equations. *Int. J. Numer. Methods Fluids* **2002**, *39*, 325–342. [CrossRef]
44. Smagorinsky, J. General Circulation Experiments with the Primitive Equations. *Mon. Weather Rev.* **1963**, *91*, 99–164. [CrossRef]
45. Hou, S.; Sterling, J.; Chen, S.; Doolen, G.D. A Lattice Boltzmann Subgrid Model for High Reynolds Number Flows. In *Pattern Formation and Lattice gas Automata*; Lawniczak, A.T., Kapral, R., Eds.; American Mathematical Society: Providence, RI, USA, 1996; pp. 151–166.
46. Zhang, S.J.; Lin, C.X. Application of Lattice Boltzmann Method in Indoor Airflow Simulation. *HVAC&R Res.* **2010**, *16*, 825–841.
47. Brimicombe, A. *GIS Environmental Modelling and Engineering*; Taylor & Francis: London, UK, 2003; p. 320.
48. Hellweger, F.L.; Maidment, D.R. Definition and Connection of Hydrologic Elements using Geographic Data. *J. Hydrol. Eng.* **1999**, *4*, 10–18. [CrossRef]
49. Olivera, F.; Valenzuela, M.; Srinivasan, R.; Choi, J.; Cho, H.; Koka, S.; Agrawal, A. ArcGIS-SWAT: A geodata model and GIS interface for SWAT. *J. Am. Water Resour. Assoc.* **2006**, *42*, 295–309. [CrossRef]
50. Huang, B.; Jiang, B. AVTOP: A full integration of TOPMODEL into GIS. *Environ. Model. Softw.* **2002**, *17*, 261–268. [CrossRef]

© 2019 by the authors. Licensee MDPI, Basel, Switzerland. This article is an open access article distributed under the terms and conditions of the Creative Commons Attribution (CC BY) license (http://creativecommons.org/licenses/by/4.0/).

applied sciences

MDPI

Article

Evaluation of Reliable Digital Elevation Model Resolution for TOPMODEL in Two Mountainous Watersheds, South Korea

Daeryong Park [1], Huan-Jung Fan [2], Jun-Jie Zhu [3], Sang-Eun Oh [4], Myoung-Jin Um [5] and Kichul Jung [1,*]

1 Department of Civil and Environmental Engineering, Konkuk University, 120 Neungdong-ro, Gwangjin-gu, Seoul 05029, Korea
2 Department of Safety, Health and Environmental Engineering, Hungkuang University, Taichung 43302, Taiwan
3 Department of Civil, Architectural and Environmental Engineering, Illinois Institute of Technology, Chicago, IL 60616-3793, USA
4 Department of Biological Environment, Kangwon National University, 192-1 Hyoja-2-dong, Gangwondo, Chuncheon 200-701, Korea
5 Department of Civil Engineering, Kyonggi University, 154-42 Gwanggyosan-ro, Yeongtong-gu, Suwon 16227, Korea
* Correspondence: jkichul11@konkuk.ac.kr

Received: 8 June 2019; Accepted: 3 September 2019; Published: 5 September 2019

Abstract: This study analyzed the result of parameter optimization using the digital elevation model (DEM) resolution in the TOPography-based hydrological MODEL (TOPMODEL). Also, this study investigated the sensitivity of the TOPMODEL efficiency by applying the varying resolution of the DEM grid cell size. This work applied TOPMODEL to two mountainous watersheds in South Korea: the Dongkok watershed in the Wicheon river basin and the Ieemokjung watershed in the Pyeongchang river basin. The DEM grid cell sizes were 5, 10, 20, 40, 80, 160, and 300 m. The effect of DEM grid cell size on the runoff was investigated by using the DEM grid cell size resolution to optimize the parameter sets. As the DEM grid cell size increased, the estimated peak discharge was found to increase based on different parameter sets. In addition, this study investigated the DEM grid cell size that was most reliable for use in runoff simulations with various parameter sets in the experimental watersheds. The results demonstrated that the TOPMODEL efficiencies in both the Dongkok and Ieemokjung watersheds rarely changed up to a DEM grid-size resolution of about 40 m, but the TOPMODEL efficiencies changed with the coarse resolution as the parameter sets were changed. This study is important for understanding and quantifying the modeling behaviors of TOPMODEL under the influence of DEM resolution based on different parameter sets.

Keywords: DEM grid cell size; efficiency; mountainous watersheds; optimized parameter set; TOPMODEL

1. Introduction

Rainfall–runoff relationships are generally simulated using a simplified conceptual model based on physical processes. Conventionally, this process involves the use of a lumped model, which considers the watershed as a homogenous element. The lumped model assumes that the watershed is a single black box that is highly dependent on the reliability of rainfall and runoff data; it suffers from the disadvantage that the temporal and spatial transformations of the rainfall to runoff are not explicitly considered. The development of Geographic Information Systems has facilitated the development of a dynamic model for analyzing the runoff phenomena and has allowed the development of a distributed parameter model that considers the spatial variability of parameters related to the runoff process. It focuses on the spatial variability of the process and explains the physical relationship between the topographic and runoff model parameters by analyzing the runoff.

To analyze the runoff event pattern using computational watershed models, the input parameters should be accurately estimated by reflecting the hydrological and hydraulic conditions such as the topographic characteristics, rainfall, and infiltration. However, accurately simulating the rainfall–runoff event is difficult because of the nonlinear characteristics of natural phenomena. One of the distributed parameter models developed to overcome these limitations is the TOPography-based hydrological MODEL (TOPMODEL), which uses the digital elevation model (DEM) [1]. TOPMODEL is semidistributed watershed model and has mainly been applied to hydrologic simulations in mountain basins [2–6]. TOPMODEL has the advantage in that it can easily separate surface runoff from basal runoff, and a runoff simulation can be performed for a long period. Similarly, the variable infiltration capacity (VIC) model can simulate rainfall–runoff processes and separates surface runoff and baseflow using the concept of three layers [7,8]. The VIC model simulates both water- and energy-balance equations and considers land-cover type as well as the concept of three soil layers. In contrast, TOPMODEL has a physical basis and focuses on hydrologic processes. Furthermore, the two models employ different infiltration estimation methods. TOPMODEL applies the Green Amp equation, whereas the VIC model uses various empirical methods [9].

Beven and Kirby [1] proposed TOPMODEL as a distributed parameter model for a runoff-contributing area. Beven and Wood [10] analyzed the effects of watershed topography on a river runoff using TOPMODEL. The influences of the spatial scale of the DEM, and the topographic characteristics on the hydrological process in TOPMODEL have been studied [11–14]. Saulnier et al. [15,16] analyzed the influence of permeability coefficient based on the DEM resolution. When the DEM resolution is larger, the permeability coefficient accordingly increases, and the variation in the coefficient is smaller.

To date, many researchers have pointed out that the topographic index in TOPMODEL is affected by the DEM resolution [17–19]. They usually demonstrated that high DEM resolutions can yield accurate topographic index estimates [20–22]. However, the effects of optimized input parameters based on the resolution of topographic data (e.g., effects of changes in DEM grid cell size), as well as the concept of reliable DEM grid cell size resolution on experimental watersheds in TOPMODEL have been rarely studied. In particular, Lin et al. [23] investigated the influence of DEM resolution in TOPMODEL in the Hanjiang River, China, and reported no evident differences in the uncertainty intervals corresponding to different DEM resolutions represented in TOPMODEL. However, uncertainty intervals corresponding to different DEM resolutions depending on changing input parameters in TOPMODEL have not yet been clarified. The objective of this study is to suggest a reliable DEM grid-size resolution that exhibits a low sensitivity to change in the input parameters during runoff simulations in experimental watersheds. To this end, this study attempted (1) to investigate the effects of DEM grid cell size on the simulated runoffs during the application of different optimized input parameters, and (2) to investigate the effects of DEM grid cell size on the simulated runoffs when a constant input parameter set is applied.

2. Geographical Setting and Morphometry of the Study Area

TOPMODEL was applied as a conceptual distributed runoff model to generate runoffs in the target watersheds. TOPMODEL considered the topographic factors estimated by DEM in the watershed. This study investigated two mountainous watersheds, namely, the Dongkok and Ieemokjung watersheds, using TOPMODEL, as shown in Figure 1. The Dongkok watershed is located upstream of the Wicheon river, a tributary of the Nakdong River, and contains one rain- and a number of streamflow-gauge stations. Similarly, the Ieemokjung watershed is a subcatchment of the Pyeongchang River, a tributary of the South Han River, and contains one rain- and one streamflow-gauge station. Figure 1 shows the locations of the rain- and streamflow-gauge stations in the Dongkok and Ieemokjung watersheds.

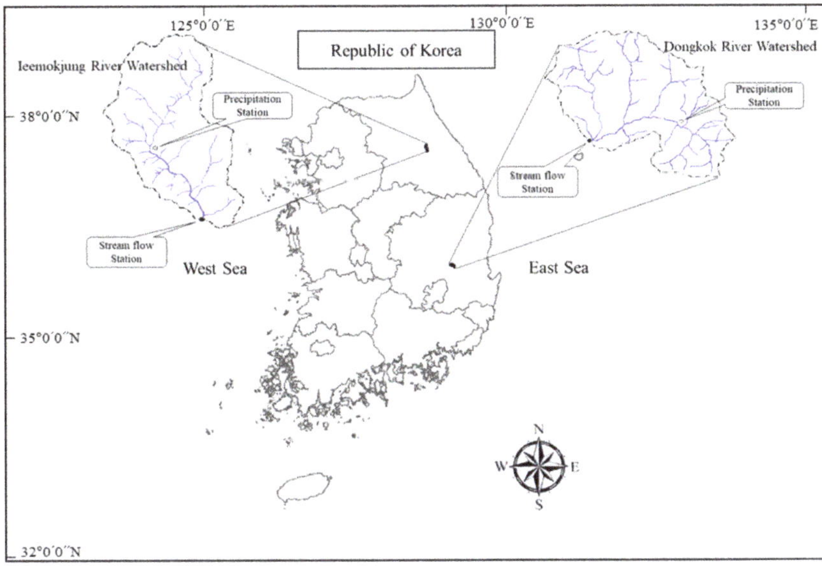

Figure 1. Location of the Dongkok (Lat: 36°07′11″ N, Lon: 128°47′46″ E) and Ieemokjung (Lat: 37°37′17″ N, Lon: 128°28′19″ E) watersheds in South Korea.

Table 1 lists the topographical and morphological characteristics of the Dongkok and Ieemokjung watersheds. The drainage area, river length, and main tributary length in the Ieemokjung watershed are larger than those in the Dongkok watershed. However, the mean width of the basin and shape factor in the Dongkok watershed is larger than that in the Ieemokjung watershed. These conditions indicate that the Dongkok watershed has a more circular shape and a wider channel than the Ieemokjung watershed. In terms of the river density and basin slope, the Dongkok watershed is slightly larger than the Ieemokjung watershed.

Table 1. Topographic characteristics of the Dongkok and Ieemokjung watersheds.

Characteristics	Dongkok Watershed	Ieemokjung Watershed
Drainage area (km^2)	33.63	55.76
River length (km)	8.00	16.55
Main tributary length (km)	39.92	53.25
Mean width of basin (km)	4.203	3.370
Shape factor	0.525	0.204
Drainage density (km/km^2)	1.188	0.955
Basin slope	0.452	0.344

The applied DEM spatial resolution values were 5, 10, 20, 40, 80, 160, 200, and 300 m, using ArcGIS. DEM data scales for Dongkok and Ieemokjung watersheds were 1:5000 and 1:25,000, respectively. The hydrological data were based on rainfall data, and the runoff data were obtained from the International Hydrological Programme (IHP) report [24]. The present study chose four rainfall–runoff events in the Dongkok watershed and seven events in the Ieemokjung watershed. The parameters of the selected rainfall–runoff events are listed in Table 2. Rainfall–runoff events in the 1990s were adopted because the applied DEM data and verified streamflow data for TOPMODEL for the target watersheds were created in the late 1990s by the National Geographic Information Institute in Korea (https://www.ngii.go.kr) [24,25].

Table 2. Selected rainfall–runoff events in the TOPMODEL application.

Watershed	Year	Rainfall–Runoff Event Time (m.dd.hh)	Durations of Events (h)	Name of Events
Dongkok	1996	6. 16. 23–6. 17. 22 8. 26. 22–8. 27. 19	16 17	DK960616 DK960826
	1997	7. 15. 18–7. 16. 21	17	DK970715
	1998	8. 15. 14–8. 17. 10	16	DK980815
Ieemokjung	1991	5. 25. 13–5. 26. 17	26	IM910525
	1992	9. 24. 6–9. 24. 24	17	IM920924
	1994	6. 29. 23–7. 1. 23 8. 1. 3–8. 2. 4	42 19	IM940629 IM940801
	1996	7. 4. 13–7. 5. 5 8. 25. 13–8. 27. 24	13 17	IM960704 IM960825
	1999	7. 28. 3–7. 29. 24	42	IM990728

3. Materials and Methods

3.1. TOPMODEL

TOPMODEL simulates the runoffs using the water balance due to precipitation, estimated infiltration (regarded as a saturation deficit), and the loss by evapotranspiration. Figure 2 shows the basic concept of the TOPMODEL scheme. P represents the precipitation, a_i is the slope area per unit contour length, q_i is the runoff per unit width, S_{rz} is the root-zone deficit, and S_{uz} is the local water storage in the unsaturated zone. q_v is the flux of water that locally enters the water table (per unit area), and Q represents the drainage of the watershed unit. The formula that serves as the basis of TOPMODEL is based on the continuity equation and Darcy's Law. The model equation is developed through several important assumptions as follows.

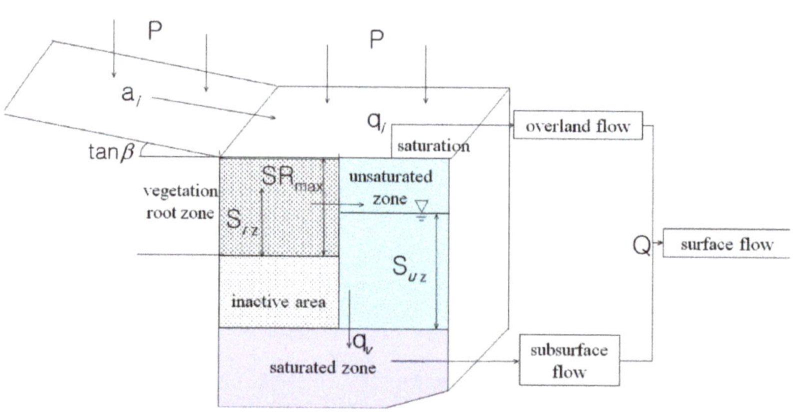

Figure 2. Basic concept of the TOPMODEL scheme (modified from [5,14,19]).

First, the flow of water in the watershed is assumed to be at a steady state.

Second, the hydraulic gradient of the saturated soil in the watershed can be expressed as tan β, which represents the surface gradient.

Third, the saturation permeability coefficient (T_0, m^2/h) exponentially decreases from the soil surface to the groundwater surface. These assumptions indicated that the most important factor on the storage and groundwater of the watershed is the topographic index, also known as topographic wetness index, i.e., ($ln(\frac{a}{tan\beta})$) [26–30].

The saturated hydraulic conductivity of soil (K_s, m/h) can be expressed as a negative exponential function with respect to depth, as follows:

$$K_s = K_0 \exp(-fz) \tag{1}$$

where z (m) is the underground water depth, K_0 (m/h) is the coefficient of permeability of the soil at the ground surface, and f (1/m) is the coefficient ratio of the K_s change depending on underground water depth z. f was assumed to be constant in the entire watershed.

The hydraulic gradient in the saturated layer can be expressed as a topographic slope in the watershed, and the behavior of the saturated layer follows Darcy's Law. Therefore, the flow rate per unit width q_i (m^2/h) at point i is expressed as follows:

$$q_i = T_i(z_i) tan\beta_i \tag{2}$$

where $tan\beta_i$ is the slope of the surface at location i, and $T_i(z_i)$ is the transmissivity coefficient. $T_i(z_i)$ in Equation (2) can be obtained by integrating Equation (1) as follows:

$$T_i(z_i) = \int_{z_i}^{z(t)} K_s(x)dx = \frac{K_0}{f}[\exp(-fz_i) - \exp(-fz)] = \frac{1}{f}[K_s(z_i) - K_s(z)] \tag{3}$$

Generally, when depth z of the groundwater is large, the saturation permeability coefficient $K_s(z)$ (m/h) is negligible, because it is very small compared with saturation permeability $K_s(z_i)$ at depth z_i. Therefore, substituting Equation (3) into Equation (2) yields the following equation:

$$q_i = \frac{K_0}{f} tan\beta_i \exp(-fz_i) = T_0 tan\beta_i \exp(-fz_i) \tag{4}$$

where $T_0 = K_0/f$ is the coefficient of permeability of the fully saturated soil. K_0 and f were assumed to be constant in the entire watershed.

In addition, the behavior of the flow in the saturation layer can be expressed as a continuous steady state as follows:

$$q_i = a_i R = T_0 tan\beta_i \exp(-fz_i) \tag{5}$$

where R (m/h) is the spatially constant recharge of the groundwater at any given time, and a_i (m) is the area of the slope per unit contour length drained through point i.

Equation (5) can be summarized based on z_i, and the average value ($\overline{z}$) of the entire watershed can be obtained as follows:

$$z_i = -\frac{1}{f}\left(ln\frac{Ra_i}{T_0 tan\beta_i}\right) \tag{6}$$

where R is spatially constant, and $ln(\frac{a}{T_0 tan\beta})$ is the soil topographic index.

Introducing average topographic indexes to Equation (6) yields the following equation:

$$z_i = \overline{z} - \frac{1}{f}\left[ln\frac{a_i}{T_0 tan\beta_i} - E\left(ln\frac{a_i}{T_0 tan\beta_i}\right)\right] \tag{7}$$

where $E(\,)$ denotes the average topographic indexes of the entire watershed [31].

Equation (7) demonstrates that depth z_i of the groundwater can be determined by factor f and topographic index ($ln(\frac{a}{tan\beta})$).

Thus, given the value of $\overline{z}$, Equation (7) can be used to estimate the spatial distribution of the topographic indexes and to predict the behavior of the surface water and groundwater levels in the entire watershed, assuming that the topographic index is spatially constant [11]. The soil topographical factors and topographic constants represent the behavior of all the flows in the watershed. All points with the same soil topographic factors and topographic constants exhibit the same flow performance. Therefore, the soil topographical factors and topographic constants could indicate the hydrological similarities between the watersheds.

The input data of TOPMODEL were divided into hydrological data, topographic data, and input parameters. The hydrological data comprised the rainfall, evapotranspiration, and observed flow data. The topographic data were obtained by extracting the DEM data from a digital map. This study applied the FORTRAN source code of TOPMODEL [1]. Figure 3 shows the calculation procedure of TOPMODEL. TOPMODEL calculated the input data, compared the calculated value with the actual value, and reduced the error generated by parameter optimization.

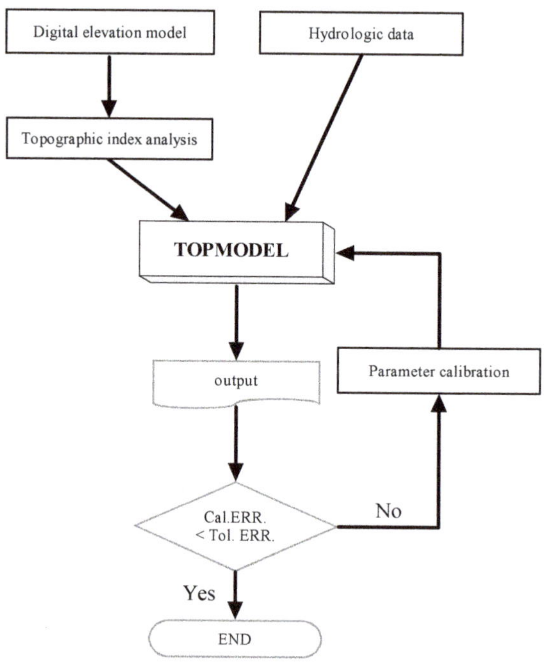

Figure 3. Flowchart of the TOPMODEL simulation. TOPMODEL: TOPography-based hydrological MODEL.

3.2. Input Parameters Optimization

TOPMODEL is a model that analyzes the runoff characteristics in a watershed by simulating the behavior of groundwater in the soil using the soil topographical factors of the watershed. The topographic parameters are the main parameters that determine the response characteristics in the subwatershed or slope, including the topographic index factors. TOPMODEL contains five calibration parameters such as the exponential storage parameter (m), log-transformed soil–water permeability coefficient ($ln(T_0)$), unsaturated lag time (T_d), maximum vegetation storage (SR_{max}), and vegetation deficiency (SR_0), as listed in Table 3. Each input parameter in TOPMODEL is constant to all DEM grid cell sizes as one parameter set. In other words, each input parameter is changed as the grid cell size or applied Rainfall–runoff events are changed, but is not changed by DEM grid cell numbers. These parameters are characterized as nominal values. The input parameters were estimated using all possible combinations of calibration ranges in Table 3 to compare the observed flows with the calculated flows. Calibration ranges of all input parameters were based on Cho [32] and Lee [33]. This study optimized the input parameters for all five parameter combinations with different segments (e.g., 0.001 for m and 0.1 for SR_{max}, 0.01 for SR_0, and 1 for $ln(T_0)$ and T_d), depending on grid cell size change and Rainfall–runoff events in Fortran. The objective function of optimization parameters in this study adopted the Nash–Sutcliffe efficiency (NSE) [34]. This study selected input parameter values that maximized the NSE estimations between the actual and simulated runoffs using the actual rainfall and runoff data from the Dongkok and Ieemokjung watersheds.

Table 3. Calibration parameters and ranges used in TOPMODEL in this study [32,33].

Parameters	Description	Units	Calibration Range
m	Exponential storage parameter	m	0.001–0.1
$ln(T_0)$	Soil-water permeability coefficient	$\ln(m^2/h)$	1–50
T_d	residence time in unsaturated zone	h	1–50
SR_{max}	Maximum vegetation storage	m	0.1–1
SR_0	Vegetation deficiency	m	0.01–1

4. Results

This study investigated the results of the topographic characteristics of the experimental watersheds, DEM size effects based on fixed parameter sets, and DEM size sensitivity based on various parameter sets. According to these results, the least sensitive resolution of the DEM grid cell size for TOPMODEL in the Dongkok and Ieemokjung watersheds was analyzed.

4.1. Frequencies of the Topographic Indexes Based on the DEM Grid Cell Size

Figures 4 and 5 show the frequency and cumulative frequency of the topographical indexes based on the grid cell size of the Dongkok and Ieemokjung watersheds, respectively. The curves in Figures 4 and 5 shifted from left to right as the grid cell size increased for both the Dongkok and Ieemokjung watersheds. These results indicate that the mean of topographic index ($ln(\frac{a}{tan\beta})$) increases as DEM cell resolution is coarser. It is because contributing area (a) increases, and the slope of watershed ($tan\ \beta$) decreases as DEM cell size increases. This analysis is coincided with Figure 6, which shows the log-transformed contributing area ($ln(a)$) and the log-transformed inverse slope of watershed ($ln(\frac{1}{tan\beta})$) increase as DEM cell resolution is coarser [13,35].

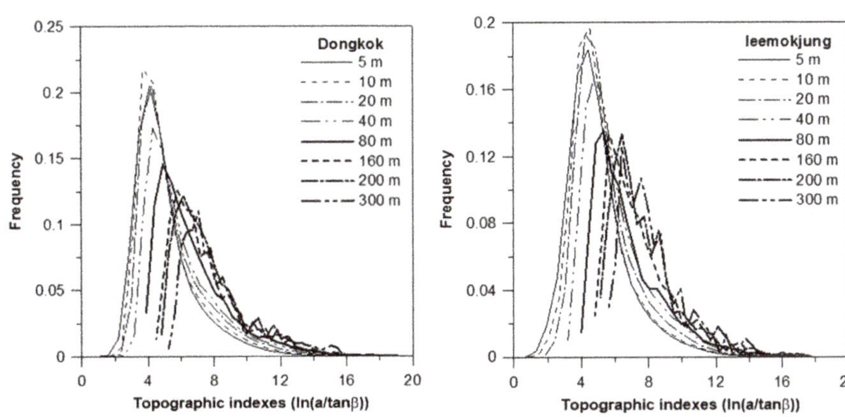

Figure 4. Frequency distribution of the topographic indexes ($ln(\frac{a}{tan\beta})$) based on the DEM grid cell size. (**a**) Dongkok watershed. (**b**) Ieemokjung watershed.

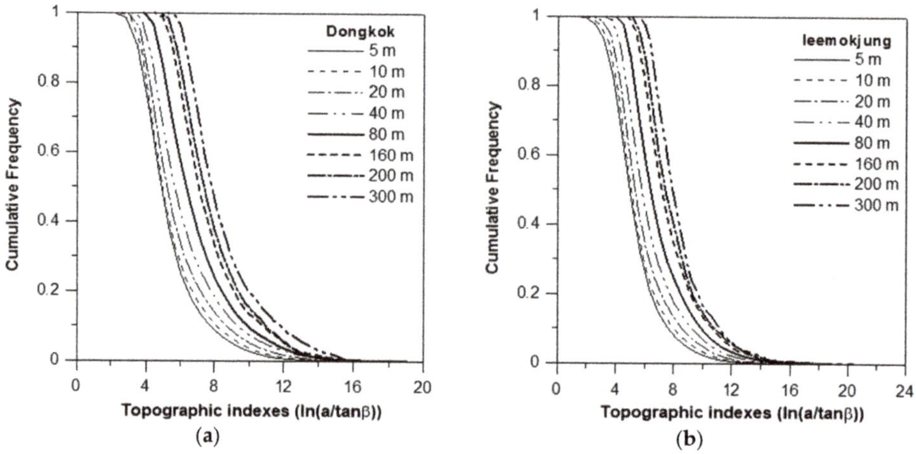

Figure 5. Cumulative frequency of the topographic indexes ($ln(\frac{a}{tan\beta})$) based on the DEM grid cell size. (**a**) Dongkok watershed. (**b**) Ieemokjung watershed.

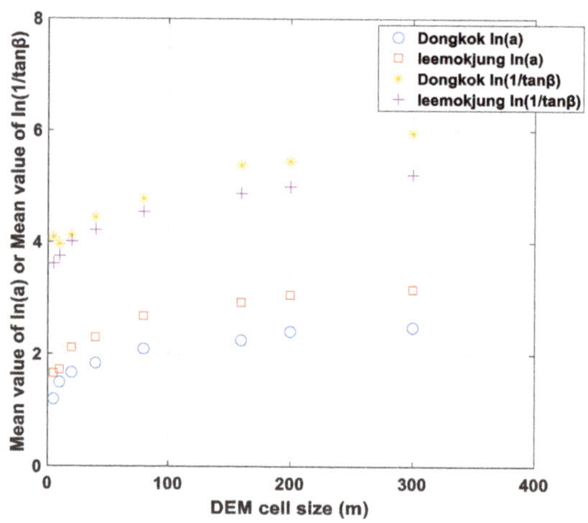

Figure 6. Relation of mean topographic indexes ($ln(\frac{a}{tan\beta})$) to mean ln(a) and mean ln(1/tanβ) values based on DEM cell size in Dongkok and Ieemokjung watersheds.

Figure 7 shows the variation in the contributing area of the grid cell size in the Dongkok and Ieemokjung watersheds, which depended on the Rainfall–runoff events. Figure 7a shows that the contributing area increased with respect to the grid cell size, but the slopes of the curves in the grid cell size in the Dongkok watershed were different depending on the Rainfall–runoff events. The slopes (ratio) of the curves in DK960826 and DK980815 were 0.239 and 0.337, respectively, and those in DK960616 and DK970715 were 0.609 and 0.803, respectively. The variation range in the slopes was between 0.239 and 0.803. Figure 7b shows the change in the slope of the curves in the grid cell size from 0.356 to 0.540 in the Rainfall–runoff events in the Ieemokjung watershed. Figure 7 shows that the contributing area in the Rainfall–runoff events and grid cell size varied. In other words, when the grid cell size was large, the contributing area increased, but the contributing area varied depending on the characteristics of the Rainfall–runoff events, even when the grid cell size was constant. The change of contributing area depending on rainfall events in Figure 6 shows a good agreement that the contributing area is not stationary but changes over both storm and seasonal timescales, reflecting changes in the average watershed water balance, topography, and soil conditions. Particularly, the contributing area differs from the mean watershed deficit related to the watershed water balance for the saturated zone [1,36,37]. It implies that rainfall events affect mean watershed deficit, and different mean watershed deficits influence the contributing area. Also, several studies found that the influence of DEM resolution was through its effect on the calculation of the contributing area [13,35]. However, more results and studies need to be provided to clarify the relation among rainfall events, contributing area, input parameters, and DEM resolution.

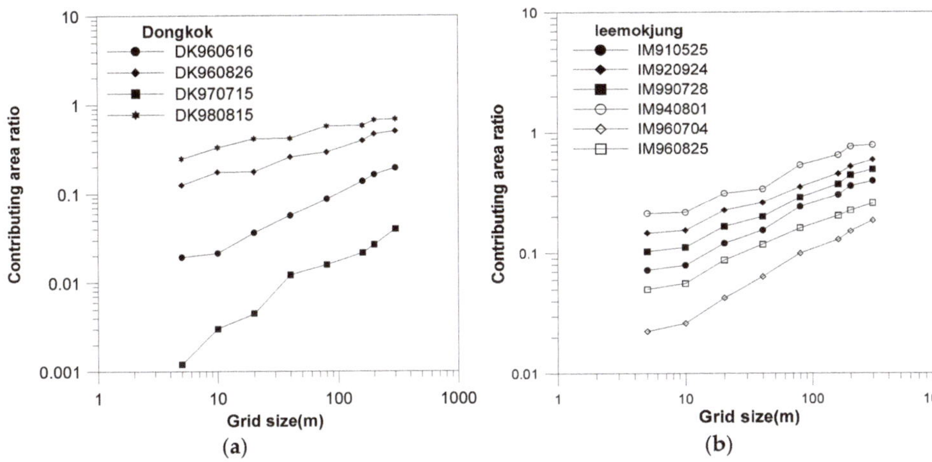

Figure 7. Change in the ratio of the contributing area based on the DEM grid cell size. (**a**) Dongkok watershed. (**b**) Ieemokjung watershed.

4.2. Resolution Effects of the Grid Cell Size with Fixed Parameter Sets

Figure 8 shows the hydrographs in the Dongkok watershed, which varied depending on the grid cell size with fixed input parameter sets. Fixed parameter sets were used to investigate the resolution effects of the grid cell size. This study chose a fixed parameter set that displayed the best fit in a 5 m grid cell size based on the Rainfall–runoff events in Table 4. Figure 8 shows that the estimated hydrographs with a smaller grid cell size were closer to the observed hydrograph. In contrast, when the grid cell size was larger, the simulated hydrograph was larger than the observed one because the grid cell size and contributing area were larger. The larger contributing area resulted in more runoffs. In particular, Figure 8b,d show that the peak rainfall matched with the estimated peak hydrographs under different grid cell sizes, which indicated that the larger the grid cell size, the larger the contributing area and the more sensitive the runoff to peak rainfall. Figure 9 shows the hydrographs whose DEM grid size varied with the application of fixed parameter sets in each hydrological event in the Ieemokjung watershed. Similarly, the input parameter set used the best fitting parameter set for the 5-m grid size in Table 5. Similar to the Dongkok watershed, the estimated runoffs well matched with the observed hydrographs because the DEM grid size was smaller in the Ieemokjung watershed. In Figures 8 and 9, the hydrographs for coarser DEM resolution provided greater surface runoffs. It is because the depth z_i of the groundwater decreases as the topographic index increases, from Equation (6) [16,35]. Particularly, negative z_i means the unsaturated zone in the DEM grid is saturated. Decreasing depth of the groundwater leads to decreasing infiltration rate and increasing surface runoff in the constant rainfall condition. In other words, the infiltration rate decreases as a topographic index increases or the DEM grid cell resolution is coarser [38,39].

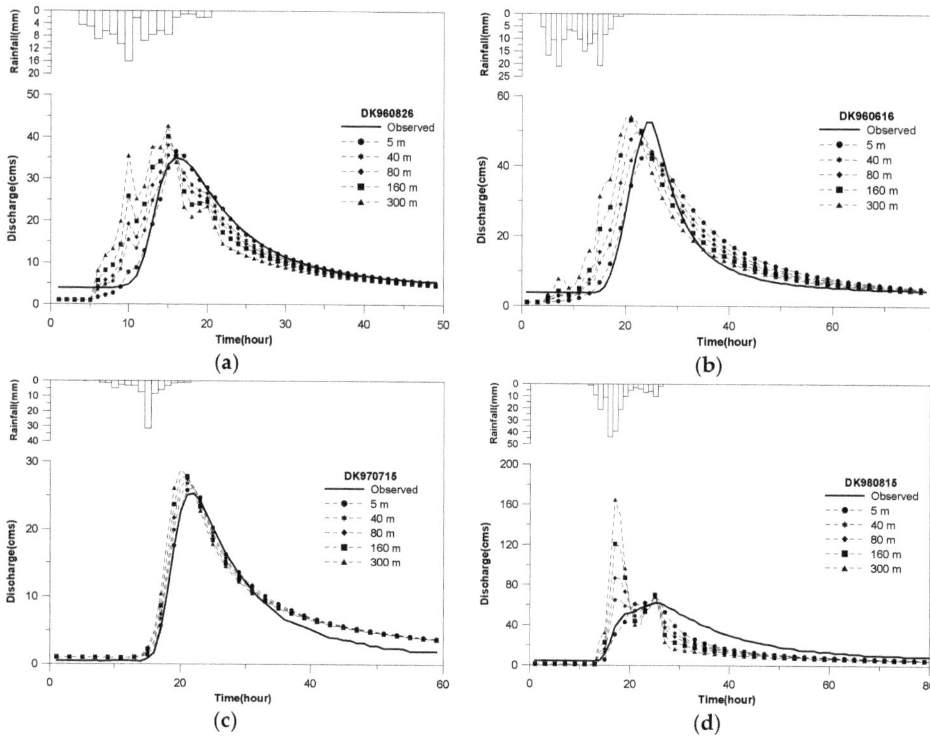

Figure 8. Rainfall–runoff events by applying a constant parameter set depending on different DEM resolution values in the Dongkok watershed. (**a**) 16 June 1996. (**b**) 26 August 1996. (**c**) 26 July 1997. (**d**) 15 August 1998.

Table 4. Applied fixed input parameter sets in the Dongkok watershed.

Hydrologic Events	Input Parameters				
	m	$ln(T_0)$	T_d	SR_{max}	SR_0
DK960616	0.025	5.0	49.0	0.1	0.01
DK960826	0.019	2.0	17.0	0.1	0.01
DK970715	0.016	8.0	15.0	0.1	0.01
DK980815	0.032	1.0	14.0	0.1	0.01

Table 5. Applied fixed input parameter sets in the Ieemokjung watershed.

Hydrologic Events	Input Parameters				
	m	$ln(T_0)$	T_d	SR_{max}	SR_0
IM910525	0.024	3.0	6.0	0.1	0.01
IM920924	0.027	1.0	1.0	0.1	0.01
IM940801	0.028	1.0	14.0	0.1	0.01
IM960704	0.016	3.0	23.0	1.0	0.01
IM960825	0.047	1.0	10.0	1.0	0.01
IM990728	0.016	3.0	40.0	1.0	0.01

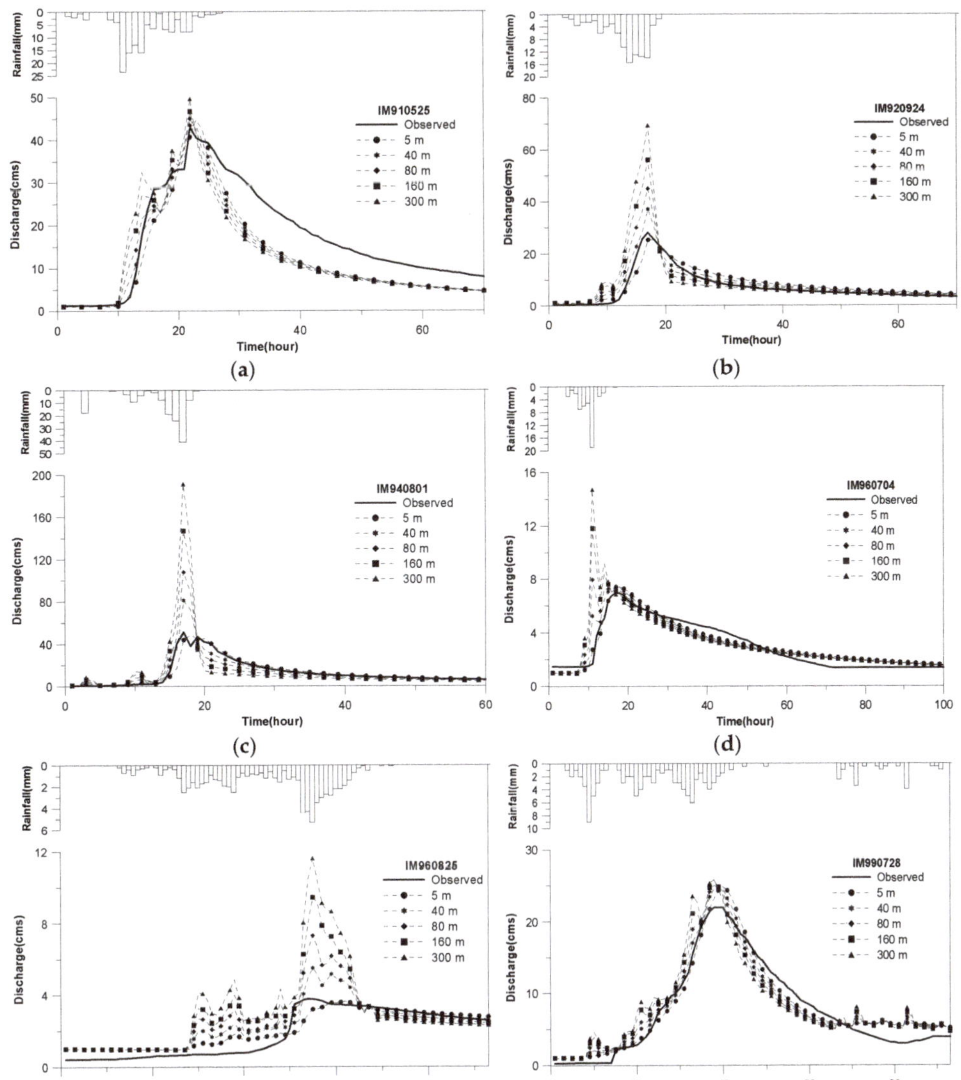

Figure 9. Rainfall–runoff events by applying a constant parameter set depending on different DEM resolution values in the Ieemokjung watershed. (**a**) 25 May 1991. (**b**) 24 September 1992. (**c**) 1 August 1994. (**d**) 4 July 1996. (**e**) 25 August 1996. (**f**) 25 July 1999.

4.3. Parameter Optimization for Each DEM Grid Cell Size

This study optimized input parameters of TOPMODEL depending on DEM resolutions and hydrologic events in the Dongkok and Ieemokjung watersheds in Figures 10 and 11. The input parameters are optimized based on the changes of DEM grid cell size and rainfall–runoff events. When the grid cell size changes, the simulated runoff shows a large change with parameter m, a small change with parameters ln (T_0) and T_d, and no change with SR_{max} and SR_0. Parameter m is the most sensitive because it is a parameter in an exponential equation, and the results in Figures 10 and 11 change significantly with changes in this parameter. The parameters ln (T_0) and T_d increase as DEM grid size is greater. It demonstrates that an ultimate infiltration capacity, the soil–water permeability coefficient (ln (T_0)) in this study, decreases as DEM grid cell resolution is coarser [38,39]. Therefore, it is necessary to increase the parameter ln (T_0) as well as the lag time in the unsaturated zone (T_d) to increase the actual infiltration capacity rate as DEM grid cell resolution decreases.

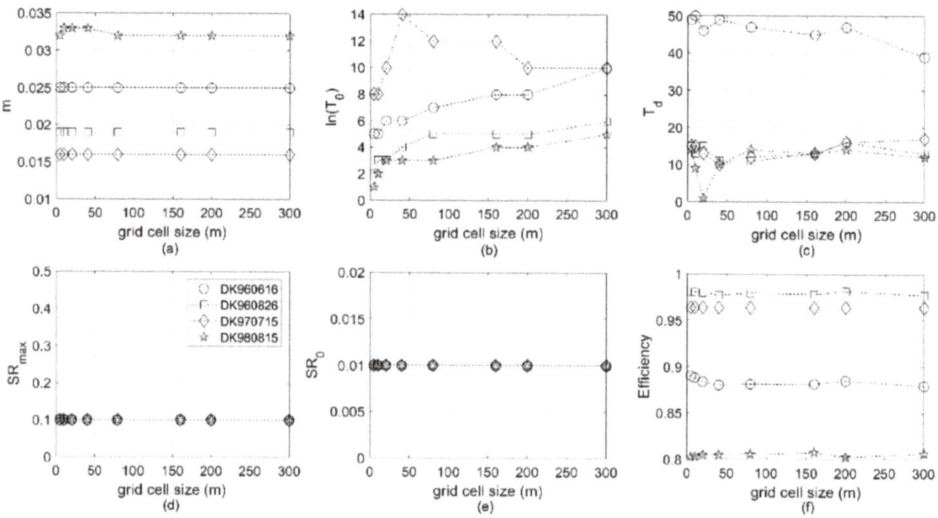

Figure 10. Optimized parameters depending on DEM resolutions in each hydrologic event, the Dongkok watershed. (**a**) exponential storage parameter (m), (**b**) soil–water permeability coefficient ($ln(T_0)$), (**c**) unsaturated lag time (T_d), (**d**) maximum vegetation storage (SR_{max}), (**e**) Vegetation deficiency (SR_0), (**f**) model efficiency.

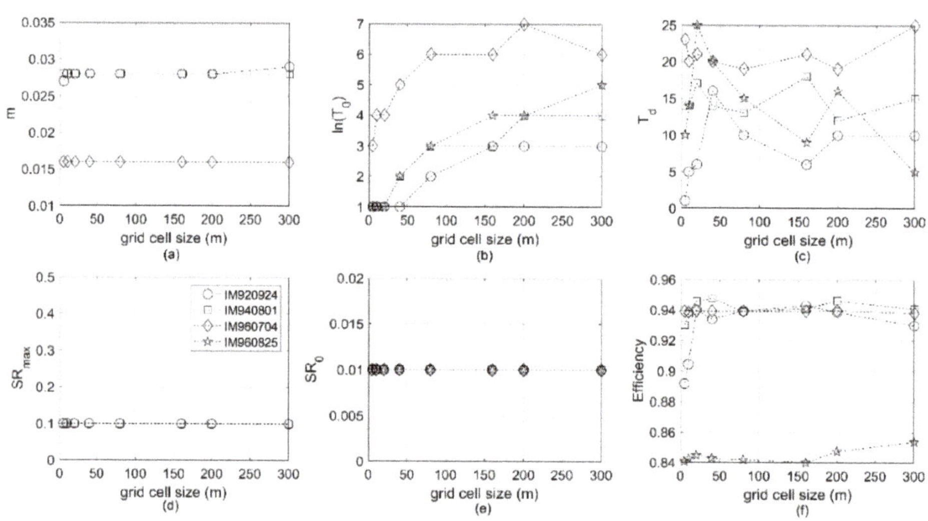

Figure 11. Optimized parameters depending on DEM resolutions in each hydrologic event, the Ieemokjung watershed. (**a**) exponential storage parameter (m), (**b**) soil–water permeability coefficient ($ln(T_0)$), (**c**) unsaturated lag time (T_d), (**d**) maximum vegetation storage (SR_{max}), (**e**) Vegetation deficiency (SR_0), (**f**) model efficiency.

Figures 12 and 13 show TOPMODEL hydrographs based on the optimization of the input parameters in the Dongkok and Ieemokjung watersheds depending on the DEM resolution. The NSEs of the Dongkok watershed shown in Figures 10 and 12a ranged from 0.88 to 0.90, 0.97 to 0.98, 0.94 to 0.96, and 0.77 to 0.80 depending on the events. In particular, in the DK960616 and DK970715 events in Figure 10, the generated peak flows were very close to observed peak flows depending on the resolution of the DEM grid cell size. Therefore, the result indicated that the TOPMODEL efficiencies were slightly different based on the DEM grid cell size resolution; however, the optimized input parameters were significantly different based on the hydrological events. The NSE of the Ieemokjung watershed shown in Figures 11 and 12b ranged from 0.89 to 0.94, 0.88 to 0.94, 0.88 to 0.93, and 0.8 to 0.86. In the IM960704 and IM960825 events, the estimated peak flow rates were very close to observed peak flows depending on the DEM resolution. In addition, similar to the Dongkok watershed, the NSE efficiencies significantly varied depending on the rainfall–runoff events instead of the DEM resolution.

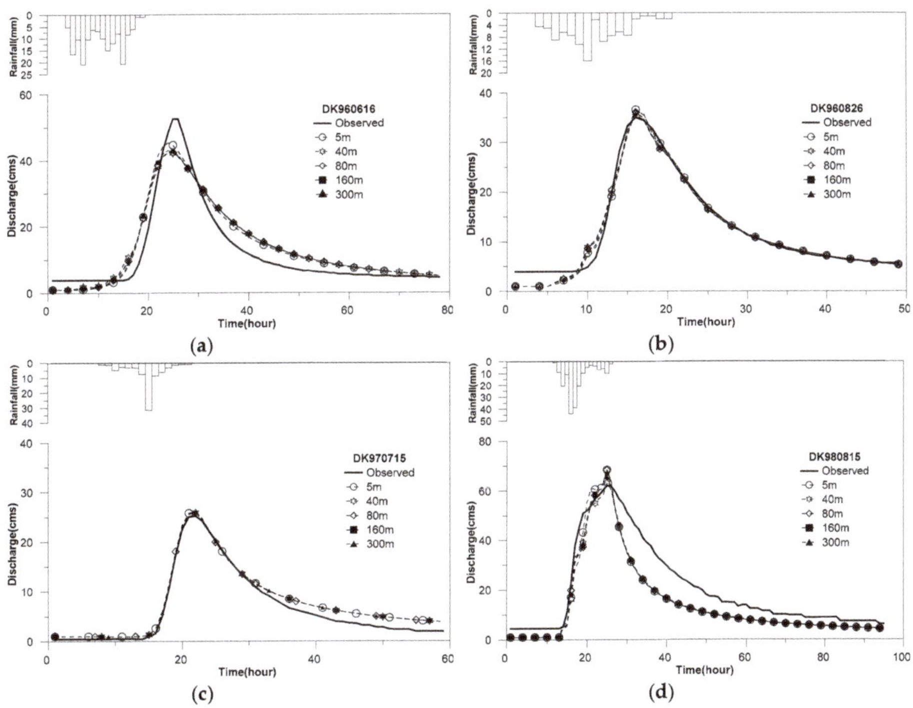

Figure 12. Rainfall–runoff events by applying optimized parameters in the Dongkok watershed depending on each DEM resolution. (**a**) 16 June 1996. (**b**) 8 August 1996. (**c**) 15 July 1997. (**d**) 15 August 1998.

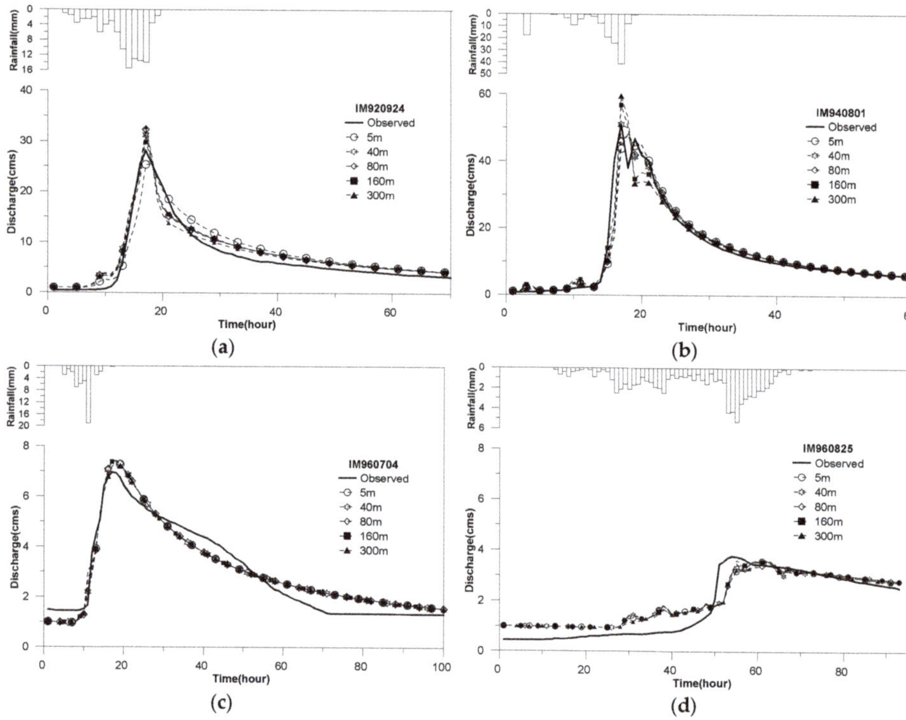

Figure 13. Rainfall–runoff events by applying optimized parameters in the Ieemokjung watershed depending on each DEM resolution. (**a**) 24 September 1992. (**b**) 1 August 1994. (**c**) 4 July 1996. (**d**) 25 August 1996.

5. Discussion

Because the watershed in the distributed watershed models was divided into multiple grid networks, the topography and Rainfall–runoff characteristics of the watershed varied depending on the DEM grid cell size. If the resolution of the DEM grid cell size is very high, it can accurately represent the characteristics of actual Rainfall–runoff events in a watershed. However, the simulation time for the runoff estimation and topography analysis in the model increase. If the resolution of the DEM grid cell size is too low, the geographical characteristics of the actual watershed are very simple, and a risk of computational errors exists [35]. Therefore, the present study proposed optimum DEM resolution on the basis of the work efficiency and runoff output accuracy in the Dongkok and Ieemokjung watersheds.

Figure 14 shows the NSE box plots in the Dongkok and Ieemokjung watersheds when the parameters were optimized in each grid for all Rainfall–runoff events. TOPMODEL hydrographs obtained using the optimized parameters in each DEM resolution were similar to the observed runoff data, which indicated that the resolution did not affect the model because the input parameters were well optimized depending on the DEM resolution.

Figure 15a shows the efficiency under a constant parameter set in the Dongkok watershed when the resolution was varied. The efficiencies of the 5 and 40 m DEM resolution slightly changed, but the efficiencies of the 80 m or higher DEM resolution significantly changed with the constant parameter set. Similarly, Figure 15b shows the efficiency under a constant parameter set in the Ieemokjung watershed depending on various DEM resolution values. Similar to Figure 15a, the variation in the efficiency was very small when the DEM resolution values were 5 and 40 m. However, the change in the efficiency was very large with the constant parameter set depending on the 80, 160, and 300 m DEM resolution values. Figure 15 shows that the combination of the input parameters and DEM resolution smaller than 40 m was not sensitive, which implied that the estimated topographic index, which was the area of the slope per unit contour length divided by the slope of the surface, was similarly smaller in the 40 m DEM grid cell size. This analysis agreed with that shown in Figures 4 and 5. The frequencies of the topographic indexes of the 40 m DEM grid cell size shown in Figures 4 and 5 were slightly different from those of the 5 and 20 m DEM grid cell size. Therefore, the result suggested that at least the 40 m resolution showed reliable results with the variation in the input parameters in the TOPMODEL application to the Dongkok and Ieemokjung watersheds. The proposed resolutions in this study were similar to the good-performance resolutions DEM that used 30 m resolution [40], greater than 10 m resolution [35], or smaller than 100 m resolution [37,41]. However, it is necessary to investigate resolutions that provide similar results as this study to the different characteristics of watersheds or to other hydrological models.

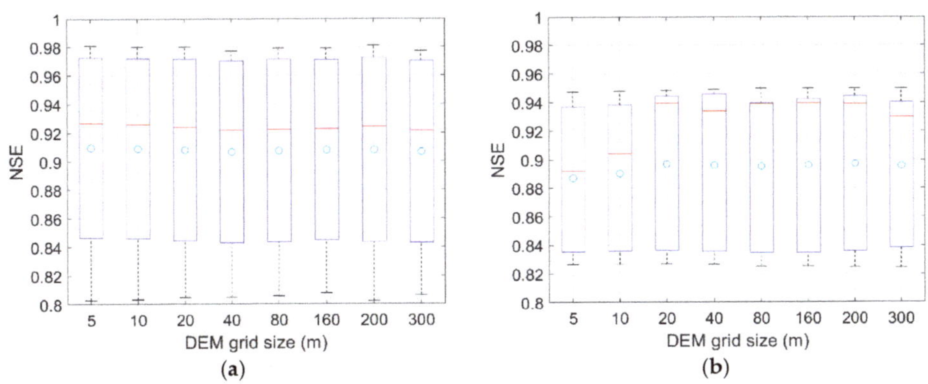

Figure 14. Results obtained using the optimized input parameters depending the DEM resolution for all Rainfall–runoff events. (**a**) Dongkok watershed. (**b**) Ieemokjung watershed.

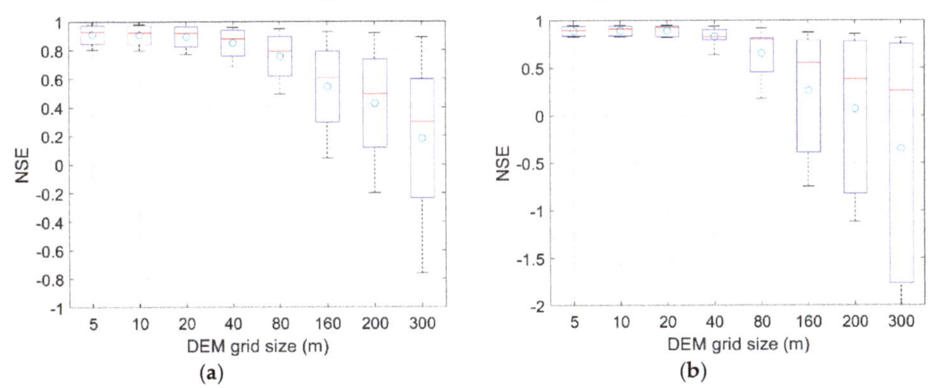

Figure 15. NSE (Nash–Sutcliffe efficiency) changes based on the resolution of the DEM grid cell size with constant parameter sets. (**a**) Dongkok watershed. (**b**) Ieemokjung watershed.

6. Conclusions

This study has applied TOPMODEL to two mountainous watersheds, namely, Dongkok and Ieemokjung watersheds, which are representative IHP watersheds in South Korea. This study investigated TOPMODEL runoff optimizations that depend on the resolution of the DEM grid cell size and analyzed the effects of DEM resolution using fixed input parameter sets. Further, this work combined the resolution of the DEM grid cell size and optimized input parameter sets and proposed the most reliable DEM grid-size resolution for TOPMODEL in the two mountainous watersheds. The optimized parameter results represent that an exponential storage parameter among calibrating input parameter is the most sensitive parameter based on TOPMODEL efficiencies.

The estimated hydrographs obtained by TOPMODEL based on well-optimized parameters that depend on each resolution of the DEM grid cell size were very similar to the observed runoff, regardless of the DEM resolution (up to 300 m in this study), which demonstrated that DEM resolution does not affect the runoff estimation when the input parameters are well optimized. In addition, this study analyzed the DEM resolution effects by applying a fixed parameter set and found that the NSE of the estimated runoff in TOPMODEL decreased as the DEM resolution increased. The estimated runoff results depending on the different optimized input parameters in both the Dongkok and Ieemokjung watersheds were similar to the observed runoff data when the DEM resolution was less than about 40 m. The results of this study imply that reliable DEM grid size depending on experimental watersheds can be evaluated in TOPMODEL. However, these results have limitations in that the reliable DEM grid cell size was evaluated using only one watershed model and two study watersheds. For future studies, it is necessary to describe more about the relations among contributing area, input parameters, and DEM resolution with rainfall events, because those relations have not been clarified. Also, it is necessary to apply different scales of watersheds and watershed models to support the relationship between reliable DEM resolution and watershed characteristics in this study.

Author Contributions: Conceptualization, D.P. and K.J.; Funding acquisition, D.P.; Investigation, D.P., H.-J.F., J.-J.Z., S.-E.O. and K.J.; Methodology, D.P. and M.-J.U.; Writing—review & editing, D.P. and K.J.

Funding: This research was supported by Basic Science Research Program through the National Research Foundation of Korea (NRF) funded by the Ministry of Education (No. 2019R1A2C1007447) and was written as part of Konkuk University's research support program for its faculty on sabbatical leave in 2019.

Conflicts of Interest: The authors declare no conflict of interest.

References

1. Beven, K.J.; Kirby, M.J. A physically based variable contributing area model of basin hydrology. *Hydrol. Sci. Bull.* **1979**, *24*, 43–69.
2. Güntner, A.; Uhlenbrook, S.; Seibert, J.; Leibundgut, C. Multi-criterial validation of TOPMODEL in a mountainous catchment. *Hydrol. Process.* **1999**, *13*, 1603–1620. [CrossRef]
3. Dolcine, L.; Andrieu, H.; Sempere-Torres, D.; Creutin, D. Flash flood forecasting with coupled precipitation model in mountainous Mediterranean basin. *J. Hydrol. Eng.* **2001**, *6*, 1–10. [CrossRef]
4. Huang, J.-C.; Lee, T.-Y.; Kao, S.-J. Simulating typhoon-induced storm hydrographs in subtropical mountainous watershed: An integrated 3-layer TOPMODEL. *Hydrol. Earth Syst. Sci.* **2008**, *5*, 1101–1135. [CrossRef]
5. Jeziorska, J.; Niedzielski, T. Applicability of TOPMODEL in the mountainous catchments in the upper Nysa Kłodzka River basin (SW Poland). *Acta Geophys.* **2018**, *66*, 203–222. [CrossRef]
6. Xue, L.; Yang, F.; Yang, C.; Wei, G.; Li, W.; He, X. Hydrological simulation and uncertainty analysis using the improved TOPMODEL in the arid Manas River basin. *China Sci. Rep.* **2018**, *8*, 452. [CrossRef]
7. Liang, X.; Lettenmaier, D.P.; Wood, E.F.; Burges, S.J. A simple hydrologically based model of land surface water and energy fluxes for general circulation models. *J. Geophys. Res. Atmos.* **1994**, *99*, 14415–14428. [CrossRef]
8. Park, D.; Markus, M. Analysis of a changing hydrologic flood regime using the Variable Infiltration Capacity model. *J. Hydrol.* **2014**, *515*, 267–280. [CrossRef]
9. Devia, G.K.; Ganasri, B.P.; Dwarakish, G.S. A review on hydrological models. *Aquatic Procedia* **2015**, *4*, 1001–1007. [CrossRef]
10. Beven, K.J.; Wood, E.F. Catchment geomorphology and the dynamics of runoff contributing areas. *J. Hydrol.* **1983**, *65*, 139–158. [CrossRef]
11. Sivapalan, M.; Wood, E.F.; Brven, K.J. On hydrologic similarity: 3, A dimensionless flood frequency model using a generalized geomorphic unit hydrograph and partial area runoff generation. *Water Resour. Res.* **1990**, *25*, 43–58. [CrossRef]

12. Wolock, D.M.; Hornberger, G.M.; Musgrove, T. Topographic controls on episodic streamwater acidification in Wales. *J. Hydrol.* **1990**, *115*, 243–259. [CrossRef]
13. Wolock, D.M.; Price, C.V. Effects of digital elevation map scale and data resolution on a topography-based watershed model. *Water Resour. Res.* **1994**, *30*, 3041–3052. [CrossRef]
14. Franchini, M.; Wending, J.; Obled, C.; Todini, E. Physical interpretation and sensitivity analysis of the TOPMODEL. *J. Hydrol.* **1996**, *175*, 293–338. [CrossRef]
15. Saulnier, G.M.; Obled, C.; Beven, K.J. Analytical compensation between DTM grid resolution and effective values of saturated hydraulic conductivity within the TOPMODEL framework. *Hydrol. Process.* **1997**, *11*, 1331–1346. [CrossRef]
16. Saulnier, G.M.; Beven, K.J.; Obled, C. Digital elevation analysis for distributed hydrological modeling: Reducing scale dependence in effective hydraulic conductivity values. *Water Resour. Res.* **1997**, *33*, 2097–2101. [CrossRef]
17. Quinn, P.F.; Beven, K.J.; Lamb, R. The ln (a/tanb) index: How to calculate it and how to use it within the TOPMODEL framework. *Hydrol. Process.* **1995**, *9*, 161–182. [CrossRef]
18. Hancock, G.R. The use of digital elevation models in the identification and characterization of catchments over different grid scales. *Hydrol. Process.* **2005**, *19*, 1727–1749. [CrossRef]
19. Nourani, V.; Roughani, A.; Gebremichael, M. TOPMODEL capability for Rainfall–runoff modeling of the Ammameh watershed at different time scales using different terrain algorithms. *J. Urban Environ. Eng.* **2011**, *5*, 1–15. [CrossRef]
20. Nourani, V.; Zanardo, S. Wavelet-based regularization of the extracted topographic index from high-resolution topography for hydro-geomorphic applications. *Hydrol. Process.* **2014**, *28*, 1345–1357. [CrossRef]
21. Sörensen, R.; Seibert, J. Effects of DEM resolution on the calculation of topographical indices: TWI and its components. *J. Hydrol.* **2007**, *347*, 79–89. [CrossRef]
22. Azizian, A.; Shokoohi, A. Investigation of the Effects of DEM Creation Methods on the Performance of a Semidistributed Model: TOPMODEL. *J. Hydrol. Eng.* **2015**, *20*, 05015005. [CrossRef]
23. Lin, K.; Zhang, Q.; Chen, X. An evaluation of impacts of DEM resolution and parameter correlation on TOPMODEL modeling uncertainty. *J. Hydrol.* **2010**, *394*, 370–383. [CrossRef]
24. Ministry of Land, Transport and Maritime Affairs (MLTMA). *International Hydrological Programme Report*; Ministry of Land, Transport and Maritime Affairs (MLTMA): Sejong, Korea, 2001. (In Korean)
25. Bae, D.-H.; Kim, J.-H.; Kwon, W.-T. A Feasibility Study of TOPMODEL for a Flood Forecasting Model on a Single Watershed. *J. Korean Water Resour. Assoc.* **2000**, *33*, 87–97. (In Korean)
26. Ambroise, B.; Beven, K.J.; Freer, J. Toward a generalization of the TOPMODEL concepts: Topographic indices of hydrological similarity. *Water Resour. Res.* **1996**, *32*, 2135–2145. [CrossRef]
27. Grabs, T.; Seibert, J.; Bishop, K.; Laudon, H. Modeling spatial patterns of saturated areas: A comparison of the topographic wetness index and a dynamic distributed model. *J. Hydrol.* **2009**, *373*, 15–23. [CrossRef]
28. Qin, C.Z.; Zhu, A.X.; Pei, T.; Li, B.L.; Scholten, T.; Behrens, T.; Zhou, C.H. An approach to computing topographic wetness index based on maximum downslope gradient. *Precis. Agric.* **2011**, *12*, 32–43. [CrossRef]
29. Buchanan, B.P.; Fleming, M.; Schneider, R.L.; Richards, B.K.; Archibald, J.; Qiu, Z.; Walter, M.T. Evaluating topographic wetness indices across central New York agricultural landscapes. *Hydrol. Earth Syst. Sci.* **2014**, *18*, 3279–3299. [CrossRef]
30. Pourali, S.H.; Arrowsmith, C.; Chrisman, N.; Matkan, A.A.; Mitchell, D. Topography wetness index application in flood-risk-based land use planning. *Appl. Spat. Anal. Policy* **2016**, *9*, 39–54. [CrossRef]
31. Tada, A.; Namihira, A.; Tanakamaru, H.; Hata, T. Application of TOPMODEL to Long-and Short-term Runoff of Small Forested Catchment. *J. Jpn. Soc. Hydrol. Water Resour.* **2002**, *15*, 399–412. (In Japanese) [CrossRef]
32. Cho, I.R. Runoff Analysis Using TOPMODEL. Master's Thesis, Ulsan University, Ulsan, Korea, 1998. (In Korean)
33. Lee, G.S. Improvement and Application of TOPMODEL to Estimate Nonpoint Source Pollution Loadings in Small Basin. Master's Thesis, Seoul National University, Seoul, Korea, 1998. (In Korean)
34. Nash, J.E.; Stutcliffe, J.V. River flow forecasting through conceptual models, Part I-a discussion of principles. *J. Hydrol.* **1970**, *10*, 282–290. [CrossRef]
35. Zhang, W.; Montgomery, D.R. Digital elevation model grid size, landscape representation, and hydrologic simulations. *Water Resour. Res.* **1994**, *30*, 1019–1028. [CrossRef]

36. Beven, K.J.; Kirkby, M.J.; Schofield, N.; Tagg, A.F. Testing a physically-based flood forecasting model (TOPMODEL) for three UK catchments. *J. Hydrol.* **1984**, *69*, 119–143. [CrossRef]
37. Brasington, J.; Richards, K. Interactions between model predictions, parameters and DTM scales for TOPMODEL. *Comput. Geosci.* **1998**, *24*, 299–314. [CrossRef]
38. Fernández-Pato, J.; Caviedes-Voullième, D.; García-Navarro, P. Rainfall/runoff simulation with 2D full shallow water equations: Sensitivity analysis and calibration of infiltration parameters. *J. Hydrol.* **2016**, *536*, 496–513.
39. Sanzana, P.; Gironás, J.; Braud, I.; Hitschfeld, N.; Branger, F.; Rodriguez, F.; Fuamba, M.; Romero, J.; Vargas, X.; Muñoz, J.F.; et al. Decomposition of 2D polygons and its effect in hydrological models. *J. Hydroinform.* **2018**, *21*, 104–122. [CrossRef]
40. Gumindoga, W.; Rwasoka, D.T.; Murwira, A. Simulation of streamflow using TOPMODEL in the Upper Save River catchment of Zimbabwe. *Phys. Chem. Earth Parts A/B/C* **2011**, *36*, 806–813. [CrossRef]
41. Manfreda, S.; Leo, M.D.; Sole, A. Detection of flood-prone areas using digital elevation models. *J. Hydrol. Eng.* **2011**, *16*, 781–790. [CrossRef]

© 2019 by the authors. Licensee MDPI, Basel, Switzerland. This article is an open access article distributed under the terms and conditions of the Creative Commons Attribution (CC BY) license (http://creativecommons.org/licenses/by/4.0/).

applied sciences

MDPI

Article

Spatiotemporal Changes of Urban Rainstorm-Related Micro-Blogging Activities in Response to Rainstorms: A Case Study in Beijing, China

Nan Wang [1,2], Yunyan Du [1,2,*], Fuyuan Liang [3], Jiawei Yi [1,2] and Huimeng Wang [1,2]

1 State Key Laboratory of Resources and Environmental Information System, Institute of Geographic Science and Natural Resources Research, Chinese Academy of Sciences, Beijing 100101, China; wangnan171@mails.ucas.edu.cn (N.W.); yijw@lreis.ac.cn (J.Y.); wanghm@lreis.ac.cn (H.W.)

2 University of Chinese Academy of Sciences, Beijing 100101, China

3 Department of Earth, Atmospheric, and Geographic Information Sciences, Western Illinois University, Macomb, IL 61455, USA; F-Liang@wiu.edu

* Correspondence: duyy@lreis.ac.cn; Tel.: +8610-64888973

Received: 18 September 2019; Accepted: 26 October 2019; Published: 31 October 2019

Featured Application: This paper examines rainstorm-related micro-blogging activities in response to rainstorms in an urban environment at fine spatial and temporal scales. Results could be used in supporting disaster assessment and mitigation decision making.

Abstract: Natural disasters cause significant casualties and losses in urban areas every year. Further, the frequency and intensity of natural disasters have increased significantly over the past couple of decades in the context of global climate change. Understanding how urban dwellers learn about and response to a natural hazard is of great significance as more and more people migrate to cities. Social media has become one of the most essential communication platforms in the virtual space for users to share their knowledge, information, and opinions about almost everything in the physical world. Geo-tagged posts published on different social media platforms contain a huge amount of information that can help us to better understand the dynamics of collective geo-tagged human activities. In this study, we investigated the spatiotemporal distribution patterns of the collective geo-tagged human activities in Beijing when it was afflicted by the "6-22" rainstorm. We used a variety of machine learning and statistical methods to examine the correlations between rainstorm-related microblogs and the rainstorm characteristics at a fine spatial and a fine temporal scale across Beijing. We also studied factors that could be used to explain the changes of the rainstorm-related blogging activities. Our results show that the human response to a disaster is very consistent, though with certain time lags, in the virtual and physical spaces at both the grid and city scales. Such a consistency varies significantly across our study area.

Keywords: social media; rainstorm event; spatiotemporal analysis; factor assessing

1. Introduction

Natural disasters such as hurricanes, floods and tornadoes can cause significant life losses, property damages, and even political instability [1–4]. In the context of global climate change, natural disasters have become more frequent and pose increasing physical, social, and economic threats to human society [5,6]. Closely monitoring how a natural disaster disturbs human activities is thus of great value [7–9], particularly in an urban environment where human-environment relationships are usually much more complicated.

Different methods and datasets have been used to study urban dwellers' response to a natural disaster. A comprehensive evaluation model was built to evaluate the macro-population vulnerability in response to an earthquake at city and county levels [10]. Questionnaire data were used to examine the impacts of bushfires on residents [11]. Qualitative studies have been conducted to examine how communities are resilient to the impacts of natural disasters using interview data [12].

Natural disasters usually pose significant threats to the dwellers being inflicted. The threats vary dynamically. Consequently, people's emotion and public opinions to such threats may also change rapidly [13]. Such dynamic changes couldn't be timely captured and examined using the afore-mentioned conventional methods and data. Unfortunately, the dynamic changes of the human response to a disaster are extremely important in disaster evaluation and mitigation.

The emerging social media data have shown their usefulness in tracking human behaviors in response to natural disasters and emergencies. Nowadays, Twitter, Facebook, Flicker and Weibo have become indispensable platforms for people to share their ideas and disseminate vital information in time [14–16], in particularly when a natural disaster, public safety event, or disease infection event occurs [17–21]. Data collected through these platforms have been used in improving situation awareness [22–24], event detection [25,26], communication analysis [27,28] and even helping governments guide the public opinions [29,30].

Social media data have also been used to detect and monitor the ongoing development of disasters such as influenza transmission, floods, typhoons, hurricanes, and terrorist attacks [31–34]. Different social media data have been examined to reveal the different development stages of a disaster and how human response to the development stages [35–37]. A research framework was constructed to extract multidimensional (time, space, content, and network) information from social data [8]. Disaster-related information derived from social media data was also used to examine the spatiotemporal impacts of a disaster on human activities and assess the actual disaster-induced damages [38–40]. A variety of machine learning methods have also been used to examine social media data to evaluate the disaster-induced damages [41,42].

Many studies have shown that public opinions in the virtual space, as reflected by social media data, is consistent with what happens in the real world at a broad geographical scale. Such consistency has not been sufficiently examined and clearly illustrated at a finer temporal and spatial scale. There is also a need to study the factors driving the human response to a disaster, particularly in an urban environment. In this study, we investigated how dwellers in a mega city react to rainstorms as revealed by blogging activities on Weibo, one of the most popular social media platforms in China. We then explored various factors that may contribute to the blogging activities in response to rainstorms.

2. Study Area and Data

2.1. Study Area

This study examined Weibo users' blogging activities in response to the rainstorms that hit Beijing from 21 to 24 June 2017, we chose the main urban area of Beijing as the study area as shown in Figure 1 (The water ponding points and major hubs in the main urban area have been marked in the Figure 1). Beijing is the capital of China. It significant grew over the past decade. The percent of the developed land in Beijing increases from 7.9% in 2005 to 16.3% in 2015. The total population has increased from 1538 million in 2005 to 2170.5 million in 2015. Beijing is also one of the most economically active cities in China with a total GDP of $975 billion in 2005 to $3270 billion in 2015.

Dwellers in Beijing use a variety of social media platforms on a daily base. Sina Weibo is one of the most popular platforms that allow people to stay in touch and share each other information about any on-going events. As of 2018, Sina Weibo has 462 million active users in China and every day an average 1.30 million words are posted online through it [42].

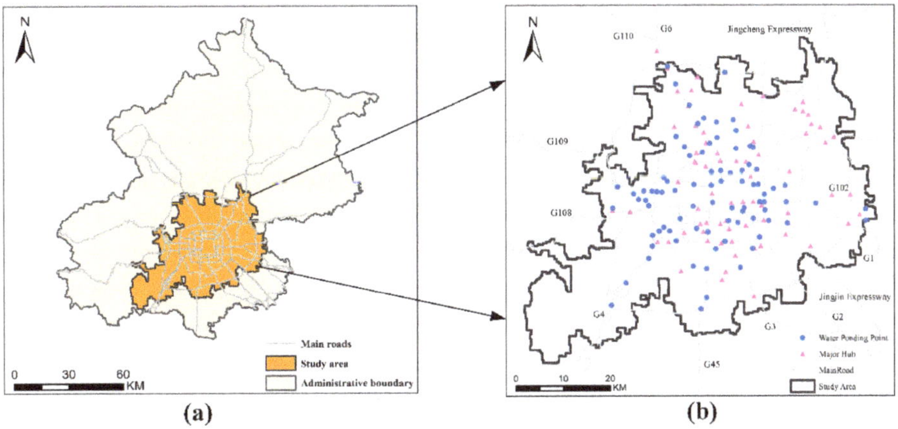

Figure 1. A map showing the boundary of our study area. Location of the study area in Beijing (**a**) and overview of the study area (**b**).

2.2. Data

A total of 3.32 million Weibo blogs geotagged with Beijing were crawled from the Sina Weibo platform. All posts were published from June to September 2017, a period that Beijing receives most of the rains all over the year. Every blog comes with the information of its user, publishing location, publishing time, and texts.

We also collected data of rainfall amount, points of interest (POIs), and water ponding sites. We used two precipitation data sets in this study. The hourly precipitation data at meteorological stations were collected from the China Meteorological Data Network (http://data.cma.cn/). The 1-h cumulative precipitation dataset was generated from the meteorological radar in Daxing, a town located 13 km south of the city. The radar covers the entire Beijing area and provides precipitation data with a spatial resolution of 1.051 × 1.051 km (After image processing and registration processing). In this paper, we unified the study of grid scale to this resolution, and the other data sources used are also processed to this resolution for further analysis and processing.

The POIs include the locations of businesses, educational institutions, residential areas, transportation facilities, open spaces, and others. The data set was produced mainly for navigation by the Beijing NavInfo Co., Ltd, Beijing, 110000. Each POI comes with its coordinates (latitude and longitude), type, name, address and flag (show its importance level). In this paper, we categorized the POIs into five classes (Table 1).

Table 1. Summary statistics of the POIs.

POI Class	Abbreviations	Total Numbers
Common Service and Education Culture	CE	1,402,310
Residential	R	977,745
Business	B	6,695,615
Scenic area and Green Open Space	S	264,065
Transportation facilities	T	299,360

3. Methods

Figure 2 shows our data processing and analysis processes. We first used ArcGIS 10.5 to aggregate geotagged Weibo posts to grids with the same resolution (1.051 × 1.051 km) as of the rainfall data. We used the support vector machine (SVM) model to classify and extract rainstorm-related microblogs. We then analyzed how city dwellers respond to the rainstorms in terms of the changes of the numbers of rainstorm-related microblogs at the grid scale and city scale. Finally, we investigated the factors that could be used to explain the city dwellers' response to the rainstorms.

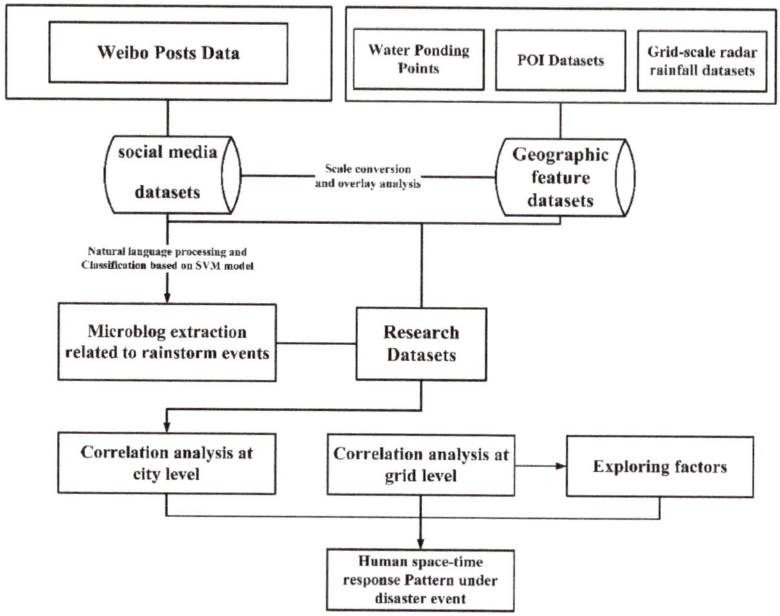

Figure 2. A flowchart showing the data processing and analysis processes.

3.1. Extraction of Rainstorm-Related Weibo Posts

In total we crawled 3.32 million Weibo posts that were published from June to September in 2017 and geo-tagged with city Beijing. We then used the keywords such as thunderbolt, storm, water, and rainfall to filter and found around 8000 posts that are possibly related to the rainstorms.

We then randomly selected 2000 out of the 8000 posts and manually checked each post. The post was labeled with "true" if it is truly related to a rainstorm otherwise "false" if it is not. The 2000 manually labeled posts were then evenly divided into two subsets, which were used to train the SVM classifier and validate the classification results, respectively.

The SVM classifier has been used in previous studies to label the microblogs either as event-related or event-independent [18,43]. It is a nonlinear classifier that was generated using the radial basis function (RBF). Essentially, it produces an optimal hyperplane that can best separate the rainstorm-related posts from those none-related. The hyperplane is defined by two parameters, C and gamma, which represent the influencing range of a single sample and the influencing degree of the support vector, respectively. The two parameters were calculated using the GridSearchCV method [44] based on the training data subset. The validation subset data was then used to evaluate the separation accuracy using the five-fold cross validation method [45]. In this study, we obtained an F-score of 0.85, which indicates that the SVM classifier could be used to identify the truly-rainstorm-related Weibo posts. The final SVM model was then used to examine all unlabeled Weibo posts. In total we found 6072 out of the 8000 posts were truly rainstorm-related during the period of June to September 2017.

3.2. Weibo Blogging Index

We used two indexes to measure the blogging activities in response to the rainstorms. The first index, the human's event response index (*HERI*), is defined as the ratio between the standardized number of the rainstorm-related Weibo posts (RRWP) to the standardized total number of the posts within a specific cell (In the data standardization process, we normalize the total number of Weibo posts and RRWP to 0–1 for each grid.).

$$HERI = \frac{\text{Standardized number of the RRWP}}{\text{Standardized total number of Weibo Posts}} \tag{1}$$

The *HERI* could be used to measure human response intensity. A higher *HERI* value would indicate city dwellers are more active in blogging the rainstorms. A very similar index was used to estimate hazard-induced damages and monitor the post-hazard recovery speed [40,46].

The *HERI* could be significantly affected by the rainfall amount. Thus, we used another index, the event normalized response relation (ENRR), to evaluate the human response to a rainstorm by eliminating the bias introduced by the variations in rainfall amount. The ENRR is expressed as the relationship between the *HERI* and the rainfall levels per cell. Both the *HERI* and rainfall amount values were first broken into three levels (high, medium, and low) using the Jenks Natural Breaks classification method, which clusters data into different classes by seeking to reduce the variance within a class and maximize the variance between classes. The different combinations of the three *HERI* and rainfall levels would reflect how dwellers response to a rainstorm which brings different rainfall amount across our study area. Figure 3 shows the nine relationships represented by ENRR. In the study, we mapped the relationship of ENRR to each grid to reflect the relationship between *HERI* and rainfall intensity in different regions.

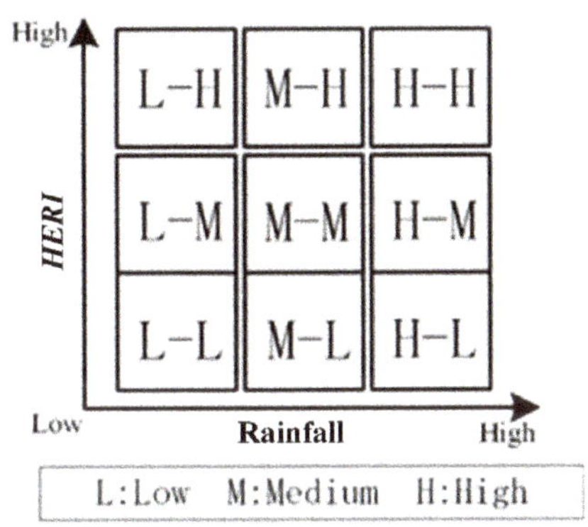

Figure 3. The 9 relationships of ENRR.

3.3. Statistical Analysis

We used a variety of conventional and spatial statistics methods to evaluate the areal difference of the blogging activities in response to the rainstorms across our study area. The hourly rainfall and the corresponding hourly RRWP were separately divided into four different groups according to their quartile levels, from which a confusion matrix was constructed. We then used the weighted Kappa coefficient to evaluate the consistency of the relationship between different levels of rainfall and the RRWP.

Quantile regression was used to estimate the conditional quantiles (0.05, 0.25, 0.50, 0.75, 0.9, and 1) of the number of posts in response to certain rainfall amount by measuring their central tendency and statistical dispersion. Quantile regression could more accurately describe the variation range of the dependent variable in response to the dependent variable. We then used the receiver operating characteristic (ROC) curves [47,48] to obtain the range within which the water ponding sites and major transportation hubs affect the blogging activities in response to the rainstorm. An optimal threshold is obtained by weighting both the sensitivity and the specificity equally, as measured by the closest distance between the points along the ROC curves and the top-left point, i.e., the perfect classification where the sensitivity and specificity both equal to 1. In addition, we also performed hotspot analysis based on the *HERI* and the ENRR.

In this paper, we examined the blogging activities in response to the rainstorms at both city and grid levels, respectively. The city extent is defined by the administrative boundary of Beijing. Within the city, the rainfall and Weibo posts were aggregated to individual grids of 1.051 km × 1.051 km. There are 2776 grids within our study area, covering a total area of 2749.23 km^2.

4. Analysis and Results

Five heavy rainstorms hit Beijing on 22 June, 6 July, 20 July, 2 August, and 22 August (Figure 4). The 22 June rainstorm brought historical record precipitation, flooded the city, and caused significant economic losses. When the city was afflicted by the 22 June rainstorm, Weibo users posted over 1000 blogs, the maximum blog number among all rainstorm events that hit Beijing in summer 2017. In this study, we mainly focus on the blogging activities in response to the 22 June rainstorm.

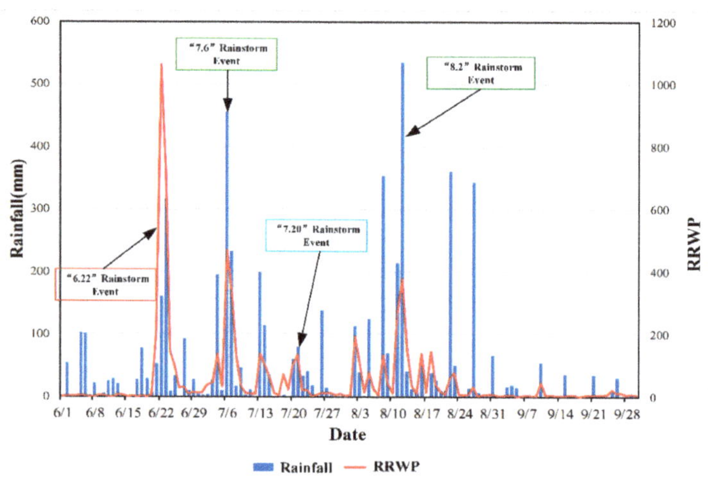

Figure 4. The time series RRWP and rainfall in Beijing from June to September 2017.

4.1. The "622" Rainstorm

At 16:00 on 21 June 2017, the Beijing Meteorological Bureau issued the first yellow lightning storm warning of a high altitude and low vortex, which later evolved into the 22 June rainstorm. The rainstorm first brought rain to the western part of Beijing from noon on 21 June and then across the whole Beijing city (Figure 5a). The storm lasted for 66 h and finally ended at 06:00 on 24 June. Heavy rains and floods were reported in Fangshan, Shunyi, Yanqing and Changping Districts. The storm flooded 131.5 hectares of agricultural land, affected 8594 people, and caused about $3.22 million direct economic losses.

Both conventional and online medias covered this rainstorm extensively and generated a large number of news reports and Weibo posts. In total, we crawled 230,125 geotagged Weibo posts during the time period (20–27 June 2017) (Figure 5b) and 2362 out of them are RRWP (Figure 5c).

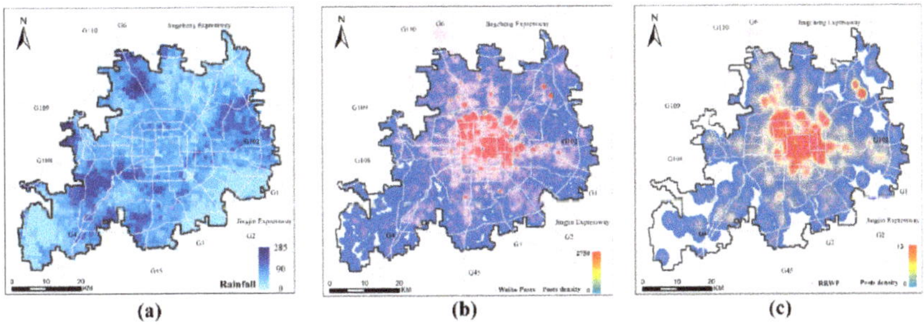

Figure 5. Spatial distribution of the rainfall amount (**a**), the number of Weibo posts (**b**), and the number of RRWP (**c**).

4.2. *Blogging Activities at City Level*

Figure 6 shows the time series hourly rainfall and the hourly number of RRWP from 20 to 28 June. No RRWP was found on the social media platform before the rainstorm hit the city. Rainstorm-related blogging activities were first detected when the first rainstorm warning was issued at 4:20 p.m. on 21 June. The blogging activities significantly intensified, particularly when the rainstorm is most intensive during the time period from 19:00 on 21 June to 04:00 on 24 June. During this time period, the city rainfall amount accounts for 90.5% of the total rainfall brought to Beijing by the "6.22" rainstorm. About 91.6% of the RRWP was posted during this time period.

Figure 6 also shows that variations in the rainfall amount are generally consistent with the blogging activities though there seems to be a 1-h time lag. The blogging activities are most intense in about 10 min before the release of the rainstorm warning. A high rainfall amount is not always accompanied by strong blogging activities, particularly when raining occurs from the late night to the early morning and the rainstorm hits suburbs with a small population flow.

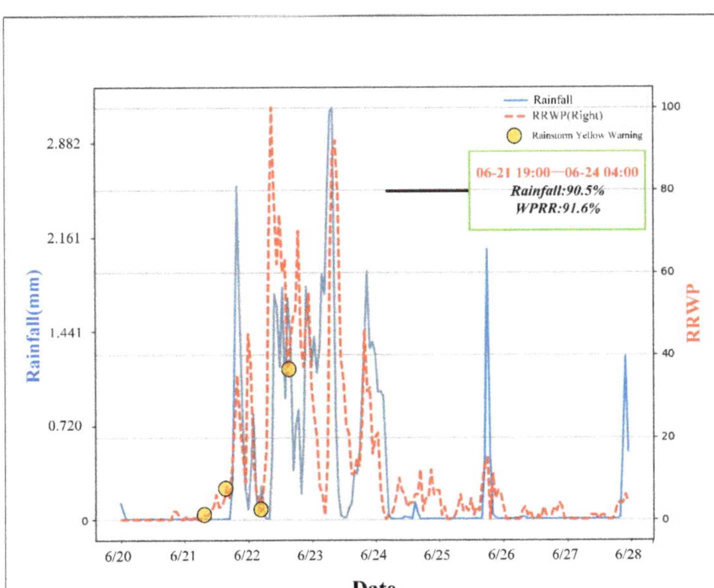

Figure 6. Time series of hourly accumulated rainfall and Weibo posts.

The confusion matrix (Figure 7) between the different levels of rainfall amount and the RRWP shows that higher rainfall levels are always associated with more RRWP. We found 38 h with higher rainfall amount and more RRWP. Lower rainfall levels are associated with fewer RRWP. A statistically significant weighted Kappa coefficient of 0.63 indicates that the levels of blogging activities are consistent with the rainfall levels across the city.

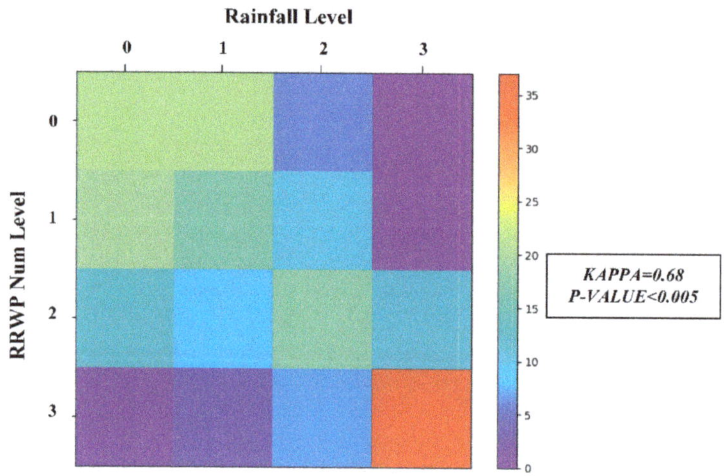

Figure 7. Consistency between rainfall and RRWP levels.

Figure 8 shows the correlation coefficients between the number of RRWP and rainfall amount with a time lag up to 6 h. With increased time lags, the coefficients drop though the correlations are statistically significant at a confidence level of 0.01. The highest coefficient 0.653 was found when the time lag is 1 h, suggesting that more RRWP were posted one hour after the rainstorm. In other words, heavy rainfall usually triggers intensified blogging activities one hour later. It seems that, after 1 h of the rainstorm, the city starts to be afflicted by issues such as waterlogging and traffic congestion. Such issues tend to intensify rainstorm-related blogging activities.

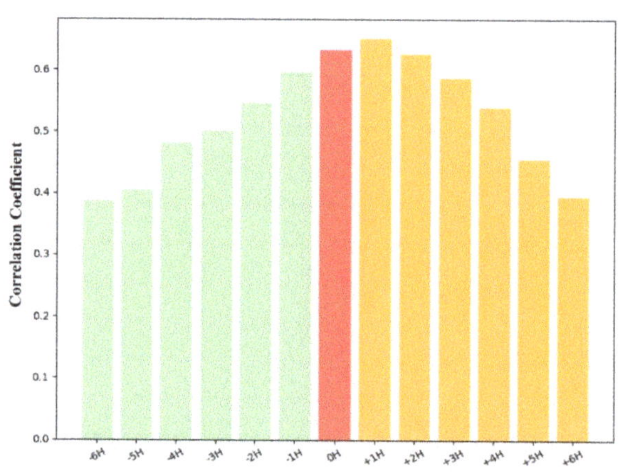

Figure 8. Correlation coefficients between the rainfall amount and the number of RRWP with different time lags.

Quantile regression analysis between the rainfall amount and the number of RRWP (Figure 9) shows a steeper slope for the higher percentile data. In other words, increased rainfall shows a more significant impact on the number of RRWP when the rainfall is heavier. By contrast, when the rainfall is less than 30th percentile (the average rainfall of the grid in the study area is 8.6), an increase in the rainfall amount shows little impacts on the change of the RRWP. Once the rainfall exceeds the 30th percentile, the RRWP starts to increase. As the rainfall percentile increases, the regression slope becomes steeper. In other words, once the rainfall is over the 30th percentile, it tends to trigger Weibo users to post much more RRWP.

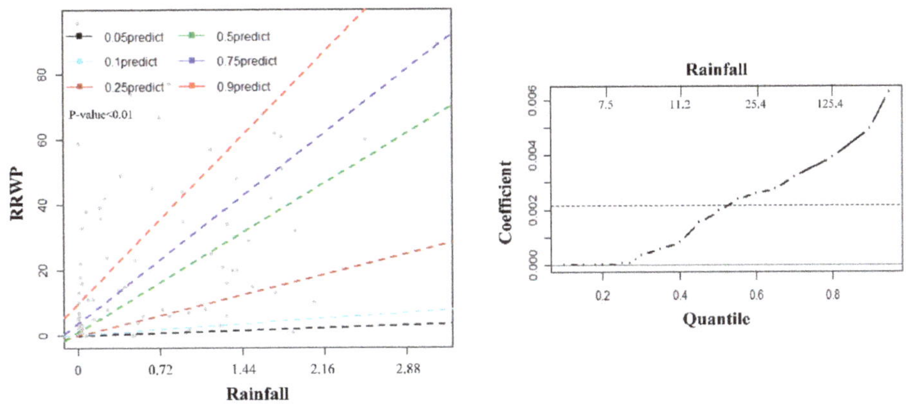

Figure 9. Quantile regression estimates of the relationship between the RRWP and the rainfall amount at the city scale.

4.3. Human Response at Grid Scale

In order to explore the differences in human response intensity of different time periods at grid scale, we first divided the whole day into four time periods (08–10, 11–16, 17–20, 21–07), and then map the RRWPs in different time periods by the 4 time periods' dot maps. Figure 10 shows the results. We can find that in the study area, the morning rush hours (08–10) and the evening rush hours (17–20) have the strongest human response intensity. During these two periods, important transportation hubs (Commercial business center, large jobs-housing area) and water ponding point areas have become regions with a high response in the main urban area of Beijing. In addition, there was a phenomenon in which dense points are distributed around the subway and along important roads. The occurrence of rainstorm event has caused great obstacles to traffic operation and delayed human's travel. The points in the second period (11–16) are mainly distributed near the traffic station and the more severely affected areas. The points in the last period are sparsely distributed in the study area. We also found the density distribution of four periods' points in major traffic stations such as airports, railway stations, and bus stations were relatively uniform, while large jobs-housing areas are densely distributed at points of the morning rush hours and the evening rush hours.

Figure 11 shows the correlation between the rainfall amount and the number of RRWP at the grid scale. The correlation coefficients vary between −0.14 and 0.86 with an average of 0.22. The negative or no correlation relationship is mainly found in suburbs, such as the Changping, Huairou and Miyun Districts. By contrast, higher correlation coefficients are mainly found in populated areas within the city, including populated residential communities, important transportation hubs, and areas significantly impacted by the rainstorms as shown in news reports.

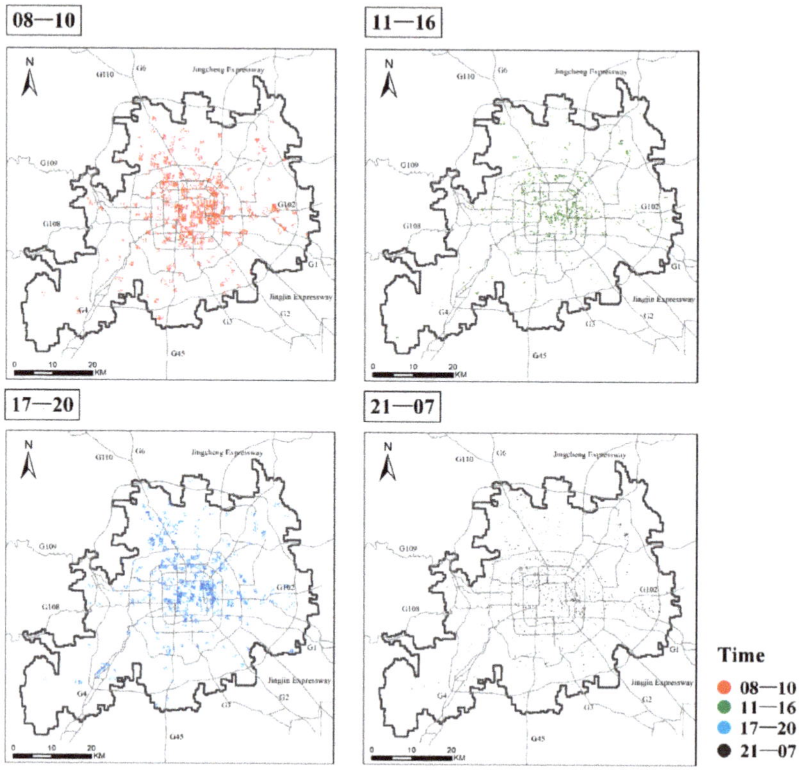

Figure 10. Dot maps of in different time periods based on grid cells.

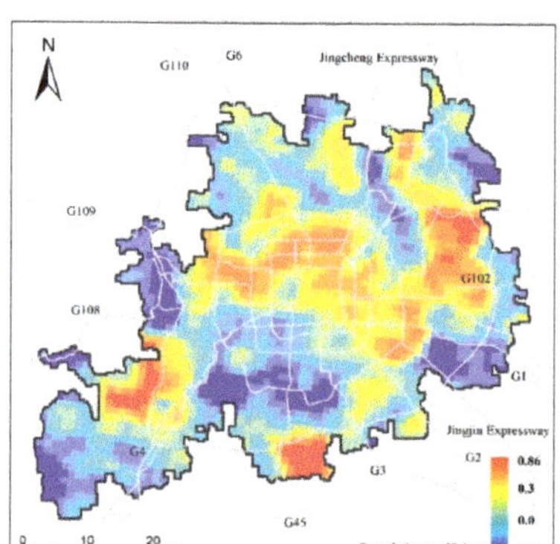

Figure 11. Correlation coefficients at grid level across our study area.

Figure 12a shows the *HERI* across our study area. Only the 2203 grids with at least 10 daily Weibo posts are selected to calculate the *HERI*. At the grid scale, *HERI* values range between 0 and 9.83 with an average value of 1.23.

The regions with a higher *HERI* value are mainly found in three places in our study area. The first are the areas with more rainstorm-induced damages, including serious house collapse, road blockage and mudslides. These areas are mainly found in the suburbs such as the Fangshan and Mentougou

Districts. The densely populated regions, including Zhongguancun and the CBD, the Tongzhou residential area, an Internet technology parks also have a higher *HERI* value. Regions with important transportation hubs also have a higher *HERI* value. The transportation hubs include subway stations, train stations, and airports.

Figure 12b shows the *HERI* hotspot analysis results. Hotspots are mainly located in densely populated areas, important residential and workplaces, such as the Tongzhou residential area, CBD districts, and IT parks. It is worth noting that a large number of hotspots are found in the urban core areas. By contrast, the cold spots are mainly found in the remote suburbs of our study area. Such areas have a low population thus limited human activities.

The proportions of POIs within each hotspot identified are shown in Figure 12c. The *HERI* hotspots in the Beijing Capital Airport, Yizhuang, and Changping-Shahe Districts have the highest percent of transportation POIs. The texts of the RRWP within these hotspots show that the rainstorms may significantly delay the commute in these regions thus stimulate users to publish more RRWP to complain the traffic. The hotspots in the Tongzhou residential area and the Mentougou District show a higher proportion of residential POIs. Hotspots in Zhongguancun, Chaoyang CBD and IT Park have a highly mixture of multiple types of residential, business, and education POIs. There is no significant difference in the proportions of the POIs in other hotspots.

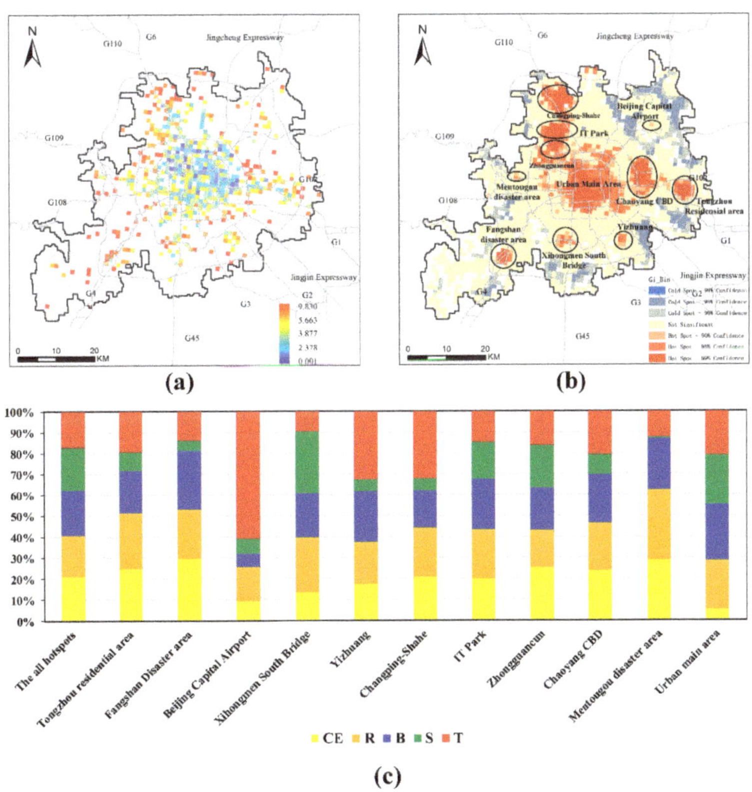

Figure 12. Spatial distribution of the *HERI* values (**a**); *HERI* hotspot analysis results (**b**); Percentage of POI types in each HERI hotspot area (**c**).

Figure 13a shows the binary relationship between the different levels of *HERI* and rainfall amounts. We classify the *HERI* and rainfall amount into three groups (high, medium, and low) using the Jenks natural breaks, respectively. In total, there would be nine combinations between the different levels of *HERI* and rainfall amounts. The HH combination (high *HERI* and more rainfall) is located in suburbs, such as Fangshan District and Xi'erqi, which were hit by heavy rains and afflicted with serious rainstorm-induced damages and losses. The combination (HL) with a higher HERI and less rainfall is mainly found in densely populated areas, including the urban core area, Tongzhou District, CBD and the Beijing Capital Airport. The texts of the RRWP show that people in these areas complain that the rainstorms caused significant traffic jams and ruined their daily routine. Passengers trapped in the Beijing Capital Airport also published more RRWP due to the significant flight delays. The LH combination (low *HERI* and high rainfall amount) is mainly in the sparsely populated regions, where few RRWP were posted due to the fewer number of the Weibo users.

Figure 13b shows the POI types in the different combinations of the *HERI* and rainfall amount levels. The areas with a higher *HERI* value tend to have more transportation POIs, no matter what the rainfall amount is. By contrast, places with fewer RRWP and higher rainfall levels are less populous and with more green space.

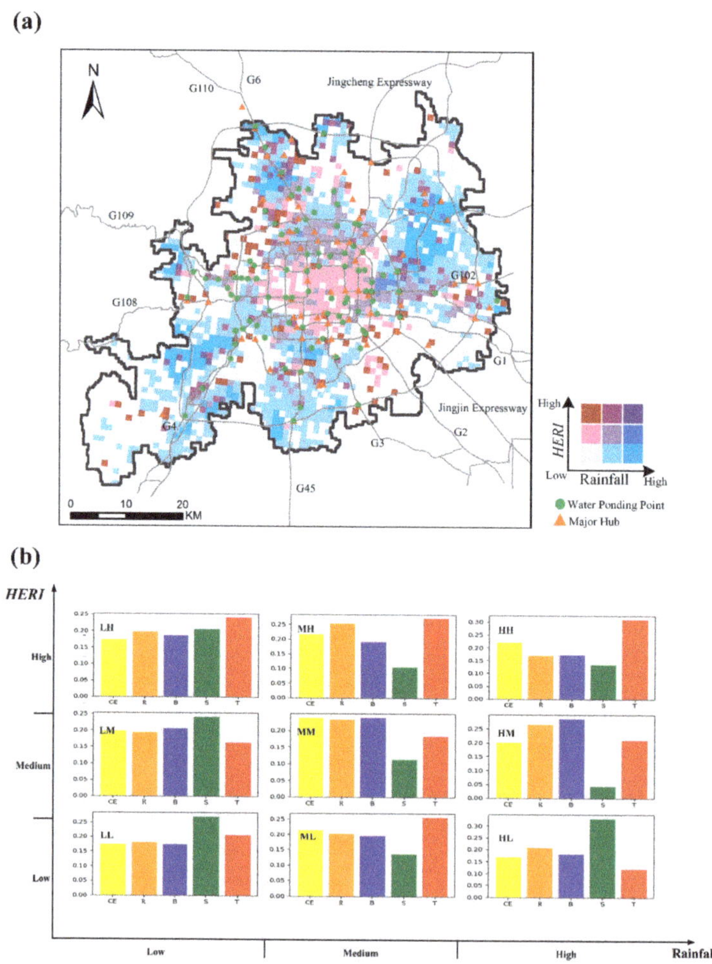

Figure 13. The ENRR values across our study area (**a**) and the proportions of different types of POIs within the areas with different ENRR values (**b**).

4.4. Factors Influencing HERI

For all grids across our study area, the AUC values on the ROC curves are 0.767 and 0.733 for the water ponding sites and the major transportation hubs, respectively (Figure 14). The most appropriate OIDF values for the afore-mentioned two factors are 3400 m and 3200 m respectively. The bilateral Welch t test results show that the *HERI* value of the areas within the OIDF distance of a water ponding site and a major transportation hub were both significantly higher than those beyond and the difference is statistically significant at the 0.01 significance level.

Table 2 shows the number and density of water ponding sites and major transportation hubs by each level combination of rainfall amount and RRWP. The level combinations with intense blogging activities are all associated with high density of water ponding sites and major transportation hubs, no matter what the rainfall amount is. By contrast, the level combinations with inactive blogging activities are associated with a low density of water ponding sites and major transportation hubs.

Table 2. The statistics of water ponding sites and transportation hubs by different ENRR categories.

ENRR	Area(km^2)	Water Ponding Points		Major Transportation Hubs	
		Number	Density	Number	Density
L_RL_P	915	35	3.8	25	2.7
ML	702	29	4.1	14	2.0
HL	202	4	2.0	10	5.0
LM	228	27	11.8	17	7.5
MM	182	15	8.2	16	8.8
HM	19.4	3	15.5	1	5.2
LH	82.5	6	7.3	9	10.9
MH	79	11	13.9	6	7.6
HH	22.8	3	13.2	2	8.8

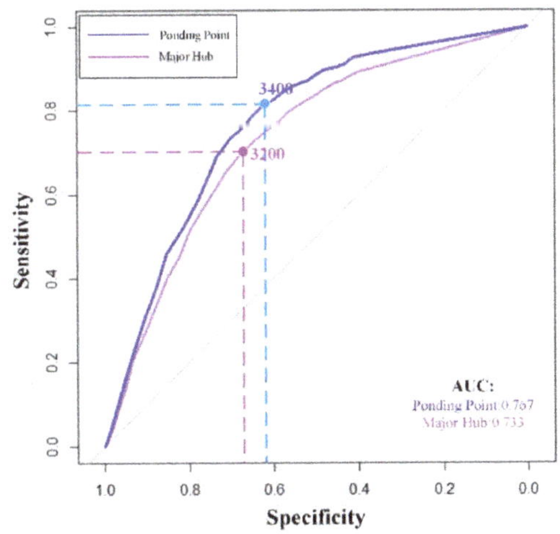

Figure 14. The ROC curves of the influencing factors.

5. Conclusions

In this study, we inferred the human activities from the rainstorm-related Weibo posts and examined how different levels of human activities are associated with different rainfall amount levels at both city and grid scales. The consistency between the rainfall amount and the human activities could be explained by the distribution of the water ponding sites and major transportation hubs. The regions with high density of water ponding sites and major transportation hubs tend to show intense human response to rainstorms in terms of the number of rainstorm-related Weibo posts. At different time periods, the intensity of human responses to rainstorm events in areas of different attributes and functions were also very different. The human response has been significantly enhanced during the early and late peak hours and is concentrated in important transportation hubs and water ponding sites. The occurrence of a rainstorm event has a huge impact on human travel. Analysis of the rainstorm-related posts suggests that there is no significant difference between the impacting ranging (~3.3 km) of a water ponding site and a major transportation hub.

We found that on the large scale, although the ground disaster space has a high consistency with the social media space, the intensity of responses at different stages, different spatial areas, and during different time periods of the disaster on social media platform were different. When looking at spatial differences on a grid scale for urban disaster events, the impacts of different types of region vary greatly due to the complexity of human–land relationships. During a rainstorm, the existence of special areas such as urban water ponding points, traffic stations, main jobs and housing areas, important line sites, and some disaster sites have led to frequent occurrence of secondary disasters and become major concentrated areas where humans respond strongly. These results show that there are time and space differences in the human response at the urban scale and grid scale under urban rainstorm events. Our research on the spatial consistency is similar to the previous research conclusions [36,39,49], but a further exploration of fine spatiotemporal process and supplementation of the factors affecting the differences in human responses give us a new understanding of the human–land relationship under the event conditions at a fine scale.

Of course, this study has some defects that can be ameliorated by additional research to improve upon our framework and further research goals. This study only examined the number of rainstorm-related Weibo posts without considering other information available in the original Weibo posts, such as the emotions, themes, and characteristics of the social media information. Other multi-source spatial data such as the nighttime lights, ambient population data, and road traffic congestion data, if successfully integrated, could provide a more comprehensive study on the human response to a natural disaster. The integration of multi-source spatial data and more comprehensive data mining methods would also significantly reduce the uncertainty of the associations between human activities in both the physical and virtual spaces.

Author Contributions: Conceptualization, N.W. and Y.D.; methodology, N.W.; validation, N.W., Y.D., H.W., and F.L.; formal analysis, N.W.; investigation, N.W.; resources, N.W and J.Y.; data curation, J.Y.; writing—original draft preparation, N.W.; writing—review and editing, F.L.; visualization, N.W.; supervision, Y.D.; project administration, Y.D.; funding acquisition, Y.D.

Funding: This research was funded by the National Key Research and Development Program of China (Grant Nos. 2017YFB0503605 and 2017YFC1503003), and the Strategic Priority Research Program of the Chinese Academy of Sciences (Grant Nos. XDA20040401 and QYZDY-SSW-DQC007-2), and the National Mountain Flood Disaster Investigation Project (SHZH-IWHR-57).

Conflicts of Interest: The authors declare no conflict of interest.

References

1. Klomp, J.; Valckx, K. Natural disasters and economic growth: A meta-analysis. *Glob. Environ. Chang.* **2014**, *26*, 183–195. [CrossRef]
2. Cole, S.A.; Healy, A.; Werker, E. Do Voters Appreciate Responsive Governments? Evidence from Indian Disaster Relief. *Soc. Sci. Electron. Publ.* **2012**, *97*, 167–181. [CrossRef]
3. Parenton, F.V.J. Disaster and Aftermath: Selected Aspects of Individual and Group Behavior in Critical Situation. *Soc. Forces* **1959**, *38*, 129–135.
4. Ramseur, J.L. *Oil Spills in U.S. Coastal Waters: Background, Governance, and Issues for Congress*; Congressional Research Service Reports; Library of Congress, Congressional Research Service: Washington, DC, USA, 2010.
5. Aalst, M.K.V. The Impacts of Climate Change on the Risk of Natural Disasters. *Disasters* **2006**, *30*, 5–18. [CrossRef] [PubMed]

6. Mendelsohn, R.; Emanuel, K.; Chonabayashi, S.; Bakkensen, L. The impact of climate change on global tropical cyclone damage. *Nat. Clim. Chang.* **2012**, *2*, 205–209. [CrossRef]
7. Goodchild, M.F. Citizens as sensors: The world of volunteered geography. *GeoJournal* **2007**, *69*, 211–221. [CrossRef]
8. Wang, Z.; Ye, X. Social media analytics for natural disaster management. *Int. J. Geogr. Inf. Sci.* **2017**, *32*, 1–24. [CrossRef]
9. Liu, Y.; Liu, X.; Gao, S.; Gong, L.; Kang, C.G.; Zhi, Y.; Chi, G.H.; Shi, L. Social Sensing: A New Approach to Understanding Our Socioeconomic Environments. *Ann. Assoc. Am. Geogr.* **2015**, *105*, 512–530. [CrossRef]
10. Linet, Ö.; Ekinci, E.; Beste, K. Emergency Logistics Planning in Natural Disasters. *Ann. Op. Res.* **2004**, *129*, 217–245.
11. De, V.P.; Guilbert, N.; Mesple-Somps, S. Impact of Natural Disasters on Education Outcomes: Evidence from the 1987-89 Locust Plague in Mali. *J. Afr. Econ.* **2015**, *24*, 57–100.
12. Nie, C.J.; Yang, L.S.; Li, H. Macro Assessment of Seismic Population Vulnerability in China. *Progress Geogr.* **2012**, *31*, 375–382.
13. Khunwishit, S. *Community Resilience in Thailand: A Case Study of Flood Response in Nakhonsawan City Municipality*; University of North Texas: Denton, Texas, 2013.
14. Morss, R.E.; Wilhelmi, O.V.; Meehl, G.A.; Dilling, L. Improving Societal Outcomes of Extreme Weather in a Changing Climate: An Integrated Perspective. *Ann. Rev. Environ. Resour.* **2011**, *36*, 1–25. [CrossRef]
15. Lazer, D.; Pentland, A.; Adamic, L. Computational Social Science. *Science* **2009**, *323*, 721–723. [CrossRef] [PubMed]
16. Vespignani, A. Predicting the Behavior of Techno-Social Systems. *Science* **2009**, *325*, 425–428. [CrossRef] [PubMed]
17. Bai, H.; Yu, G.A. Weibo-based approach to disaster informatics: Incidents monitor in post-disaster situation via Weibo text negative sentiment analysis. *Nat. Hazards* **2016**, *83*, 1177–1196. [CrossRef]
18. Li, Z.; Wang, C.; Emrich, C.T. A novel approach to leveraging social media for rapid flood mapping: A case study of the 2015 South Carolina floods. *Cartogr. Geogr. Inf. Sci.* **2017**, *45*, 97–110. [CrossRef]
19. Neppalli, V.K.; Caragea, C.; Squicciarini, A.; Tapia, A.; Stehle, S. Sentiment analysis during Hurricane Sandy in emergency response. *Int. J. Disaster Risk Reduct.* **2017**, *21*, 213–222. [CrossRef]
20. Cvetojevic, S.; Hochmair, H.H. Analyzing the spread of tweets in response to Paris attacks. *Comput. Environ. Urban Syst.* **2018**, *71*, 14–26. [CrossRef]
21. Chunara, R.; Andrews, J.R.; Brownstein, J.S. Social and News Media Enable Estimation of Epidemiological Patterns Early in the 2010 Haitian Cholera Outbreak. *Am. J. Trop. Med. Hyg.* **2012**, *86*, 39–45. [CrossRef]
22. Vieweg, S. Microblogging during two natural hazards events: What twitter may contribute to situational awareness. In *SIGCHI Conference on Human Factors in Computing Systems*; ACM: Museum, Singapore, 2010.
23. Caragea, C.; McNeese, N.; Jaiswal, A.; Traylor, G.; Kim, H.-W. Classifying text messages for the Haiti earthquake. In Proceedings of the 8th International Conference on Information Systems for Crisis Response and Management, New York, NY, USA, 20–23 May 2011.
24. Freeman, M. *Fire, Wind and Water: Social Networks in Natural Disasters*; IGI Global: Pennsylvania, NY, USA, 2011.
25. Stewart, K.; Glanville, J.L.; Bennett, D.A. Exploring Spatiotemporal and Social Network Factors in Community Response to a Major Flood Disaster. *Prof. Geogr.* **2014**, *66*, 421–435. [CrossRef]
26. Chae, J.; Thom, D.; Bosch, H. Spatiotemporal Social Media Analytics for Abnormal Event Detection using Seasonal-Trend Decomposition. In Proceedings of the IEEE Conference on Visual Analytics Science & Technology, Seattle, WA, USA, 14–19 October 2012; IEEE Computer Society: Washington, DC, USA, 2012.
27. Dong, X.; Mavroeidis, D.; Calabrese, F. Multiscale event detection in social media. *Data Min. Knowl. Discov.* **2015**, *29*, 1374–1405. [CrossRef]
28. Brennan, S.; Sadilek, A.; Kautz, H. Towards understanding global spread of disease from everyday interpersonal interactions. In Proceedings of the twenty-third international joint conference on artificial intelligence, Beijing, China, 3–19 August 2013; AAAI Press: Beijing, China, 2013; pp. 2783–2789.
29. Magdy, W.; Darwish, K.; Abokhodair, N. Quantifying Public Response towards Islam on Twitter after Paris Attacks. *arXiv* **2015**, arXiv:1512.04570.
30. Berger, L. Local, National and Global Islam: Religious Guidance and European Muslim Public Opinion on Political Radicalism and Social Conservatism. *West Eur. Politics* **2016**, *39*, 205–228. [CrossRef]

31. Kryvasheyeu, Y.; Chen, H.; Obradovich, N. Rapid assessment of disaster damage using social media activity. *Sci. Adv.* **2016**, *2*, e1500779. [CrossRef] [PubMed]
32. Fohringer, J.; Dransch, D.; Kreibich, H. Social media as an information source for rapid flood inundation mapping. *Nat. Hazards Earth Syst. Sci.* **2015**, *15*, 2725–2738. [CrossRef]
33. Takahashi, B.; Tandoc, E.C.; Carmichael, C. Communicating on Twitter during a disaster: An analysis of tweets during Typhoon Haiyan in the Philippines. *Comput. Hum. Behav.* **2015**, *50*, 392–398. [CrossRef]
34. Rosser, J.F.; Leibovici, D.G.; Jackson, M.J. Rapid flood inundation mapping using social media, remote sensing and topographic data. *Nat. Hazards* **2017**, *87*, 103–120. [CrossRef]
35. Houston, J.B.; Hawthorne, J.; Perreault, M.F. Social media and disasters: A functional framework for social media use in disaster planning, response, and research. *Disasters* **2014**, *39*, 1–22. [CrossRef]
36. Kryvasheyeu, Y.; Chen, H.; Moro, E. Performance of Social Network Sensors during Hurricane Sandy. *PLoS ONE* **2015**, *10*, e0117288. [CrossRef]
37. Zou, L.; Lam, N.S.N.; Cai, H. Mining Twitter Data for Improved Understanding of Disaster Resilience. *Ann. Am. Assoc. Geogr.* **2018**, *108*, 1–20. [CrossRef]
38. Wu, D.; Cui, Y. Disaster early warning and damage assessment analysis using social media data and geo-location information. *Decis. Support Syst.* **2018**, *111*, 48–59. [CrossRef]
39. Yago, M.; Li, Z.; Cutter, S.L. Leveraging Twitter to gauge evacuation compliance: Spatiotemporal analysis of Hurricane Matthew. *PLoS ONE* **2017**, *12*, e0181701.
40. Resch, B.; Florian, U.; Havas, C. Combining machine-learning topic models and spatiotemporal analysis of social media data for disaster footprint and damage assessment. *Cartogr. Geogr. Inf. Sci.* **2017**, *45*, 362–376. [CrossRef]
41. Weibo User Development Report for 2018. Available online: https://data.weibo.com/report/reportDetail?id=433 (accessed on 15 May 2019).
42. Zahra, K.; Ostermann, F.O.; Purves, R.S. Geographic variability of Twitter usage characteristics during disaster events. *Geo-sp. Inf. Sci.* **2017**, *20*, 231–240. [CrossRef]
43. Aramaki, E.; Maskawa, S.; Morita, M. *Twitter Catches the Flu: Detecting Influenza Epidemics Using Twitter*; Emnlp: AsiaWorld-Expo, Hong Kong, China, 2011.
44. Dilrukshi, I.; Zoysa, K.D.; Caldera, A. Twitter news classification using SVM. In Proceedings of the International Conference on Computer Science & Education, Kuala Lumpur, Malaysia, 29 November–1 December 2013; IEEE: Piscataway, NJ, USA, 2013.
45. Zhang, H.; Chen, L.; Qu, Y. Support Vector Regression Based on Grid-Search Method for Short-Term Wind Power Forecasting. *J. Appl. Math.* **2014**, *2014*, 1–11. [CrossRef]
46. Gao, Y.; Wang, S.; Padmanabhan, A. Mapping spatiotemporal patterns of events using social media: A case study of influenza trends. *Int. J. Geogra. Inf. Sci.* **2018**, *32*, 425–449. [CrossRef]
47. Hanley, J.A.; Mcneil, B.J. The meaning and use of the area under a receiver operating characteristic (ROC) curve. *Radiology* **1982**, *143*, 29–36. [CrossRef]
48. Cook, N.R. Use and Misuse of the Receiver Operating Characteristic Curve in Risk Prediction. *Circulation* **2007**, *115*, 928–935. [CrossRef]
49. Wang, Y.; Wang, T.; Tsou, M.-H. Mapping Dynamic Urban Land Use Patterns with Crowdsourced Geo-Tagged Social Media (Sina-Weibo) and Commercial Points of Interest Collections in Beijing, China. *Sustainability* **2016**, *8*, 1202. [CrossRef]

© 2019 by the authors. Licensee MDPI, Basel, Switzerland. This article is an open access article distributed under the terms and conditions of the Creative Commons Attribution (CC BY) license (http://creativecommons.org/licenses/by/4.0/).

applied sciences

MDPI

Review

Rainfall Induced Landslide Studies in Indian Himalayan Region: A Critical Review

Abhirup Dikshit [1], Raju Sarkar [2], Biswajeet Pradhan [1,3,*], Samuele Segoni [4] and Abdullah M. Alamri [5]

[1] Centre for Advanced Modeling and Geospatial Information Systems (CAMGIS), University of Technology, Sydney, NSW 2007, Australia; abhirup.dikshit@student.uts.edu.au
[2] Department of Civil Engineering, Delhi Technological University, Shahbad Daulatpur, Delhi 110042, India; rajusarkar@dce.ac.in
[3] Department of Energy and Mineral Resources Engineering, Sejong University, Choongmu-gwan, 209 Neungdong-ro, Gwangjin-gu, Seoul 05006, Korea
[4] Department of Earth Sciences, University of Firenze, Via Giorgio La Pira, 4, 50121 Florence, Italy; samuele.segoni@unifi.it
[5] Department of Geology & Geophysics, College of Science, King Saud University, Riyadh 11451, Saudi Arabia; amsamri@ksu.edu.sa
* Correspondence: Biswajeet.pradhan@uts.edu.au

Received: 25 February 2020; Accepted: 31 March 2020; Published: 3 April 2020

Abstract: Landslides are one of the most devastating and recurring natural disasters and have affected several mountainous regions across the globe. The Indian Himalayan region is no exception to landslide incidences affecting key economic sectors such as transportation and agriculture and often leading to loss of lives. As reflected in the global landslide dataset, most of the landslides in this region are rainfall triggered. The region is prone to 15% of the global rainfall-induced landslides, and thereby a review of the studies in the region is inevitable. The high exposure to landslide risk has made the Indian Himalayas receive growing attention by the landslides community. A review of landslides studies conducted in this region is therefore important to provide a general picture of the state-of-the-art, a reference point for researchers and practitioners working in this region for the first time, and a summary of the improvements most urgently needed to better address landslide hazard research and management. This article focuses on various studies ranging from forecasting and monitoring to hazard and susceptibility analysis. The various factors used to analyze landslide are also studied for various landslide zones in the region. The analysis reveals that there are several avenues where significant research work is needed such as the inclusion of climate change factors or the acquisition of basic data of highest quality to be used as input data for computational models. In addition, the review reveals that, despite the entire region being highly landslide prone, most of the studies have focused on few regions and large areas have been neglected. The aim of the review is to provide a reference for stakeholders and researchers who are currently or looking to work in the Indian Himalayas, to highlight the shortcomings and the points of strength of the research being conducted, and to provide a contribution in addressing the future developments most urgently needed to obtain a consistent advance in landslide risk reduction of the area.

Keywords: Indian Himalayas; landslides; GIS; remote sensing

1. Introduction

Landslides are the most frequent naturally occurring hazards that affect people and their livelihood worldwide. The frequency of occurrence in the Himalayan context is very large when compared with global events [1]. This review paper is an attempt to understand the research being undertaken to understand, assess, and mitigate landslide scenarios in the Indian Himalayan region. The need for such a review was raised from the compilation of the global landslide disaster database by Froude and Petley, [1]. In their database, a total of 5318 non-seismic landslides occurred from 2004 to 2017, of which 3285 landslides were triggered by rainfall. In the context of the Indian Himalayas, during the same period, 580 landslides occurred with 477 triggered by rainfall, thereby contributing 14.52% of the global landslides. These number could be even higher; for instance, based on NASA GLC data, the number of landslides in the Indian Himalaya during 2007–2015 is 691 with 6306 casualties (https://data.nasa.gov/Earth-Science/Global-Landslide-Catalog/h9d8-neg4).

This region covers more than 12% of India's landmass and is very prone to landslides due to the fragile lithology, the complex geological setting, the high energy of the relief with steep slopes, and the high topographic roughness. Moreover, most of the area is seismically active and subject to extreme precipitations, and the situation has been further worsened with the increase in anthropogenic activities and the advent of climate change. Since it is a well-established fact that most of the landslides in this area have been primarily triggered by rainfall [2–4], the focus of the present review is only on the studies considering rainfall triggered landslides. Moreover, one of the main practical purposes of this review is to serve as a starting point for future projects which consider implementing territorial landslide early warning systems, and rainfall triggering landslides are the only ones that at present can be forecasted with a certain confidence. Studies on earthquake-induced landslides, snow avalanches, and Glacier Lake Outburst Flood (GLOF) were not considered in this review. This work also covers the various methodologies being adopted for landslide monitoring and analysis, as well as reports on various mitigation measures being undertaken along with the role of Geographic Information Systems (GIS), remote sensing, and the recent use of computational techniques.

2. Materials and Methods

2.1. Study Region

The Indian Himalayan region is one of the most diverse and heterogeneous area in terms of geology, lithology, rainfall distribution, land use/land cover, soil properties, and road and stream networks, which makes it highly prone to landslides. The region covers 16.2% (~500,000 km^2) of India's landmass as well as 10 of its states (Ministry of Environment, Forest and Climate Change, Government of India). The region extends 26°20′–35°40′ N and 74°50′–95°40′ E and covers the states Jammu and Kashmir, Himachal Pradesh, Uttarakhand, Sikkim, Arunachal Pradesh, Meghalaya, Nagaland, Manipur, and Mizoram, as well as the hill stations of Assam and West Bengal.

The region has been divided both horizontally and vertically into four and three divisions, respectively. In horizontal context, it has been categorized as Jammu and Kashmir Himalaya, Himachal Himalaya, Uttarakhand Himalaya, and Eastern Himalaya (Figure 1). Uttarakhand Himalaya is further divided into the Garhwal and Kumaon Himalayas. In terms of vertical divisions, it has been divided into Great Himalaya, Middle Himalaya, and Lesser Himalaya or Shivalik Ranges [5]. Geologically, the Himalayas are broadly divided into four areas across its length: (i) Foothill or Outer Himalaya; (ii) Lesser Himalaya; (iii) Higher Himalaya; and (iv) Tethyan or Trans-Himalaya [6]. The major rivers that originate from the great Himalayan Mountain Ranges are the Indus, Ganges, and their various tributaries. The region exposes diverse geology with different rock types representing the complete spectrum ranging in age from Archaean metamorphites/granitoids to the youngest Quaternary alluvium.

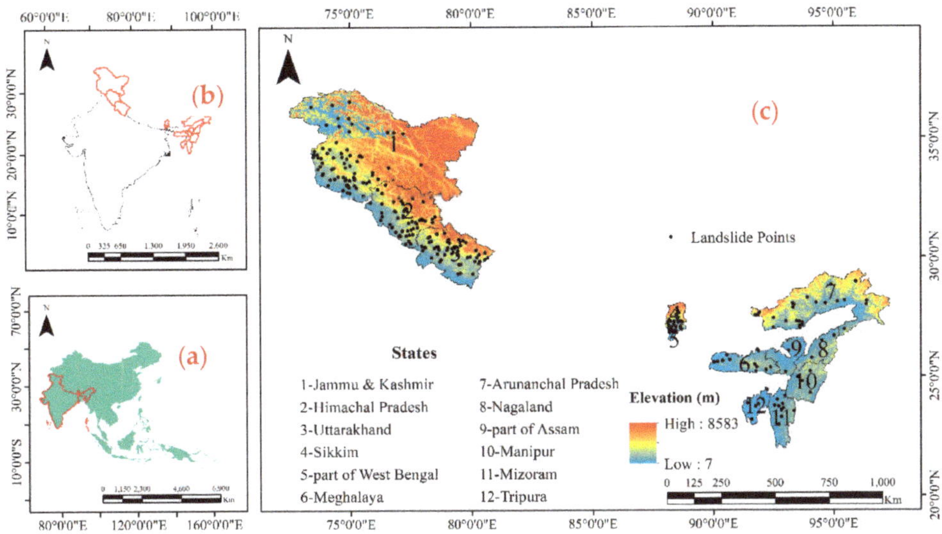

Figure 1. (**a**) Southeast Asia highlighting India; (**b**) geographical boundaries of India highlighting the Indian Himalayan region; and (**c**) location of rainfall-induced landslides in the Indian Himalayan region [1] (DEM Source: SRTM, resolution 30 m).

The region is tectonically and physiographically divided into three broad domains: the Peninsular India, the Extra-Peninsular India, and the Indo-Gangetic Plain [7]. The Indo-Gangetic plain is sandwiched between the shield area of the Peninsular India and the highly deformed suites of the Himalaya of Extra-Peninsular India, comprising essentially the younger meta-sediments. The tectonic trough (foreland basin) sandwiched between peninsular shield in the south and Himalayan Mountains in the north formed due to upliftment of the latter has been filled up by sediments derived from both sides, especially from the Himalayas. Structurally, the Himalayan Mountain chain occurring all along northern part of India can be divided into four contrasting longitudinal litho cum morphotectonic belts from south to north: (i) foothill belt; (ii) Main Himalayan belt; (iii) Indus-Shyok belt; and (iv) Karakoram belt.

The foothill Himalaya is a 10–50 km wide Miocene to Lower Pleistocene Molasse sequence represented by Siwalik, Murree, and Subathu Group of rocks. The belt is a domain of active tectonics, having participated in the terminal phase of the Himalayan Orogeny. This is followed to the north by the Lesser and Higher Himalayas, represented by geological sequences of Proterozoic age with a Phanerozoic cover of varying thicknesses in different parts. The foothill Himalaya is overlain by alluvium and separated from the Lesser Himalaya by the north-dipping fault commonly known as the Main Boundary Fault (MBF) or the Main Boundary Thrust (MBT) in Garhwal, Kumaon, Darjeeling, and Arunachal Pradesh Himalaya. The Main Frontal Thrust (MFT) limits the margins of the Siwalik Zone against the Ganga Plains [8].

The Lesser Himalaya is 60–80 km wide and is a discontinuous belt stretching between the MBT in the south and the Main Central Thrust (MCT) in the north [9]. It consists of autochthonous Late Proterozoic sediments, thrust over by three vast nappes that are built up successively of Palaeozoic sediments, Precambrian epimetamorphics, and mesograde metasediments [10]. The epi-metamorphic and meso-metamorphic nappes throughout their extent are characterized by Early Proterozoic (= 1900 Ma) and Early Palaeozoic granitic bodies of large dimension. The MCT separates the Lesser Himalaya from the Higher Himalaya to its north. The Higher Himalaya marks the region of the highest peaks of the Himalaya (Nunkun, Leopargial, Kedarnath, Badrinath, Nanda Devi, Api, Dhaulagiri, Mt. Everest, and Kanchanjunga), made up of 10–15 km thick Precambrian crystallines exhumed up and intruded by granites, some of which are Tertiary in age. The Indus Shyok belt/the Tethys Himalaya extend to the south of the Trans-Himalayan Karakoram belt and comprise ophiolite melange (Indus ophiolite and associated formation) and plutonic rocks (Ladakh Granitoid Complex) of the Indus Shyok belt [11]. These predominantly fossiliferous sediments range in age from Late Proterozoic to Eocene. Sporadic occurrences of chromite have been reported from the ultrabasic rocks associated with Dras volcanics from Ophiolite-Melange zone. Karakoram belt, the northernmost zone, comprises Palaeozoic and Mesozoic sedimentary sequence in a metamorphic basement of an unknown age. This Trans-Himalayan belt lies to the north of the Indus suture Zonein Ladakh region and extends eastward into Tibet. No important mineral occurrence is known from this belt. Figure 2 represents the geological map of the Indian Himalayan region.

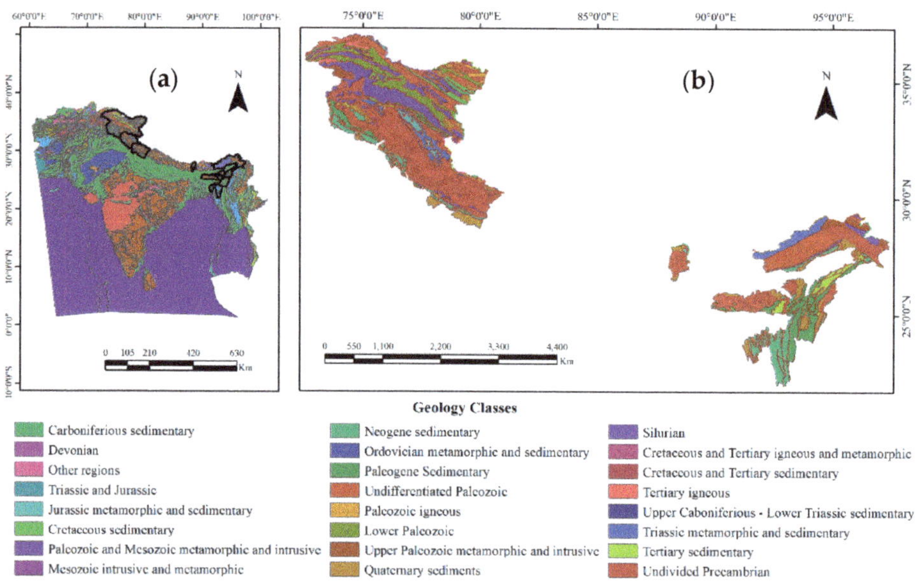

Figure 2. Geology of: (**a**) South Asia highlighting the Indian Himalayan region; and (**b**) the study region (Source: USGS World Geologic Map).

2.2. Data Collection

The review was undertaken by carrying out a bibliographic search on "Web of Science" Database (1990–October 2019) for a combination of keywords: "Landslide*", "Himalaya*", and "India". For the current analysis, only peer-reviewed journal articles written in English were considered, as peer-reviewed journals are considered as having the best quality articles and because the English language ensures that these contributions could be fully understandable for the whole international scientific community. The filtering of the articles was carried out by the number of citations (a) articles published before 2008 with 10 or more citations; (b) articles from 2008 to 2010 with 5 or more citations; and (c) all articles thereafter independent of the number of citations received (following the approach adopted from Reichenbach et al. [12]. After performing a screening of the relevant studies, the number of selected articles was narrowed down to 226 (Figure 3).

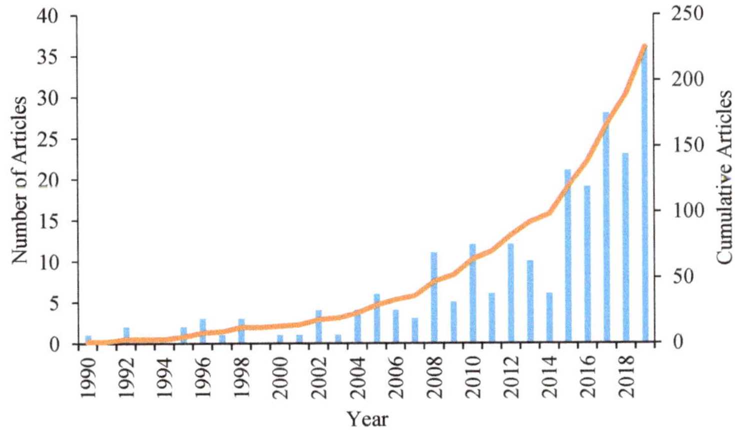

Figure 3. Analysis of the literature database during 1990–October 2019 from Web of Science. The left ordinate axis represents the number of articles per year and the right ordinate axis depicts the cumulative number of articles during the entire analysis period.

The articles were published across 75 different journals, with nearly 60% of the works published in 11 journals (Figure 4). The articles were initially majorly published in Engineering Geology and Geomorphology. The focus has shifted towards other journals such as Natural Hazards, Landslides, and Geomatics, Natural Hazards and Risk. Generally, after a major landslide event, the preliminary report is published in Current Science. However, extensive analysis of landslide study misses in such journals and is mostly focused on post-landslide studies.

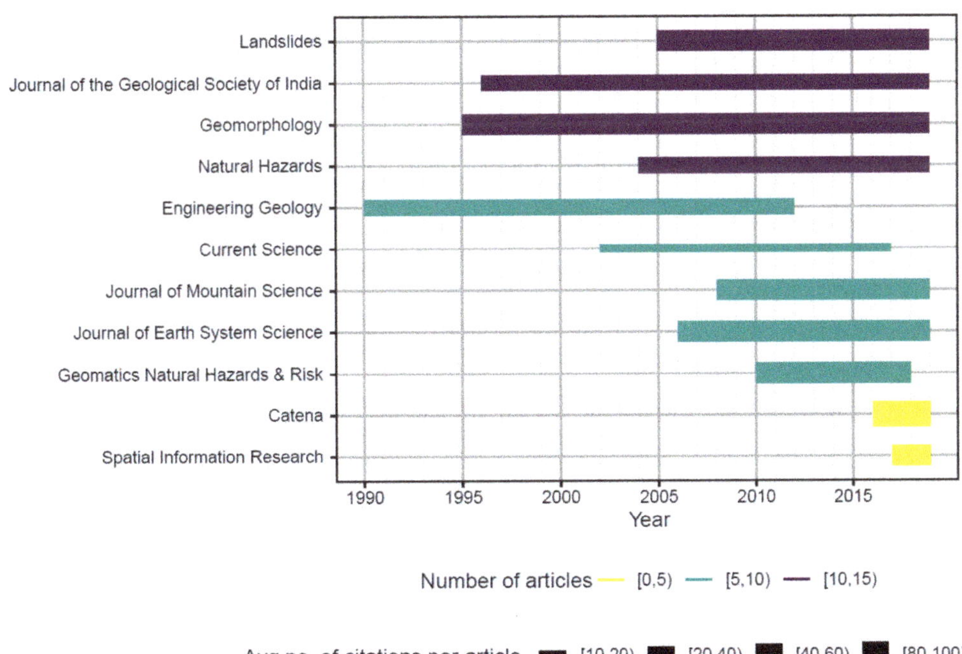

Figure 4. Distribution of the number of articles across the top 11 journals (out of 75). Horizontal bars depict the number of articles and its thickness denotes the average number of citations per article.

3. Results

The analysis reveals that the landslide studies in the region are quite biased towards Uttarakhand region, while there are few to no studies in the northeast Indian region (Figure 5).

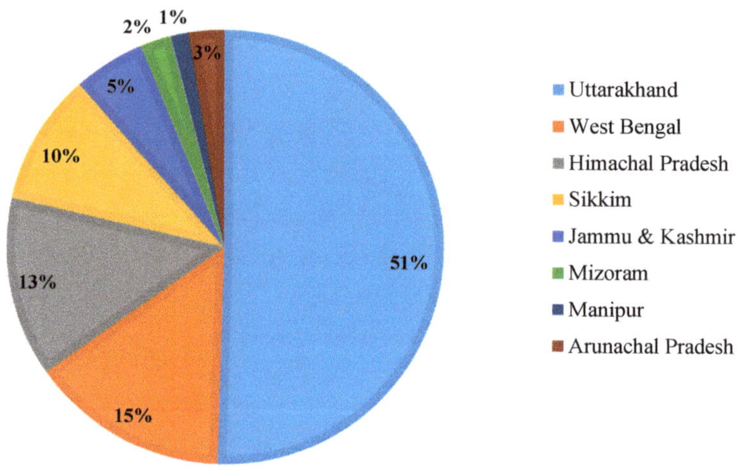

Figure 5. Distribution of landslide studies across the states in Indian Himalayan region.

This bias can be attributed to the high population density in the western sector of the Indian Himalaya, leading to a larger number of casualties and a generally higher level of landslide risk. Pham et al. [13] calculated the number of casualties due to landslide incidences during 2007–2015, which came out to be 5228, a staggering 82.9% of all disaster casualties in the region. To understand the type of studies being conducted, we divided them into seven broad categories: description of main events; identification; forecasting; monitoring and investigation; river damming; extreme events and climate change; and susceptibility mapping.

3.1. Main Landslide Events

The Himalayan region has been affected by several landslides, and a few of the most researched and damaging landslides are mentioned here. Most studies have been carried out in the Uttarakhand region. Some major landslide events in the Uttarakhand region occurred in the Okhimath area in Mandakini Valley, which was significantly damaged by the August 1998 landslide event, triggered by heavy rainfall. In total, 466 landslides were triggered, leading to 103 deaths and damaging 47 villages. Thereafter, cloudburst on July 2001 in a part of Mandakini valley led to more than 200 landslides causing 27 fatalities and affecting almost 4000 people [14]. The region also suffered from a heavy rainfall of more than 200 mm during 13–14 September 2012, which led to the death of 51 people and caused 473 landslides [15]. June 2013 witnessed another heavy rainfall of 350 mm within a period of five days in the Chamoli and Rudraprayag districts of Uttarakhand [16]. Other major landslides in the Uttarakhand region were the 2009 landslide in Pittorgarh region with a loss of 43 people.

Another major landslide affected region is Himachal Pradesh, of which the area of Katropi has often been studied [17,18]. The area has a history of recurring landslides since 1977 and recently it suffered from major landside in August 2017 claiming 46 lives [17]. The Pawari landslide zone has also been an active landslide zone located in the southeastern part of the state, which expanded 7% during 2005–2014 [19]. The Luggarbhati landslide in 1995 and Dharla landslide event in 2007 claimed 65 and 62 lives, respectively. The Jammu and Kashmir region has suffered from major cloudbursts such as the Leh Nalla cloudburst (2005, 2006), Leh cloudburst (2010), and Baggar cloudburst (2011) [20].

In the northeast part of the region, most reported landslides have been in the Darjeeling area of West Bengal and Sikkim. The Darjeeling region has a history of several landslides with the first major recorded occurrence in 1899, which led to the death of 72 people. Thereafter, landslides happened in 1950 (127 people died) and 1968 (667 people died) [21]. The region has also suffered significant landslides in the late 20th and 21st century (1991, 1993, 2004, 2005, 2006, 2009, and 2015) [22]. Similarly, the Sikkim region faced numerous landslides such as the Lanta Khola landslide [23], south Sikkim area [24]. The prominent unstable sections in the northeast region of the Indian Himalaya are National Highway (NH)-40 [25] and NH-44 [26] in Meghalaya. In Mizoram, NH-44A has suffered several landslides, in 2011, 2013 and 2017 [27].

3.2. Landslide Identification

The identification, visualization and standardized classification is an important step for pre- and post-hazard landslide analysis, which could assist in assessing rescue and relief operations [28]. Historically, landslide damage assessment was carried out using field visits and categorizing it according to the block diagrams depicted in [29]. However, understanding landslide scenario using such simplistic illustrations fails to consider the surrounding morphometry and its contextual relationship. With the advancement of remote sensing technologies and availability of high spatial and temporal resolution data, researchers have attempted to map landslides using image classification techniques. In the case of Indian Himalaya, the identification of landslides after an event has been mostly carried out by scientists at National Remote Sensing Centre (NRSC). Vinod Kumar et al. [30] studied Varunawat Parbhat landslide in Uttarakhand following the September 2003 landslide event using post-landslide Stereoscopic Earth Observation data on a 1:10,000 scale. The availability of high-quality images enables researchers to identify the loss of vegetation and exposure of fresh rock and soil after the landslide event [31]. The methodologies to assess landslide damages using pre- and post-landslide event images can be categorized as pixel based (PB) (medium resolution images) and object based (OB) (high-resolution images). Pixel based method involves the use of pre- and post-event images, and landslides are identified using spectral information. The identification of the landslides is mainly based on change detection and image fusion techniques. However, the use of pixel-based identification approach has some limitations. Firstly, the availability of images (pre- and post-event) may not be available from the same sensor, including variation in the atmospheric conditions and bandwidth of spectral values. Secondly, the representations of landslides are based on pixel values. On the other hand, object-based approaches use spectral, shape, spatial, and contextual properties and the identification is carried out using an object rather than a pixel, which makes more sense for natural events such as landslides as the damages are of irregular size and shape. It has the potential to add morphometric information derived from Digital Elevation Model (DEM).

The work conducted in the Indian Himalayan region has been primarily conducted using an object-based image classification technique. The use of OB image classification is dependent on the data type and methodology being used. The methodology involves detection, segmentation, and classification of images. The detection procedure can further be categorized into: (i) direct object change detection; and (ii) classified object change detection. The first detection technique involves the comparison of two images for changes in geometrical and spectral information acquired at different times. Such approach is usually used when comparison is made between images acquired from same satellite sensors and generation of full change matrix is not an absolute requirement. In the latter approach, first segmentation and classification are performed independently using time series images, and then the change matrix is derived using classified images. The second approach is mostly used in land use/land cover change analysis as well as when time series data are acquired by different satellite sensors [32]. The studies conducted in the region have mainly used the first approach as a single feature, i.e., the landslide is analyzed and the changes are estimated from images acquired from the same sensor.

The capability of detection using object-based analysis depends on the segmentation technique being used, which has the ability to excerpt objects that accurately describe the relevant properties [32]. The most popular segmentation technique for classification of images is the multi-resolution segmentation (MRS). This segmentation approach is a bottom-up region-merging technique, wherein small objects are merged into bigger ones in subsequent steps, and involves three parameters: scale, shape, and compactness. The scale parameter handles the size of the image object size, whereas the shape parameter determines the degree to which shape influences segmentation vs. spectral homogeneity. The compactness defines the weight of the compactness criteria. The higher the value is, more compact the objects will be [33]. Several methods exist to select an optimal scale parameter, of which plateau objective function [31,32] and optimal scale parameter selector [34] have mostly been used. However, obtaining a desired scale for all the landslide types in an area is difficult, thus over-segmentation is preferred to under-segmentation [35].

The use of the OB technique has been used for landslide detection immediately after a landslide event, as well as for creating a landslide inventory database from historical images. Martha et al. [35] prepared landslide inventory data for the 1998 Uttarakhand landslides in Okhimath region, Uttarakhand and compared with the field data acquired after the landslide event. The study identified 73 landslides using Resourcesat-1 LISS-IV multispectral data (5.8 m) and a 10-m Cartosat-1 derived DEM. This semi-automatic approach resulted in achieving an accuracy of 76% for recognition and 69% for classification in terms of number of landslides. Thereafter, Martha et al. [31] used historical panchromatic image dataset (1998–2006) for the same region to prepare a landslide inventory database. The images were acquired from Cartosat-1 (2.5 m) and IRS-1D (5.8 m) and depicted the use of texture in cases of missing spectral information. Martha et al. [36] analyzed various time series images (1997–2009) for Okhimath region, Uttarakhand using Cartosat-1 (2.5 m), Resourcesat-1 (5.8 m), and IRS-1D panchromatic (5.8 m) data. For the entire period, the accuracy of landslide detection varied from 60% to 89%, whereas it varied from 71% to 97% in terms of landslide extent.

The identification and classification of landslides is available for two severe landslide events occurring in Uttarakhand in September 2012 [15] and June 2013 [16,32]. Martha and Vinod Kumar [15] used very-high resolution (VHR) satellite data (Resourcesat-2 (5.8 m), Cartosat-2 (1 m), Kompsat-2 (1 m), GeoEye-1 (0.5 m), and Cartosat-1 DEM (10 m)) covering the September 12 Uttarakhand landslide event and identified 473 landslides. Martha et al. [16] compared pre-disaster images (Resourcecat-2 and GeoEye-1) with post-disaster images (Resourcesat-2 and Cartosat-2a (1 m)) after the June 2013 Uttarakhand event. The study identified 6031 landslides and classified those as new (57.74%), old (18.92%), and reactivated (23.34%). Martha et al. [32] identified new landslides by comparing the pre- and post-landslide Resourcesat-2 images with a detection accuracy of 81%. Mohan Vamsee et al. [34] improved the scale component of MRS technique and developed the optimal scale parameter selector (OSPS) tool, which was applied to Uttarakhand region using Resourcesat-2 images.

However, the accuracy of the above-mentioned techniques largely depends on the resolution of the satellite images that are costly to be acquired. Therefore, free and high-resolution Google Earth (GE) images have also been used for landslide mapping [19,37]. The ability of such images to provide a 3D view and its free availability can serve researchers to exploit imagery for landslide detection and mapping [38]. Kumar et al. [37] utilized GE images for landslide dimension mapping along Satluj Valley in Northwest Himalaya. Further, Kumar et al. [19] conducted a study on a relatively smaller area (Pawari landslide) in the same region and analyzed the changes using GE images of 2005, 2012, and 2014.

As landslide mapping is the first key step towards conducting any landslide study or to set up recovery attempts after a landslide event, focus should be on the use of high temporal resolution dataset. The focus has primarily been on the use of semi-automatic identification approach and gradually improving the algorithms used to detect and classify landslides [16,28]. The classification and segmentation techniques need to be improved using computational techniques such as Machine Learning (ML) and Artificial Neural Networks (ANN) [39]. In addition, the studies have largely been concentrated in a single area and need to be applied to other Himalayan regions to understand their applicability and reliability.

3.3. Landslide Forecasting

Landslide forecasting is a key element for disaster risk reduction and is also the most challenging. Most of the landslides in the Indian Himalayan region are shallow in nature, of which rainfall is the primary triggering factor [3,4]. The analysis of precipitation for landslide occurrence can be performed by estimating minimum rainfall conditions, sub surface monitoring, or slope stability analysis. The minimum rainfall conditions, also known as thresholds, can be classified as empirical and physical approaches [40–42]. Physical models assess the relationship between rainfall conditions and hydrological conditions of the soil, which affects slope stability. The model analyzes the spatial variation of several factors such as geotechnical parameters, soil depth, volumetric water content, geology, and topography to determine the pore water pressure change and estimate the factor-of-safety. Such models need a large dataset of many parameters in spatial and temporal context, which is usually not available for the Indian Himalaya. Empirical methods analyze the rainfall conditions using statistical methods to determine threshold levels of precipitation. Such models are simpler to apply because they require only the spatial and temporal dataset of precipitation and landslide events [43,44]. The threshold values can be greatly affected by the data quantity and quality, the rain gauge density, and the methodology used. The thresholds are usually determined by drawing (usually with statistical techniques) a lower line to the precipitation conditions corresponding to landslide events in log-log, semi-log, or Cartesian coordinate system [3,41]. Segoni et al. [43] reviewed the type of thresholds, data collected, and other important information for thresholds determined in the global context for 2008–2017.

The studies on rainfall thresholds for the Indian Himalayan region have been very limited. The use of empirical approach has been explored for different Himalayan regions. Sengupta et al. [45] proposed the use of EMAP for Sikkim Himalayas, which is the ratio of cumulative event rainfall (E) to mean annual rainfall (MAP) (E_{MAP} = E/MAP), instead of rainfall event-duration (ED) or rainfall intensity-duration (ID) thresholds. Kanungo and Sharma [3] determined ID thresholds for Garhwal Himalayas using best fit of the lower boundary of ID plane. Dikshit and Satyam [4] developed the ID thresholds for Kalimpong region using frequentist statistical approach. The use of probabilistic [22,46] and semi-automated algorithmic approach [47] has only been attempted for Kalimpong region. Harilal et al. [48] developed both regional (Sikkim) and local (Gangtok, Sikkim) ID thresholds using statistical approach. The main concerns with these studies are the coarse spatial distribution of rain gauges, all studies being performed using a single rain gauge, and the availability of only daily rainfall data. In this regard, Gariano et al. [49] highlighted that a coarse temporal resolution dataset of precipitation could lead to an underestimation of rainfall thresholds, which culminates in a higher number of false alarms when using thresholds for operational early warning system. In addition, the thresholds determined are rainfall intensity-duration thresholds, which should be avoided, while the determination of event rainfall-duration (ED) thresholds should be encouraged. Mathew et al. [50] established ID thresholds along the Rishikesh–Mana pilgrimage route for Garhwal Himalaya using Tropical Rainfall Measuring Mission (TRMM)-based Multi-satellite Precipitation Analysis (TMPA) data. Despite using different methods to determine various aspects of rainfall characteristics, all studies were unified in the effect of antecedent rainfall in the region. Mathew et al. [50] also suggested that antecedent rainfall ranging from 15 to 30 days plays an influential role for destabilizing slopes in the Himalayan region, which leads the subsequent rainfall of short duration (24–72 h) to trigger landslides. The studies on rainfall thresholds in the Indian Himalayan region have been very minimal and more work should be conducted on the calculation and analysis of regional and local thresholds. Kumar et al. [51] highlighted that the mean rainfall threshold intensity of NW Himalayas (excluding Sikkim Himalayas) is roughly 290 mm/day. At present, defining rainfall thresholds at regional scale or at a state administrative level would be highly desirable but is hard to accomplish. The variation in the thresholds determined also ascertains that local thresholds would perform be better than regional thresholds in setting up an operational landslide early warning system. This could be due to the heterogeneous rainfall pattern at local scales in various Himalayan pockets. Table 1 lists the various thresholds generated for different regions of the study area. Figure 6 illustrates the thresholds in ID plane.

Table 1. Threshold equations generated for various Himalayan regions using several methods (I is Rainfall Intensity (mm/h), while D is Duration (h)).

Threshold Equation	Region	Methodologies
$I = 1.82\,D^{-0.23}$	Chamoli, Uttarakhand	Empirical thresholds
$I = 58.7\,D^{-1.12}$	NH 58, Uttarakhand	Empirical thresholds
$I = 3.52\,D^{-0.41}$	Kalimpong, West Bengal	Empirical thresholds
$E = (4.2 \pm 1.3)\,D^{(0.56 \pm 0.05)}$	Kalimpong, West Bengal	Semi-automated algorithm approach
$I = 100\,D^{-0.92}$	Gangtok, Sikkim	Empirical thresholds
$I = 43.26\,D^{-0.78}$	Sikkim	Empirical thresholds

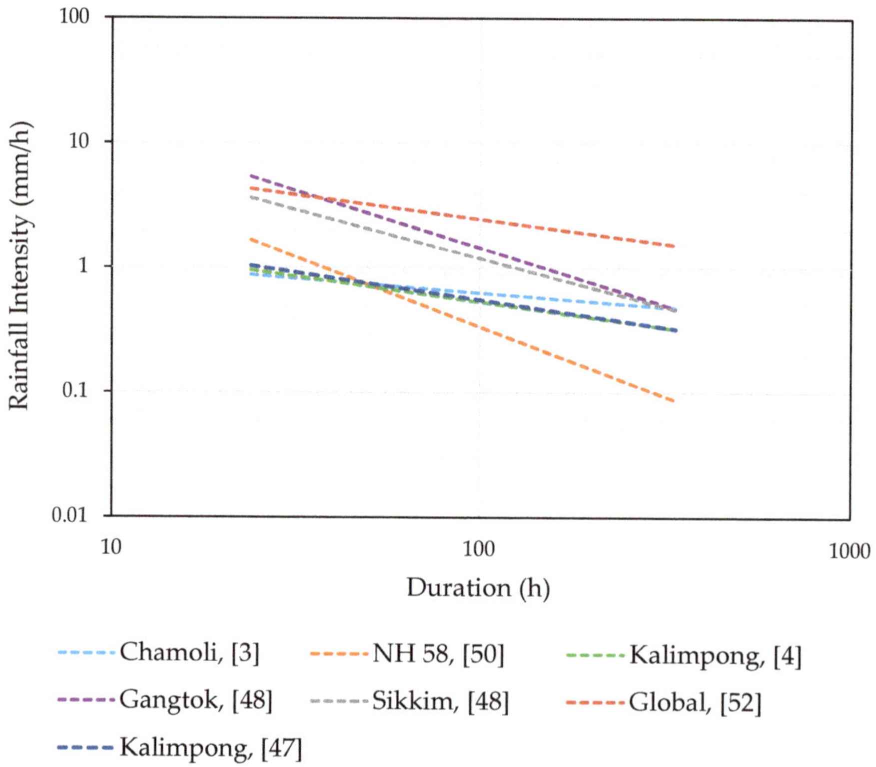

Figure 6. Comparison of various ID thresholds developed for various Himalayan regions with global thresholds defined by Caine [52].

3.4. Landslide Monitoring and Investigation

Landslide monitoring is a very significant aspect of landslide assessment, especially in the Indian Himalayan region where the slopes are generally creepy in nature, which could immediately fail during an abrupt rainfall event or seismic occurrence. Landslide monitoring is generally categorized into three types [53]:

1. Observing the changes of topography and cracks on the surface during the site investigation. This is a traditional monitoring technique used by geologists at regular time periods. The major limitation of using such a technique is that it does not provide the variations for a short time interval and it is difficult to accurately determine the time and location of future landslide incidences.
2. Remote sensing techniques such as satellite imagery analysis, GPS synthetic aperture radar (SAR) interferometry, and light detection and ranging (LiDAR). These methods can be helpful as they can measure slope displacement over a large area independent of the weather conditions.
3. In situ ground-based observation of slope movement using various instruments (extensometers, inclinometers, and tiltmeters) and installing rain gauges to accumulate local rainfall data. For example, Dikshit et al. [54] used tilt sensors for Darjeeling Himalayas, while Falae et al. [55] used Electrical resistivity tomography (ERT) for Garhwal Himalayas.

Of the various monitoring techniques, the use of remote sensing (RS) data along with traditional monitoring instruments would prove to be very helpful especially in understanding long-term deformation and the advancement in RS technology would gradually overcome the use of traditional equipment. In the case of Indian Himalaya, monitoring studies include remote sensing techniques and ground based observation. Yhokha et al. [56] used the Persistent Scatterer Interferometry (PSI) technique using ENVISAT satellite for Lesser Himalaya, Nainital. PSI is based on InSAR technique, which utilizes several SAR images and has proved to be successful in identifying the creeping zones in Nainital. Martha et al. [57] monitored the landslide dammed lake in Zanskar Himalayas for five months (January–May 2015) using multi-temporal high-resolution satellite images after the landslide event (December 2014) and depicted the variation in the dimension of the impounded lake during the monitoring period. Dikshit et al. [54] and Dikshit and Satyam, [22] used tilt sensors at shallow depths to analyze the variation in tilting angle of the instrument, which is related to the lateral displacement of the slope. The study also validated the empirical models thereby encouraging similar studies to be conducted in other Northeast Himalayan regions, which could help in setting up a preliminary early warning system.

The investigation of unstable slopes in the Indian Himalaya has been conducted using Ground Penetrating Radar (GPR) or 2D Electrical Resistivity Tomography (ERT). Mondal et al. [58] conducted ERT investigation for Naitwar Bazar landslide in Uttarakhand for six sites, which was active after the 2004 event. Kannaujiya et al. [59] compared slide dimensions observed from satellite (IRS LISS-IV) with geophysical investigation and determined the depth and slip surface geometry using 2D ERT and GPR) for Kaliganga river valley in Uttarakhand. Falae et al. [55] used ERT to understand the subsurface movement to determine the probable failure plane of the Pakhi landslide, Uttarakhand. Sharma et al. [60] conducted a geophysical study at Lanta Khola Landslide, Sikkim using very low-frequency (VLF) electromagnetic survey to understand the subsurface structure. The results reveal the presence of a water-saturated zone at the subsurface level of the slide.

The literature review reveals that only a handful of studies have been carried out for landslide monitoring and investigation; considering the increased quantity of the available remote sensing data sources, the focus needs to shift towards the use of high spatial and temporal resolution data, which would lead to near real-time monitoring results. The framework using sub surface investigation is ideal at a local scale and would fulfill the demands of a community but would be difficult to manage in long term due to the high installation and operational costs. However, in sections where the landslide problem is immense, which is slide specific, such as Paglajhora landslide (Darjeeling), Singtam landslide (Sikkim), or Kotropi landslide (Himachal Pradesh), monitoring using instruments would still be required as the region is highly vulnerable and satellite images at currently available resolution may not be able to accurately identify failure planar sections.

3.5. Lake Damming Landslides

Landslide lakes or dams are temporary lakes in the river valleys formed after a landslide blocks the river course. Landslide dammed lakes and their outburst floods (LLOFs) are common in the Indian

Himalaya. The breaching of such temporary lakes with a huge amount of accumulated water and sediments can create devastating floods in the downstream areas [61]. In the Himalayas, landslide dams commonly form in high mountains because of different mass movement types such as rock and debris avalanches, rock and soil slides, mud–debris–earth flows. The oldest recorded landslide lake was the Gohna Tal, which was formed in September 1893 as a result of heavy precipitation blocking the Birahi Ganga River situated in the Kumaon Himalaya. Gradually, the lake expanded to an approximate 4000 m length, 340 m width, and 300 m depth. On 26 August 1894, the dam collapsed, and a devastating flood hit downstream of Birahi Ganga [62].

The techniques used for assessment of damming processes are either based on geomorphic characteristics, hydraulic properties, and velocity measurement of dam material. The assessments of the damming process are dependent on the analysis of three significant factors: (a) pre-dam formation mechanism; (b) dimensional characteristics of the dam; and (c) stability analysis of dam [63]. The pre-dam formation mechanism largely depends on the slope stability analysis and landslide triggering factors, whereas dimensional characteristics are dependent on geometry of the area and landslide volume. In terms of slope stability, the techniques can be categorized as discontinuum modeling and continuum modeling [63]. Discontinuum modeling is majorly used for rock slope stability analysis, whereas continuum modeling is used for debris flows as well as rock slopes dependent on the material and geometry of the slope [19,64].

Most of the studies have focused on geomorphic analysis [15,16,57,63] using different satellite images to understand the spatiotemporal landslide changes. Gupta and Sah [65] catalogued the LLOFs developed in the Trans-Himalayan region between 2000 and 2005 and studied its impact on the stability in the region. Martha et al. [57] conducted an extensive investigation of Phutkal River landslide dammed lake for landslide occurrence on 31 December 2014 using multi-temporal Cartosat-2 images of 1 m spatial resolution and calculated slide volume using pre- and post-event datasets. Kumar et al. [63] used GE images of 1.5 m resolution and estimated landslide volume along with slope stability analysis for Urni landslide in Himachal Pradesh.

3.6. Extreme Events and Climate Change

The effect of global warming and the corresponding changes to climate and geohazards is expected to affect landslide events [66]. However, forecasting and understanding the impact of climate change on landslide activity still poses a challenge. Gariano and Guzzetti [66] in their review article on climate change studies related to landslides emphasized the need for more studies as large parts of the world suffer from a few to no studies. Although it is predicted that the Indian Himalaya region will be affected by climate change at the time of compiling this article, we could not find any article in which the effect of climate change has been considered in the Indian Himalaya for any type of rainfall-induced landslide study. However, some works exist that describe some of the most recent extreme rainfall events: Jammu and Kashmir Himalaya for the 2014 rainfall event [51], Uttarakhand Himalaya for the 2009 event [67], and Leh region for the 2010 event [68]. One of the main reasons for the lack of inclusion of climatic variation in landslide studies for the Indian Himalayan region is the unavailability of data both in spatial and temporal context, which has been highlighted by several researchers. Kumar et al. [51] estimated that the pattern of rainfall intensity along the Himalayas varies from west to east, thereby precipitation patterns being affected by western disturbances and summer monsoon. Such lack of studies in a region such as the Himalayas, which is prone to severe landslides and other geohazards, is a matter of concern. Hereon, studies should include climate aspects using remote sensing products, which could reconstruct events and provide a better understanding of climate impact.

3.7. Landslide Susceptibility

Landslide hazard is defined as the probability of landslide occurrence of a specific type ("when" or "how frequent") and magnitude ("how large") [12]. The spatial probability of landslide hazard is assessed by carrying out landslide susceptibility mapping or popularly known as landslide hazard zonation mapping. Landslide susceptibility assessment determines potential of landslide event considering several predisposing factors and investigating their spatial distribution [12]. The spatial occurrence of landslides is controlled by several and sometimes inter-playing factors such as geological setting, rainfall, morphology, soil, and vegetation conditions, thus landslide susceptibility assessment is not a straightforward task, and many methodologies have been developed to assist the susceptibility analysis. The classification techniques used for landslide susceptibility models has also changed over time. We categorized them into four groups: Qualitative, Semi-quantitative, Quantitative, and Deterministic (Figure 7). In this section, we explore the models, the parameters used, and the trend in susceptibility analysis.

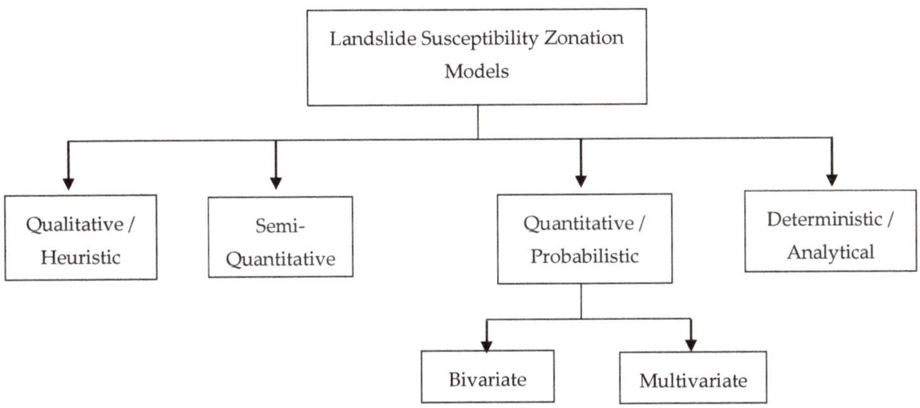

Figure 7. Types of modeling techniques used in landslide susceptibility studies.

Initially, studies were mostly based on qualitative methods wherein values are ascertained for landslide conditioning factors based on the knowledge about the study area and are quite subjective [69,70]. Semi-quantitative methods are logical tools and emphasize the significant factors by assigning weights or values. These include techniques such as Analytic Hierarchy Process (AHP), danger pixel approach, and a weighted linear combination [71].

The deterministic analysis involves an analysis of physical and mechanical soil properties and determines the susceptibility in the form of a factor of safety (FS) [72]. FS is the ratio between factors affecting landslide to factors preventing landslide incidences. This involves the utilization of several factors such as infiltration, soil cohesion, groundwater table, pore water pressure, geotechnical soil parameters, etc. Several deterministic models are available to determine slope stability such as SHALSTAB (Shallow Landsliding Stability model), SINMAP (Stability Index Mapping), SHETRAN, TRIGRS (Transient Rainfall Infiltration and Grid-based Regional Slope Stability) model, etc. In the case of the Indian Himalayas, Sarkar et al. [72] used SHALSTAB model to determine the critical steady-state rainfall for slope stability in Darjeeling Himalayas. A landslide inventory map was developed using various satellite images (Resourcesat -1, Cartosat -1, and Google Earth) and field surveys, soil parameters (depth, saturated soil density, engineering properties, and soil transmissivity), and slope parameters (angle, contour length, and upslope contributing area). Mathew et al. [73] determined FS for terrain stability for Garigaon watershed area of Uttarakhand by coupling an infinite slope stability model with a steady-state hydrological model (LIDA) using the spatial analysis in GIS environment. The landslide locations were mapped using LISS IV and Cartosat-1 PAN data and the input parameters used were soil texture, index properties, porosity, LULC, and terrain properties. The results indicate an increase in the unstable areas of more than 45% for rainfall intensity variation from 50 to 100 mm/day.

Quantitative models depend on the landslide density under each influencing factor and can be further classified as a bivariate and multivariate [74]. Both bivariate and multivariate statistical models compute weights; however, multivariate techniques depend on the collective effect of parameters [75,76]. Bivariate statistical methods include weights of evidence, frequency ratio, information value, and the combination of frequency ratio and fuzzy methods [77–79]. The most popular multivariate model used is logistic regression [80]. The analysis shows that most studies have used quantitative techniques, of which logistic regression has been most used [81–83]. Thereafter, frequency ratio and information value were used 16% and 9% of the studies, respectively [74,84–88]. In terms of semi-quantitative models, AHP has been the most used method (Figure 8).

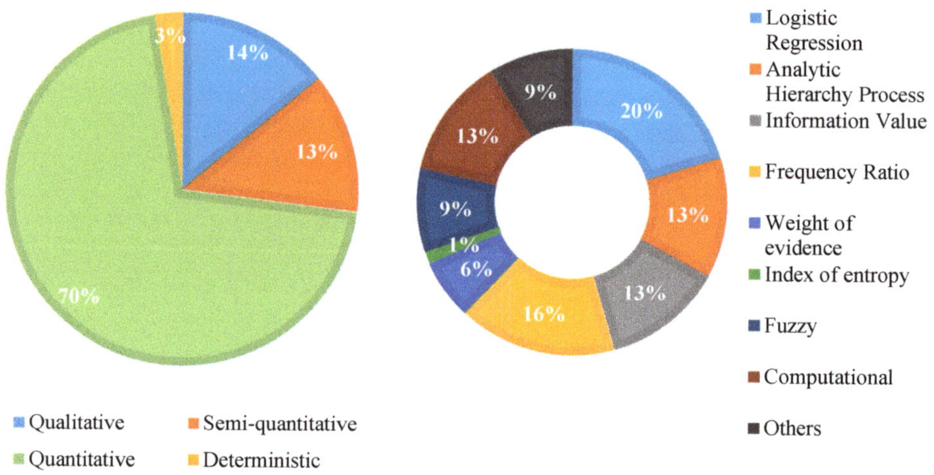

Figure 8. Model types and the techniques used in hazard and susceptibility studies.

After understanding the various models being used to determine susceptibility, the analysis of the parameters being used was investigated. The use of parameters for susceptibility analysis would affect the results; however, they usually depend on the local factors along with data availability. In addition, the factors depend on the type of movement, scale of study and the methodology used [89]. Based on the analysis of the articles, the factors influencing landslide susceptibility can be divided into four groups: (i) geological (lithology, geological structures); (ii) geomorphological (drainage, relative relief, slope, and slope aspect); (iii) environmental (land cover); and (iv) anthropogenic (roads). In the 81 studies, 36 factors were used, where some factors such as "gradient" and "slope" were clubbed together, and a word map of the factors is illustrated in Figure 9. The most used factors were slope angle (93.8%), land use/land cover (LULC) (92.5%), aspect (80.2%), and lithology (62.9%). Apart from these factors, another set of parameters, namely drainage density, curvature, topographic wetness index (TWI), stream power index (SPI), relative relief, lineament distance, and lineament density, was also utilized. Seismic factors, fault buffer, and road buffer have been less frequently used. The trend over the years in the number of factors to be used has increased with an average of six factors being used until 2010 and an average of eight factors thereafter. This trend can be ascertained to the increase in the availability of data and remote sensing products.

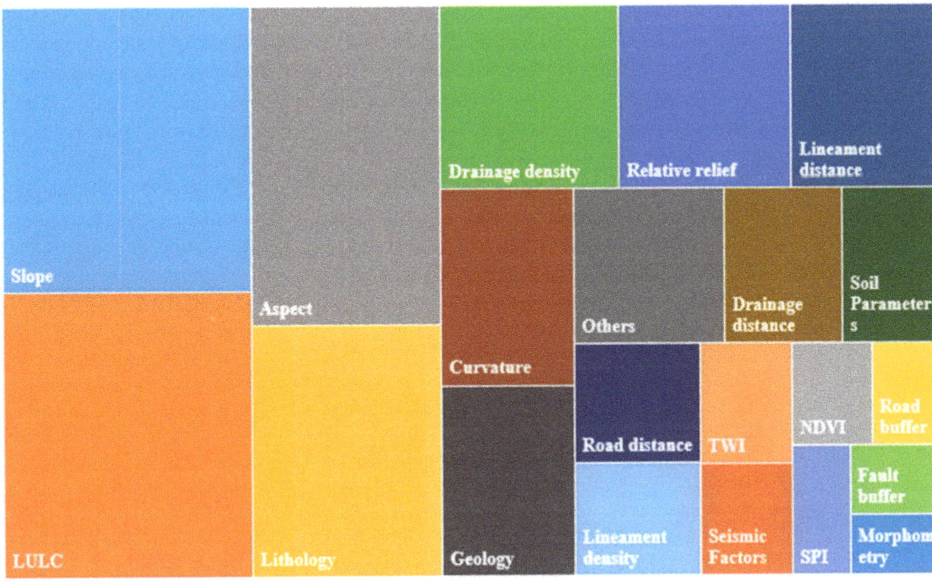

Figure 9. Tree map of the factors used for determining landslide hazard and susceptibility.

In terms of the spatial resolution of the dataset used to determine susceptibility, the most commonly used Digital Elevation Model (DEM) is the one released by USGS, with a spatial resolution of 30 m (46%). Only 28% of the studies used a spatial resolution higher than 10 m, which was obtained from Cartosat-1 images, whereas 26% of studies used the dataset with a spatial resolution of 12.5 m from ALOS-PALSAR DEM. The scale of geological maps has generally been 1:250,000 (62%) while the remaining 38% studies were conducted using 1:50,000 scale. Anbalgan [69] formulated the guidelines for landslide hazard assessment based on Landslide Hazard Evaluation factor (LHEF) using factors such as structure, relative relief, lithology, slope morphometry, land use, and groundwater conditions. The LHEF rating system was based on determining the factors and assigning values based on the understanding/knowledge about the study region. The national standard body of the country, Bureau of Indian Standards (BIS), prepared its guidelines (BIS, 1998) based on Anbalgan's study using the same factors excluding groundwater conditions [90]. BIS recommends an indirect method for medium-scale (1:25,000–1:50,000) landslide susceptibility mapping. Such technique is a generalized method that could be applied over large regions regardless of the relationship between causative factors and landslide types, often leading to moderate to poor predictions [91].

Following this, the nodal agency, Geological Survey of India (GSI), developed its own guidelines (GSI, 2005) considering 10 factors, which is a modified version of BIS (1998). Singh et al. [90] compared the zonation maps prepared using BIS and GSI guidelines for Arunachal Pradesh and found that GSI guidelines were better than BIS guidelines. Das et al. [92] studied homogenous susceptible units (HSU)-based landslide hazard utilizing spatiotemporal data and landslide size, which could better represent homogeneous susceptible regions. The study was conducted using high-resolution data for Northern Uttarakhand region. Martha et al. [36] applied the weight of evidence model to determine the susceptibility using a semi-automatic derived landslide inventories. Sarkar et al. [93] was the first to introduce landslide intensity as a parameter for hazard determination, which was applied for Garhwal Himalayas. The intensity was determined by analyzing the volume and velocity, which depends on the landslide area, debris thickness and the types of failures.

However, lately, the rise has been in the use of computational techniques such as Machine Learning (ML), Support Vector machine (SVM), and Artificial Neural Network (ANN), which have proved to outperform traditional approaches [13,94–96]. However, the studies have largely focused on landsides in Uttarakhand Himalayas. Ramakrishnan et al. [97] used backpropagation neural network and showed prediction capability of 80%. Pham et al. [98] used various ML ensemble models using Multilayer Perceptron (MLP) Neural Networks and six different ensemble techniques, thereby depicting better capability using ensemble framework. Pham et al. [99] analyzed susceptibility for an area of 323 km^2 in Uttarakhand using various ML models (SMO-SVM, VFI, and LR), FT, MLP-neural networks, and NB. In the first case, SMO-SVM performed best, whereas, in the latter case, MLP neural networks and FT models provided similar accuracy compared to NB. Pham et al. [13] used four hybrid machine learning models for Uttarakhand Himalayas. Pham et al. [13,95,100] extensively analyzed the Pittorgarh region (242 km^2) in Uttarakhand using various hybrid ML models (Table 2).

Table 2. Various computational models used for landslide susceptibility in Uttarakhand Himalayas.

Models	No of landslides	No of Factors	Test Site Area/Pixel Size	AUC/Accuracy	Articles
ANN	154	7	600 km^2/5 m × 5 m	0.84	Chauhan et al. [101]
Back propagation neural network	63	6	~116 km^2/30 m × 30 m	0.8	Ramakrishna et al. [97]
SVM, Proximal SVM, L2-SVM-Modified Finite Newton	2009 pixels	8	1625 km^2/30 m × 30 m	0.807	Kumar et al. [102]
Multilayer Perceptron (MLP) Neural Networks	930	15	1325.47 km^2/20 m × 20 m	0.886	Pham et al. [94]
sequential minimal optimization (SMO) SVM, vote feature interval (VFI), Logistic Regression (LR)	430	11	323.82 km^2/20 m × 20 m	0.891	Pham et al. [98]
Functional Trees (FT), Multilayer Perceptron (MLP) Neural Networks, Naïve Bayes (NB)	430	11	0.32 km^2/20 m × 20 m	0.850	Pham et al. [99]
Rotation Forest based Radial Basis Function (RFRBF) neural network	930	15	0.13 km^2/20 m × 20 m	0.891	Pham et al. [103]
Aggregating One-Dependence Estimators, SVM, ANN-RBF, LR, NB	1295	16	561 km^2/30 m × 30 m	0.968	Pham et al. [104]
Ensemble decision tree	103	10	242 km^2/20 m × 20 m	0.883	Pham et al. [13]
Rotation Forest Ensemble Model	103	10	242 km^2/30 m × 30 m	0.741	Pham et al. [95]
Hybrid Reduced Error Pruning Trees	103	10	242 km^2/30 m × 30 m	0.989	Pham et al. [100]
Hybrid model of Multi Boost ensemble and SVM	391	16	270 km2/10 m × 10 m	0.972	Pham et al. [105]
Neuro-fuzzy inference system	391	16	270 km^2/30 m × 30 m	0.95	Jaafari et al. [106]
Fuzzy Expert System, Extreme Learning Machine	49	8	43 km^2/15 m × 15 m	0.844	Peethambaran et al. [96]

Kanungo et al. [107] compared susceptibility map for Darjeeling Himalayas using traditional, ANN, fuzzy, and combined neural fuzzy weighing schemes and found the combined neural fuzzy model to provide more accurate results. Following this, two different models (combined neural certainty and fuzzy) were compared with the neural fuzzy model for the same region and the latter was found to be better [108]. Chawla et al. [109] used Genetic Programming (GP) and Particle Swarm Optimization (PSO) SVM model for the Darjeeling Himalayan region and found GP method to perform better. Meena et al. [110] conducted susceptibility analysis using a hybrid spatial multicriteria evaluation model for Kullu, Himachal Pradesh and found it to produce better results when compared to FR and AHP.

4. Discussion

The results of the review provide researchers a comprehensive insight to understand the studies being conducted in various avenues of the Indian Himalaya, thus providing a general picture of the state-of-the-art in this region and a reference point for those (researchers and practitioners) starting to work in this area for the first time. Moreover, the analysis of the results can be useful in identifying some existing research problem and the most important research questions to be addressed in the near future.

The first gap to be urgently filled is the lack of studies in regions such as the Northeast and Jammu and Kashmir Himalayas. Since research programs have focused mainly on Uttarakhand region, which is understandable given the high population density and risk exposure, more efforts need to be made to analyze other regions as well and to help policy makers in formulating a general guideline towards mitigation. Another issue is the urgent need for high quality data, as many of the reviewed studies suffer from the coarse dataset being employed. As an instance, rainfall thresholds have been conducted using a very sparse rain gauge network that measures rainfall only at daily time steps; and most of the susceptibility studies make use of a very coarse geological map at the 1:250,000 scale. It is well established that the poor quality of input data has a negative impact on the quality of the results of any scientific model; therefore, priority should be given to the acquisition of high-resolution basic data for further research.

In the Indian Himalaya, many studies have been conducted that explore different disciplines of landslide research in various regions, and a precise framework is not evident. One of the next steps of research in the region could be the combination of existing studies in a more comprehensive and coherent framework aimed at assessing and reducing landslide risk. For instance, susceptibility maps could be used to define the most hazardous areas, rainfall thresholds to provide temporal forecasts especially for rapid movements, regional scale remote sensing systems to get a general overview of slow-moving landslides, and ground based monitoring systems could be employed for management of the hot-spots at highest risk level. Similar programs should also consider landslide hazard evolution as influenced by climate change and human modifications to the territorial setting.

Another significant gap highlighted by the review is that, even though landslides are a well-known source of risk in the area, a quantitative risk assessment has never been carried out, thus all past and present risk reduction strategies are implemented without a comprehensive framework that could optimize the efforts made. At present, landslide hazard studies are also quite limited: examples exist that encompass temporal probability of occurrence (by means of rainfall thresholds) or spatial probability of occurrence (by landslide susceptibility maps). A spatiotemporal hazard assessment would be desirable and, to obtain it the future, two options could be explored: the application of distributed physically-based models (which, in turn, require the acquisition of many input parameters with high spatial density—see, e.g., Tofani et al. [111]) or the adoption of simpler approaches based on the dynamic combination of rainfall thresholds and susceptibility maps [112].

Several techniques have been used to understand various aspects; however, the use of computational techniques has been limited to susceptibility and hazard analysis. It is encouraging to see the use of advanced machine learning techniques (ANN, SVM, and Random Forest) for landslide mitigation. However, the use of deep learning and artificial intelligence, which shows great promise in geohazards and climatic studies, is yet to be performed for the Indian Himalayan region. The use of advanced computational techniques (deep learning) especially for landslide identification is one aspect which researchers should look to exploit.

5. Conclusions

This review paper is an attempt to understand the studies being conducted in the Indian Himalayan region, which contributes almost 15% of the global rainfall-induced landslides. The article focuses only on landslide activity which was triggered due to rainfall as it reflects a significant portion to landslides in the Indian Himalaya and it is the only typology that can be managed by early warnings and forecasting models. The review reveals that there are several avenues where significant progress is required, especially in the aspect of climate change, use of high spatial and temporal resolution data, and new methodologies such as physical methods and computational approaches. The study highlights key topics in which the research has focused: susceptibility mapping, identification, and slope stability analysis. The conclusions from each sub-section are as follows:

1. Landslide assessment (including identification, threshold estimation, and monitoring of landslides). For landslide identification and mapping, focus needs to shift in three key directions:
    a. Use of automated approaches involving the use of computational techniques.
    b. Use of higher temporal resolution datasets and assessment of their reliability.
    c. Application of the current techniques towards other significant landslide prone Himalayan regions. In terms of rainfall threshold studies, the number of articles were less than 10 with most of the work focused on the use of statistical models to define a single threshold, and majorly the Eastern Himalayas have been covered. The thresholds developed show large differences when calculated for regional and local scale, therefore it is suggested to develop thresholds at local scale to improve the understanding of the region and help in setting up an operational landslide early warning system. In general, the thresholds are very low if compared with other literature thresholds, thus confirming the basic assumption that the India Himalaya is very susceptible to landsliding. Additional research needs to be conducted on the use of physical models, including campaigns aimed at gathering input data for more complex models; moreover, efforts need to be made on the combination of empirical and physical models for a better understanding of landslide initiation.
2. Landslide monitoring has been performed using both ground instrumentation and satellite data, however a general multi-scale approach that extensively covers the whole region is missing.
3. The focus needs to shift on include climatic factors for landslide assessment as climate change is unequivocal. The use of climate models needs to be conducted with caution, especially when downscale projections are considered. As the climatic conditions are quite varied from west to east, focus should also be on the use of appropriate down-scaling models.
4. The studies on landslide susceptibility was found to have a regional bias and more research needs to be conducted in the Jammu and Kashmir Himalayas and the northeastern belt. Emphasis has primarily been on specific states and regions such as Uttarakhand, Darjeeling, and some areas of Himachal Pradesh. In modeling aspects, the use of computational approaches needs to be emphasized as it has proved to be better than traditional methods. The analysis should start focusing on the use of hybrid models and big data analytics for regional to site specific analysis, thereby understanding the heterogeneity and uncertainty of the region.
5. There is a serious lack of ground-based rainfall data in large parts of the Himalayan region which has been highlighted by several works. A solution to this flaw is to start using remote sensing data (e.g., satellite radar rainfall estimates) to compare and find the best dataset to be used for individual sections in the Himalayan region.

Author Contributions: Conceptualization, A.D. and R.S.; formal analysis, A.D. and S.S.; data curation, R.S.; writing—original draft preparation, A.D. and S.S.; writing—review and editing, B.P. and A.M.A.; and supervision, B.P. and R.S. All authors have read and agreed to the published version of the manuscript.

Funding: This research was supported by the Centre for Advanced Modelling and Geospatial Information Systems (CAMGIS) in the University of Technology Sydney (UTS) under Grants 321740.2232335, 321740.2232357,

321740.2232424, and 321740.2232452. This research was also funded by International Science Council under Grant ISCROAP/IRDR/SG/2019/009. This research was also supported by Researchers Supporting Project number RSP-2019/14, King Saud University, Riyadh, Saudi Arabia. The research was also funded by Università degli Studi di Firenze under the Grant 58517_INTERNAZIONALIZZAZIONE.

Conflicts of Interest: The authors declare no conflict of interest.

References

1. Froude, M.J.; Petley, D. Global fatal landslide occurrence from 2004 to 2016. *Nat. Hazards Earth Syst. Sci.* **2018**, *18*, 2161–2181. [CrossRef]
2. Dubey, C.S.; Chaudhry, M.; Sharma, B.K.; Pandey, A.C.; Singh, B. Visualization of 3-D digital elevation model for landslide assessment and prediction in mountainous terrain: A case study of Chandmari landslide, Sikkim, eastern Himalayas. *Geosci. J.* **2005**, *9*, 363–373. [CrossRef]
3. Kanungo, D.P.; Sharma, S. Rainfall thresholds for prediction of shallow landslides around Chamoli-Joshimath region, Garhwal Himalayas, India. *Landslides* **2014**, *11*, 629–638. [CrossRef]
4. Dikshit, A.; Satyam, D.N. Estimation of rainfall thresholds for landslide occurrences in Kalimpong, India. *Innov. Infrastruct. Solut.* **2018**, 3. [CrossRef]
5. Sati, V.P. *Towards Sustainable Livelihoods and Ecosystems in Mountain Regions*; Springer: Berlin, Germany, 2014.
6. Chakrabarti, B.K. Chapter 1—Lithotectonic Subdivisions of the Himalaya. In *Geology of the Himalayan Belt*; Chakrabarti, B.K., Ed.; Elsevier: Amsterdam, The Netherlands, 2016; pp. 1–9. [CrossRef]
7. Roy, A.B.; Purohit, R. Chapter 2—Indian Subcontinent: Geomorphic and Geophysical Traits. In *Indian Shield*; Roy, A.B., Purohit, R., Eds.; Elsevier: Amsterdam, The Netherlands, 2018; pp. 13–30. [CrossRef]
8. Mukhopadhyay, D.K.; Mishra, P. The Main Frontal Thrust (MFT), northwestern Himalayas: Thrust trajectory and hanging wall fold geometry from balanced cross sections. *J. Geol. Soc. India* **2004**, *64*, 739–746.
9. Pradhan, B.; Singh, R.; Buchroithner, M. Estimation of stress and its use in evaluation of landslide prone regions using remote sensing data. *Adv. Space Res.* **2006**, *37*, 698–709. [CrossRef]
10. DeCelles, P.G.; Carrapa, B.; Gehrels, G.E.; Chakraborty, T.; Ghosh, P. Along-strike continuity of structure, stratigraphy, and kinematic history in the Himalayan thrust belt: The view from Northeastern India. *Tectonics* **2016**, *35*, 2995–3027. [CrossRef]
11. Thakur, V.C. Plate tectonic interpretation of the western Himalaya. *Tectonophysics* **1987**, *134*, 91–102. [CrossRef]
12. Reichenbach, P.; Rossi, M.; Malamud, B.D.; Mihir, M.; Guzzetti, F. A review of statistically-based landslide susceptibility models. *Earth-Sci. Rev.* **2018**, *180*, 60–91. [CrossRef]
13. Pham, B.T.; Shirzadi, A.; Shahabi, H.; Omidvar, E.; Singh, S.K.; Sahana, M.; Asl, D.T.; Bin Ahmad, B.; Quoc, N.K.; Lee, S. Landslide Susceptibility Assessment by Novel Hybrid Machine Learning Algorithms. *Sustainability* **2019**, 4386. [CrossRef]
14. Naithani, A.K.; Kumar, D.; Prasad, C. The catastrophic landslide of 16 July 2001 in Phata Byung area, Rudraprayag district, Garhwal Himalaya, India. *Curr. Sci.* **2002**, *82*, 921–923.
15. Martha, T.R.; Vinod Kumar, K. September, 2012 landslide events in Okhimath, India—An assessment of landslide consequences using very high resolution satellite data. *Landslides* **2013**, *10*, 469–479. [CrossRef]
16. Martha, T.R.; Roy, P.; Govindharaj, K.B.; Kumar, K.V.; Diwakar, P.G.; Dadhwal, V.K. Landslides triggered by the June 2013 extreme rainfall event in parts of Uttarakhand state, India. *Landslides* **2015**, *12*, 135–146. [CrossRef]
17. Roy, P.; Martha, T.R.; Jain, N.; Kumar, K.V. Reactivation of minor scars to major landslides–a satellite-based analysis of Kotropi landslide (13 August 2017) in Himachal Pradesh, India. *Curr. Sci.* **2018**, *115*, 395. [CrossRef]
18. Pradhan, S.P.; Panda, S.D.; Roul, A.R.; Thakur, M. Insights into the recent Kotropi landslide of August 2017, India: A geological investigation and slope stability analysis. *Landslides* **2019**, *16*, 1529–1537. [CrossRef]
19. Kumar, V.; Gupta, V.; Jamir, I. Hazard evaluation of progressive Pawari landslide zone, Satluj valley, Himachal Pradesh, India. *Nat. Hazards* **2018**, *93*, 1029–1047. [CrossRef]
20. Banerjee, A.; Dimri, A.P. Comparative analysis of two rainfall retrieval algorithms during extreme rainfall event: A case study on cloudburst, 2010 over Ladakh (Leh), Jammu and Kashmir. *Nat. Hazards* **2019**, *97*, 1357–1374. [CrossRef]

21. Mondal, S.; Mandal, S. Landslide susceptibility mapping of Darjeeling Himalaya, India using index of entropy (IOE) model. *Appl. Geomat.* **2019**, *11*, 129–146. [CrossRef]
22. Dikshit, A.; Satyam, N. Probabilistic rainfall thresholds in Chibo, India: Estimation and validation using monitoring system. *J. Mt. Sci.* **2019**, *16*, 870–883. [CrossRef]
23. Anbarasu, K.; Sengupta, A.; Gupta, S.; Sharma, S.P. Mechanism of activation of the Lanta Khola landslide in Sikkim Himalayas. *Landslides* **2010**, *7*, 135–147. [CrossRef]
24. Bera, A.; Mukhopadhyay, B.P.; Das, D. Landslide hazard zonation mapping using multi-criteria analysis with the help of GIS techniques: A case study from Eastern Himalayas, Namchi, South Sikkim. *Nat. Hazards* **2019**, *96*, 935–959. [CrossRef]
25. Umrao, R.K.; Singh, R.; Sharma, L.; Singh, T. Soil slope instability along a strategic road corridor in Meghalaya, north-eastern India. *Arab. J. Geosci.* **2017**, *10*, 260. [CrossRef]
26. Sarkar, K.; Buragohain, B.; Singh, T. Rock slope stability analysis along NH-44 in Sonapur area, Jaintia hills district, Meghalaya. *J. Geol. Soc. India* **2016**, *87*, 317–322. [CrossRef]
27. Sardana, S.; Verma, A.; Singh, A. Comparative analysis of rockmass characterization techniques for the stability prediction of road cut slopes along NH-44A, Mizoram, India. *Bull. Eng. Geol. Environ.* **2019**, *78*, 5977–5989. [CrossRef]
28. Martha, T.R.; Kerle, N.; Jetten, V.; van Westen, C.J.; Kumar, K.V. Landslide Volumetric Analysis Using Cartosat-1-Derived DEMs. *Ieee Geosci. Remote Sens. Lett.* **2010**, *7*, 582–586. [CrossRef]
29. Cruden, D.M.; Varnes, D.J. Landslides: Investigation and mitigation. In *National Research Council Transportation Research Board Special Report (Book 247)*; Turner, A.K., Schuster, R.L., Eds.; Transportation Research Board: Washington, DC, USA, 1996; pp. 36–75.
30. Vinod kumar, K.; Lakhera, R.C.; Martha, T.R.; Chatterjee, R.S.; Bhattacharya, A. Analysis of the 2003 Varunawat Landslide, Uttarkashi, India using Earth Observation data. *Environ. Geol.* **2008**, *55*, 789–799. [CrossRef]
31. Martha, T.R.; Kerle, N.; van Westen, C.J.; Jetten, V.; Kumar, K.V. Object-oriented analysis of multi-temporal panchromatic images for creation of historical landslide inventories. *Isprs J. Photogramm. Remote Sens.* **2012**, *67*, 105–119. [CrossRef]
32. Martha, T.R.; Kamala, P.; Jose, J.; Kumar, K.V.; Sankar, G.J. Identification of new Landslides from High Resolution Satellite Data Covering a Large Area Using Object-Based Change Detection Methods. *J. Indian Soc. Remote Sens.* **2016**, *44*, 515–524. [CrossRef]
33. Amatya, P.; Kirschbaum, D.; Stanley, T. Use of Very High-Resolution Optical Data for Landslide Mapping and Susceptibility Analysis along the Karnali Highway, Nepal. *Remote Sens.* **2019**, *11*, 2284. [CrossRef]
34. Vamsee, A.M.; Kamala, P.; Martha, T.R.; Kumar, K.V.; Amminedu, E. A tool assessing optimal multi-scale image segmentation. *J. Indian Soc. Remote Sens.* **2018**, *46*, 31–41. [CrossRef]
35. Martha, T.R.; Kerle, N.; Jetten, V.; van Westen, C.J.; Kumar, K.V. Characterising spectral, spatial and morphometric properties of landslides for semi-automatic detection using object-oriented methods. *Geomorphology* **2010**, *116*, 24–36. [CrossRef]
36. Martha, T.R.; van Westen, C.J.; Kerle, N.; Jetten, V.; Kumar, K.V. Landslide hazard and risk assessment using semi-automatically created landslide inventories. *Geomorphology* **2013**, *184*, 139–150. [CrossRef]
37. Kumar, V.; Gupta, V.; Sundriyal, Y.P. Spatial interrelationship of landslides, litho-tectonics, and climate regime, Satluj valley, Northwest Himalaya. *Geol. J.* **2019**, *54*, 537–551. [CrossRef]
38. Guzzetti, F.; Mondini, A.C.; Cardinali, M.; Fiorucci, F.; Santangelo, M.; Chang, K.-T. Landslide inventory maps: New tools for an old problem. *Earth-Sci. Rev.* **2012**, *112*, 42–66. [CrossRef]
39. Ghorbanzadeh, O.; Blaschke, T.; Gholamnia, K.; Meena, S.R.; Tiede, D.; Aryal, J. Evaluation of different machine learning methods and deep-learning convolutional neural networks for landslide detection. *Remote Sens.* **2019**, *11*, 196. [CrossRef]
40. Guzzetti, F.; Peruccacci, S.; Rossi, M.; Stark, C.P. Rainfall thresholds for the initiation of landslides in central and southern Europe. *Meteorol. Atmos. Phys.* **2007**, *98*, 239–267. [CrossRef]
41. Dikshit, A.; Sarkar, R.; Pradhan, B.; Acharya, S.; Dorji, K. Estimating rainfall thresholds for landslide occurrence in the Bhutan Himalayas. *Water* **2019**, *11*, 1616. [CrossRef]
42. Dikshit, A.; Sarkar, R.; Pradhan, B.; Jena, R.; Drukpa, D.; Alamri, A.M. Temporal Probability Assessment and Its Use in Landslide Susceptibility Mapping for Eastern Bhutan. *Water* **2020**, *12*, 267. [CrossRef]

43. Segoni, S.; Piciullo, L.; Gariano, S.L. A review of the recent literature on rainfall thresholds for landslide occurrence. *Landslides* **2018**, *15*, 1483–1501. [CrossRef]
44. Gariano, S.L.; Sarkar, R.; Dikshit, A.; Dorji, K.; Brunetti, M.T.; Peruccacci, S.; Melillo, M. Automatic calculation of rainfall thresholds for landslide occurrence in Chukha Dzongkhag, Bhutan. *Bull. Eng. Geol. Environ.* **2019**, *78*, 4325–4332. [CrossRef]
45. Sengupta, A.; Gupta, S.; Anbarasu, K. Rainfall thresholds for the initiation of landslide at Lanta Khola in north Sikkim, India. *Nat. Hazards* **2010**, *52*, 31–42. [CrossRef]
46. Dikshit, A.; Sarkar, R.; Satyam, N. Probabilistic approach toward Darjeeling Himalayas landslides-A case study. *Cogent Eng.* **2018**, 5. [CrossRef]
47. Teja, T.S.; Dikshit, A.; Satyam, N. Determination of Rainfall Thresholds for Landslide Prediction Using an Algorithm-Based Approach: Case Study in the Darjeeling Himalayas, India. *Geosciences* **2019**, 302. [CrossRef]
48. Harilal, G.T.; Madhu, D.; Ramesh, M.V.; Pullarkatt, D. Towards establishing rainfall thresholds for a real-time landslide early warning system in Sikkim, India. *Landslides* **2019**, *16*, 2395–2408. [CrossRef]
49. Gariano, S.L.; Melillo, M.; Peruccacci, S.; Brunetti, M.T. How much does the rainfall temporal resolution affect rainfall thresholds for landslide triggering? *Nat. Hazards* **2019**, 1–16. [CrossRef]
50. Mathew, J.; Babu, D.G.; Kundu, S.; Kumar, K.V.; Pant, C.C. Integrating intensity-duration-based rainfall threshold and antecedent rainfall-based probability estimate towards generating early warning for rainfall-induced landslides in parts of the Garhwal Himalaya, India. *Landslides* **2014**, *11*, 575–588. [CrossRef]
51. Kumar, A.; Asthana, A.L.; Priyanka, R.S.; Jayangondaperumal, R.; Gupta, A.K.; Bhakuni, S.S. Assessment of landslide hazards induced by extreme rainfall event in Jammu and Kashmir Himalaya, northwest India. *Geomorphology* **2017**, *284*, 72–87. [CrossRef]
52. Caine, N. The rainfall intensity-duration control of shallow landslides and debris flows. *Geogr. Ann. Ser. APhys. Geogr.* **1980**, *62*, 23–27.
53. Chae, B.-G.; Park, H.-J.; Catani, F.; Simoni, A.; Berti, M. Landslide prediction, monitoring and early warning: A concise review of state-of-the-art. *Geosci. J.* **2017**, *21*, 1033–1070. [CrossRef]
54. Dikshit, A.; Satyam, D.N.; Towhata, I. Early warning system using tilt sensors in Chibo, Kalimpong, Darjeeling Himalayas, India. *Nat. Hazards* **2018**, *94*, 727–741. [CrossRef]
55. Falae, P.O.; Kanungo, D.P.; Chauhan, P.K.S.; Dash, R.K. Electrical resistivity tomography (ERT) based subsurface characterisation of Pakhi Landslide, Garhwal Himalayas, India. *Environ. Earth Sci.* **2019**, 78. [CrossRef]
56. Yhokha, A.; Goswami, P.K.; Chang, C.P.; Yen, J.Y.; Ching, K.E.; Aruche, K.M. Application of Persistent Scatterer Interferometry (PSI) in monitoring slope movements in Nainital, Uttarakhand Lesser Himalaya, India. *J. Earth Syst. Sci.* **2018**, 127. [CrossRef]
57. Martha, T.R.; Reddy, P.S.; Bhatt, C.M.; Raj, K.B.G.; Nalini, J.; Padmanabha, E.A.; Narender, B.; Kumar, K.V.; Muralikrishnan, S.; Rao, G.S.; et al. Debris volume estimation and monitoring of Phuktal river landslide-dammed lake in the Zanskar Himalayas, India using Cartosat-2 images. *Landslides* **2017**, *14*, 373–383. [CrossRef]
58. Mondal, S.K.; Sastry, R.G.; Pachauri, A.K.; Gautam, P.K. High resolution 2D electrical resistivity tomography to characterize active Naitwar Bazar landslide, Garhwal Himalaya, India. *Curr. Sci.* **2008**, *94*, 871–875.
59. Kannaujiya, S.; Chattoraj, S.L.; Jayalath, D.; Ray, P.K.C.; Bajaj, K.; Podali, S.; Bisht, M.P.S. Integration of satellite remote sensing and geophysical techniques (electrical resistivity tomography and ground penetrating radar) for landslide characterization at Kunjethi (Kalimath), Garhwal Himalaya, India. *Nat. Hazards* **2019**, *97*, 1191–1208. [CrossRef]
60. Sharma, S.P.; Anbarasu, K.; Gupta, S.; Sengupta, A. Integrated very low-frequency EM, electrical resistivity, and geological studies on the Lanta Khola landslide, North Sikkim, India. *Landslides* **2010**, *7*, 43–53. [CrossRef]
61. Raj, K.; Govindha, B.; Martha, T.R.; Kumar, K.V. A bird's-eye view of landslide dammed lakes in Zanskar Himalaya, India. *Curr. Sci.* **2017**, *112*, 1109–1112.
62. Weidinger, J.T. Case history and hazard analysis of two lake-damming landslides in the Himalayas. *J. Asian Earth Sci.* **1998**, *16*, 323–331. [CrossRef]
63. Kumar, V.; Gupta, V.; Jamir, I.; Chattoraj, S.L. Evaluation of potential landslide damming: Case study of Urni landslide, Kinnaur, Satluj valley, India. *Geosci. Front.* **2019**, *10*, 753–767. [CrossRef]

64. Gupta, V.; Bhasin, R.K.; Kaynia, A.M.; Kumar, V.; Saini, A.S.; Tandon, R.S.; Pabst, T. Finite element analysis of failed slope by shear strength reduction technique: A case study for Surabhi Resort Landslide, Mussoorie township, Garhwal Himalaya. *Geomat. Nat. Hazards Risk* **2016**, *7*, 1677–1690. [CrossRef]
65. Gupta, V.; Sah, M. Impact of the trans-Himalayan landslide lake outburst flood (LLOF) in the Satluj catchment, Himachal Pradesh, India. *Nat. Hazards* **2008**, *45*, 379–390. [CrossRef]
66. Gariano, S.L.; Guzzetti, F. Landslides in a changing climate. *Earth-Sci. Rev.* **2016**, *162*, 227–252. [CrossRef]
67. Sarkar, S.; Kanungo, D. Landslide Disaster on Berinag-Munsiyari Road, Pithoragarh District, Uttarakhand. *Curr. Sci.* **2010**, *98*, 2.
68. Bhan, S.; Devrani, A.; Sinha, V. An analysis of monthly rainfall and the meteorological conditions associated with cloudburst over the dry region of Leh (Ladakh), India. *Mausam* **2015**, *66*, 107–122.
69. Anbalagan, R. Landslide hazard evaluation and zonation mapping in mountainous terrain. *Eng. Geol.* **1992**, *32*, 269–277. [CrossRef]
70. Gupta, P.; Anbalagan, R. Slope stability of Tehri Dam Reservoir Area, India, using landslide hazard zonation (LHZ) mapping. *Q. J. Eng. Geol.* **1997**, *30*, 27–36. [CrossRef]
71. Kanungo, D.; Arora, M.; Gupta, R.; Sarkar, S. Landslide risk assessment using concepts of danger pixels and fuzzy set theory in Darjeeling Himalayas. *Landslides* **2008**, *5*, 407–416. [CrossRef]
72. Sarkar, S.; Roy, A.K.; Raha, P. Deterministic approach for susceptibility assessment of shallow debris slide in the Darjeeling Himalayas, India. *Catena* **2016**, *142*, 36–46. [CrossRef]
73. Mathew, J.; Kundu, S.; Kumar, K.V.; Pant, C.C. Hydrologically complemented deterministic slope stability analysis in part of Indian Lesser Himalaya. *Geomat. Nat. Hazards Risk* **2016**, *7*, 1557–1576. [CrossRef]
74. Kumar, R.; Anbalagan, R. Landslide susceptibility zonation in part of Tehri reservoir region using frequency ratio, fuzzy logic and GIS. *J. Earth Syst. Sci.* **2015**, *124*, 431–448. [CrossRef]
75. Ghosh, S.; Carranza, E.J.M.; van Westen, C.J.; Jetten, V.G.; Bhattacharya, D.N. Selecting and weighting spatial predictors for empirical modeling of landslide susceptibility in the Darjeeling Himalayas (India). *Geomorphology* **2011**, *131*, 35–56. [CrossRef]
76. Mandal, S.; Mandal, K. Modeling and mapping landslide susceptibility zones using GIS based multivariate binary logistic regression (LR) model in the Rorachu river basin of eastern Sikkim Himalaya, India. *Modeling Earth Syst. Environ.* **2018**, *4*, 69–88. [CrossRef]
77. Mathew, J.; Jha, V.K.; Rawat, G.S. Application of binary logistic regression analysis and its validation for landslide susceptibility mapping in part of Garhwal Himalaya, India. *Int. J. Remote Sens.* **2007**, *28*, 2257–2275. [CrossRef]
78. Pourghasemi, H.R.; Yansari, Z.T.; Panagos, P.; Pradhan, B. Analysis and evaluation of landslide susceptibility: A review on articles published during 2005–2016 (periods of 2005–2012 and 2013–2016). *Arab. J. Geosci.* **2018**, *11*, 193. [CrossRef]
79. Sharma, S.; Mahajan, A.K. A comparative assessment of information value, frequency ratio and analytical hierarchy process models for landslide susceptibility mapping of a Himalayan watershed, India. *Bull. Eng. Geol. Environ.* **2019**, *78*, 2431–2448. [CrossRef]
80. Das, I.; Stein, A.; Kerle, N.; Dadhwal, V.K. Landslide susceptibility mapping along road corridors in the Indian Himalayas using Bayesian logistic regression models. *Geomorphology* **2012**, *179*, 116–125. [CrossRef]
81. Mathew, J.; Jha, V.K.; Rawat, G.S. Landslide susceptibility zonation mapping and its validation in part of Garhwal Lesser Himalaya, India, using binary logistic regression analysis and receiver operating characteristic curve method. *Landslides* **2009**, *6*, 17–26. [CrossRef]
82. Chauhan, S.; Mukta, S.; Arora, M.K. Landslide susceptibility zonation of the Chamoli region, Garhwal Himalayas, using logistic regression model. *Landslides* **2010**, *7*, 411–423. [CrossRef]
83. Sahana, M.; Sajjad, H. Evaluating effectiveness of frequency ratio, fuzzy logic and logistic regression models in assessing landslide susceptibility: A case from Rudraprayag district, India. *J. Mt. Sci.* **2017**, *14*, 2150–2167. [CrossRef]
84. Saha, A.K.; Gupta, R.P.; Sarkar, I.; Arora, M.K.; Csaplovics, E. An approach for GIS-based statistical landslide susceptibility zonation - with a case study in the Himalayas. *Landslides* **2005**, *2*, 61–69. [CrossRef]
85. Sarkar, S.; Anbalagan, R. Landslide hazard zonation mapping and comparative analysis of hazard zonation maps. *J. Mt. Sci.* **2008**, *5*, 232–240. [CrossRef]

86. Mondal, S.; Maiti, R. Integrating the Analytical Hierarchy Process (AHP) and the Frequency Ratio (FR) Model in Landslide Susceptibility Mapping of Shiv-khola Watershed, Darjeeling Himalaya. *Int. J. Disaster Risk Sci.* **2013**, *4*, 200–212. [CrossRef]
87. Sharma, L.P.; Patel, N.; Ghose, M.K.; Debnath, P. Application of frequency ratio and likelihood ratio model for geo-spatial modelling of landslide hazard vulnerability assessment and zonation: A case study from the Sikkim Himalayas in India. *Geocarto Int.* **2014**, *29*, 128–146. [CrossRef]
88. Balamurugan, G.; Ramesh, V.; Touthang, M. Landslide susceptibility zonation mapping using frequency ratio and fuzzy gamma operator models in part of NH-39, Manipur, India. *Nat. Hazards* **2016**, *84*, 465–488. [CrossRef]
89. Manzo, G.; Tofani, V.; Segoni, S.; Battistini, A.; Catani, F. GIS techniques for regional-scale landslide susceptibility assessment: The Sicily (Italy) case study. *Int. J. Geogr. Inf. Sci.* **2013**, *27*, 1433–1452. [CrossRef]
90. Singh, C.; Kohli, A.; Kumar, P. Comparison of results of BIS and GSI guidelines on macrolevel landslide hazard zonation—A case study along highway from Bhalukpong to Bomdila, West Kameng district, Arunachal Pradesh. *J. Geol. Soc. India* **2014**, *83*, 688–696. [CrossRef]
91. Ghosh, S.; van Westen, C.J.; Carranza, E.J.M.; Ghoshal, T.B.; Sarkar, N.K.; Surendranath, M. A quantitative approach for improving the BIS (Indian) method of medium-scale landslide susceptibility. *J. Geol. Soc. India* **2009**, *74*, 625–638. [CrossRef]
92. Das, I.; Stein, A.; Kerle, N.; Dadhwal, V.K. Probabilistic landslide hazard assessment using homogeneous susceptible units (HSU) along a national highway corridor in the northern Himalayas, India. *Landslides* **2011**, *8*, 293–308. [CrossRef]
93. Sarkar, S.; Kanungo, D.P.; Sharma, S. Landslide hazard assessment in the upper Alaknanda valley of Indian Himalayas. *Geomat. Nat. Hazards Risk* **2015**, *6*, 308–325. [CrossRef]
94. Pham, B.T.; Bui, D.T.; Prakash, I.; Dholakia, M.B. Hybrid integration of Multilayer Perceptron Neural Networks and machine learning ensembles for landslide susceptibility assessment at Himalayan area (India) using GIS. *Catena* **2017**, *149*, 52–63. [CrossRef]
95. Pham, B.T.; Prakash, I.; Dou, J.; Singh, S.K.; Trinh, P.T.; Tran, H.T.; Le, T.M.; Phong, T.V.; Khoi, D.K.; Shirzadi, A.; et al. A novel hybrid approach of landslide susceptibility modelling using rotation forest ensemble and different base classifiers. *Geocarto Int.* **2019**. [CrossRef]
96. Peethambaran, B.; Anbalagan, R.; Shihabudheen, K.V.; Goswami, A. Robustness evaluation of fuzzy expert system and extreme learning machine for geographic information system-based landslide susceptibility zonation: A case study from Indian Himalaya. *Environ. Earth Sci.* **2019**, 78. [CrossRef]
97. Ramakrishnan, D.; Singh, T.N.; Verma, A.K.; Gulati, A.; Tiwari, K.C. Soft computing and GIS for landslide susceptibility assessment in Tawaghat area, Kumaon Himalaya, India. *Nat. Hazards* **2013**, *65*, 315–330. [CrossRef]
98. Pham, B.T.; Bui, D.T.; Prakash, I.; Nguyen, L.H.; Dholakia, M.B. A comparative study of sequential minimal optimization-based support vector machines, vote feature intervals, and logistic regression in landslide susceptibility assessment using GIS. *Environ. Earth Sci.* **2017**, 76. [CrossRef]
99. Pham, B.T.; Bui, D.T.; Pourghasemi, H.R.; Indra, P.; Dholakia, M. Landslide susceptibility assesssment in the Uttarakhand area (India) using GIS: A comparison study of prediction capability of naïve bayes, multilayer perceptron neural networks, and functional trees methods. *Theor. Appl. Climatol.* **2017**, *128*, 255–273. [CrossRef]
100. Pham, B.T.; Prakash, I.; Singh, S.K.; Shirzadi, A.; Shahabi, H.; Tran, T.T.G.; Buig, D.T. Landslide susceptibility modeling using Reduced Error Pruning Trees and different ensemble techniques: Hybrid machine learning approaches. *Catena* **2019**, *175*, 203–218. [CrossRef]
101. Chauhan, S.; Sharma, M.; Arora, M.K.; Gupta, N.K. Landslide Susceptibility Zonation through ratings derived from Artificial Neural Network. *Int. J. Appl. Earth Obs. Geoinf.* **2010**, *12*, 340–350. [CrossRef]
102. Kumar, D.; Thakur, M.; Dubey, C.S.; Shukla, D.P. Landslide susceptibility mapping & prediction using Support Vector Machine for Mandakini River Basin, Garhwal Himalaya, India. *Geomorphology* **2017**, *295*, 115–125. [CrossRef]
103. Pham, B.T.; Shirzadi, A.; Bui, D.T.; Prakash, I.; Dholakia, M. A hybrid machine learning ensemble approach based on a radial basis function neural network and rotation forest for landslide susceptibility modeling: A case study in the Himalayan area, India. *Int. J. Sediment Res.* **2018**, *33*, 157–170. [CrossRef]

104. Pham, B.T.; Prakash, I.; Jaafari, A.; Bui, D.T. Spatial prediction of rainfall-induced landslides using aggregating one-dependence estimators classifier. *J. Indian Soc. Remote Sens.* **2018**, *46*, 1457–1470. [CrossRef]
105. Pham, B.T.; Jaafari, A.; Prakash, I.; Bui, D.T. A novel hybrid intelligent model of support vector machines and the MultiBoost ensemble for landslide susceptibility modeling. *Bull. Eng. Geol. Environ.* **2019**, *78*, 2865–2886. [CrossRef]
106. Jaafari, A.; Panahi, M.; Pham, B.T.; Shahabi, H.; Bui, D.T.; Rezaie, F.; Lee, S. Meta optimization of an adaptive neuro-fuzzy inference system with grey wolf optimizer and biogeography-based optimization algorithms for spatial prediction of landslide susceptibility. *Catena* **2019**, *175*, 430–445. [CrossRef]
107. Kanungo, D.P.; Arora, M.K.; Sarkar, S.; Gupta, R.P. A comparative study of conventional, ANN black box, fuzzy and combined neural and fuzzy weighting procedures for landslide susceptibility zonation in Darjeeling Himalayas. *Eng. Geol.* **2006**, *85*, 347–366. [CrossRef]
108. Kanungo, D.P.; Sarkar, S.; Sharma, S. Combining neural network with fuzzy, certainty factor and likelihood ratio concepts for spatial prediction of landslides. *Nat. Hazards* **2011**, *59*, 1491–1512. [CrossRef]
109. Chawla, A.; Chawla, S.; Pasupuleti, S.; Rao, A.C.S.; Sarkar, K.; Dwivedi, R. Landslide Susceptibility Mapping in Darjeeling Himalayas, India. *Adv. Civ. Eng.* **2018**. [CrossRef]
110. Meena, S.R.; Mishra, B.K.; Piralilou, S.T. A Hybrid Spatial Multi-Criteria Evaluation Method for Mapping Landslide Susceptible Areas in Kullu Valley, Himalayas. *Geosciences* **2019**, 156. [CrossRef]
111. Tofani, V.; Bicocchi, G.; Rossi, G.; Segoni, S.; D'Ambrosio, M.; Casagli, N.; Catani, F. Soil characterization for shallow landslides modeling: A case study in the Northern Apennines (Central Italy). *Landslides* **2017**, *14*, 755–770. [CrossRef]
112. Segoni, S.; Tofani, V.; Rosi, A.; Catani, F.; Casagli, N. Combination of rainfall thresholds and susceptibility maps for dynamic landslide hazard assessment at regional scale. *Front. Earth Sci.* **2018**, *6*, 85. [CrossRef]

© 2020 by the authors. Licensee MDPI, Basel, Switzerland. This article is an open access article distributed under the terms and conditions of the Creative Commons Attribution (CC BY) license (http://creativecommons.org/licenses/by/4.0/).

applied sciences

MDPI

Article

GIS-Based Evaluation of Landslide Susceptibility Models Using Certainty Factors and Functional Trees-Based Ensemble Techniques

Xia Zhao [1,2] **and Wei Chen** [1,2,*]

[1] College of Geology & Environment, Xi'an University of Science and Technology, Xi'an 710054, China; 19209071022@stu.xust.edu.cn

[2] Key Laboratory of Coal Resources Exploration and Comprehensive Utilization, Ministry of Natural Resources, Xi'an 710021, China

* Correspondence: chenwei0930@xust.edu.cn

Received: 15 October 2019; Accepted: 25 November 2019; Published: 18 December 2019

Abstract: The main purpose of this paper is to use ensembles techniques of functional tree-based bagging, rotation forest, and dagging (functional trees (FT), bagging-functional trees (BFT), rotation forest-functional trees (RFFT), dagging-functional trees (DFT)) for landslide susceptibility modeling in Zichang County, China. Firstly, 263 landslides were identified, and the landslide inventory map was established, and the landslide locations were randomly divided into 70% (training data) and 30% (validation data). Then, 14 landslide conditioning factors were selected. Furthermore, the correlation analysis between conditioning factors and landslides was applied using the certainty factor method. Hereafter, four models were applied for landslide susceptibility modeling and zoning. Finally, the receiver operating characteristic (ROC) curve and statistical parameters were used to evaluate and compare the overall performance of the four models. The results showed that the area under the curve (AUC) for the four models was larger than 0.74. Among them, the BFT model is better than the other three models. In addition, this study also illustrated that the integrated model is not necessarily more effective than a single model. The ensemble data mining technology used in this study can be used as an effective tool for future land planning and monitoring.

Keywords: landslide susceptibility mapping; ensemble techniques; functional trees; bagging; rotation forest; dagging

1. Introduction

A landslide is a complex natural phenomenon [1]. It is influenced by many geological environmental factors, such as topography, landform, geology, land use, and vegetation [2]. A landslide is one of the most familiar and disastrous geological hazards with great destructiveness, which always poses a serious threat to human life, property, and living environment, and restricts human progress and development, especially when geological environments are increasingly affected by human engineering activities [3]. Therefore, landslide prediction is of great significance for landslide prevention and control [4,5]. One of the greatest tasks of landslide disaster and risk mitigation is to prepare landslide susceptibility maps [6].

With the development and progress of the geographic information system (GIS), its application in spatial analysis of landslides is becoming more and more popular. With proper use of GIS, most of the landslide susceptibility mapping methods can realize the automation of evaluation and standardization of data management technology, and enable us to build more efficient and accurate maps [7,8]. This is because these technologies can obtain, query, store, analyze, manipulate, and display a set of spatial and non-spatial data about landslide conditioning factors [8–10]. Landslide susceptibility zoning mapping technology includes a variety of statistical techniques and statistical methods, including Dempster–Shafer [11–13], entropy [14–16], logistic regression [17–19], certainty factors [20–22], statistical index [23,24], analytic hierarchy process [25–27], frequency ratio [20,28], weight of evidence [29–32], index of entropy [20,33], multivariate adaptive regression spline [34–36], and evidential belief function [37–39].

Landslide susceptibility mapping is a typical complex nonlinear problem in a large area of a landslide research area [5]. Thus, the results obtained by statistical techniques and statistical methods may not be able to achieve satisfactory accuracy [5,40]. Later, many researchers proposed a large number of machine learning techniques for evaluating the susceptibility of landslides, which usually have high prediction accuracy and better performance in data-driven models, such as naive Bayes [41–43], random forests [2,44–46], artificial neural networks [47–50], kernel logistic regression [51,52], support vector machine [53,54], and decision trees [55,56]. However, the performance of machine learning methods is generally influenced by the quality and quantity of training data, and the dependence on modeling parameters is very high [5,57]. So far, it is not clear which method is most suitable for landslide susceptibility mapping [5].

In recent years, hybrid technology is considered to be more effective than single technology [58]. In order to explore more reasonable and perfect research results, a variety of integrated algorithms have been developed for landslide susceptibility modeling [6], such as adaptive neuro-fuzzy inference system [59,60], artificial neural networks-Bayes analysis [61], and Evidential Belief Function-fuzzy logic [62]. The important capability of the integrated model is that the method is more accurate in identification and greatly improves the prediction ability compared with the single machine learning model [6].

The purpose of this study is to propose and validate the ability and effect of ensemble techniques in landslide susceptibility modeling, and functional trees are selected as the base classifier to ensemble with bagging, rotation forest, and dagging models in Zichang County (China). Receiver operating characteristics (ROCs) and statistical parameters were used to evaluate and compare the overall performance of the four models.

2. Study Area and Data Used

2.1. The Study Area

Zichang County is located in the north of Yan'an City, Shaanxi Province, China, between longitudes 109°11′58″ E and 110°01′22″ E and between latitudes 36°59′30″ N and 37°30′00″ N, with a total area of 2405 km^2 (Figure 1). Zichang County is a typical hilly and gully region of the Loess Plateau. The terrain is tilted from northwest to southeast, with an elevation of 933 to 1574 m. Zichang County prevails a warm temperate semi-arid continental monsoon climate, with low temperature and large temperature difference. The annual average temperature within the territory is 9.1 °C, the annual average precipitation is 514.7 mm. The rivers in the territory belong to the Yellow River system, which is divided into three tributaries: Qingjian River, Wuding River, and Yanhe River.

2.2. Data Preparation

The quality of landslide inventory is very significant for landslide susceptibility modeling, and an accurate landslide inventory map is the foundation of landslide susceptibility modeling [63,64]. In this study, three techniques were used to improve the reliability and accuracy of the landslide inventory map: historical report, aerial photo interpretation, and field survey using Global Navigation Satellite Systems (GNSS). According to the landslide inventory map in this area, 263 landslides were identified, and 184 landslide locations (70%) were randomly sampled as the training data and the other 79 landslide locations (30%) were used to validate models.

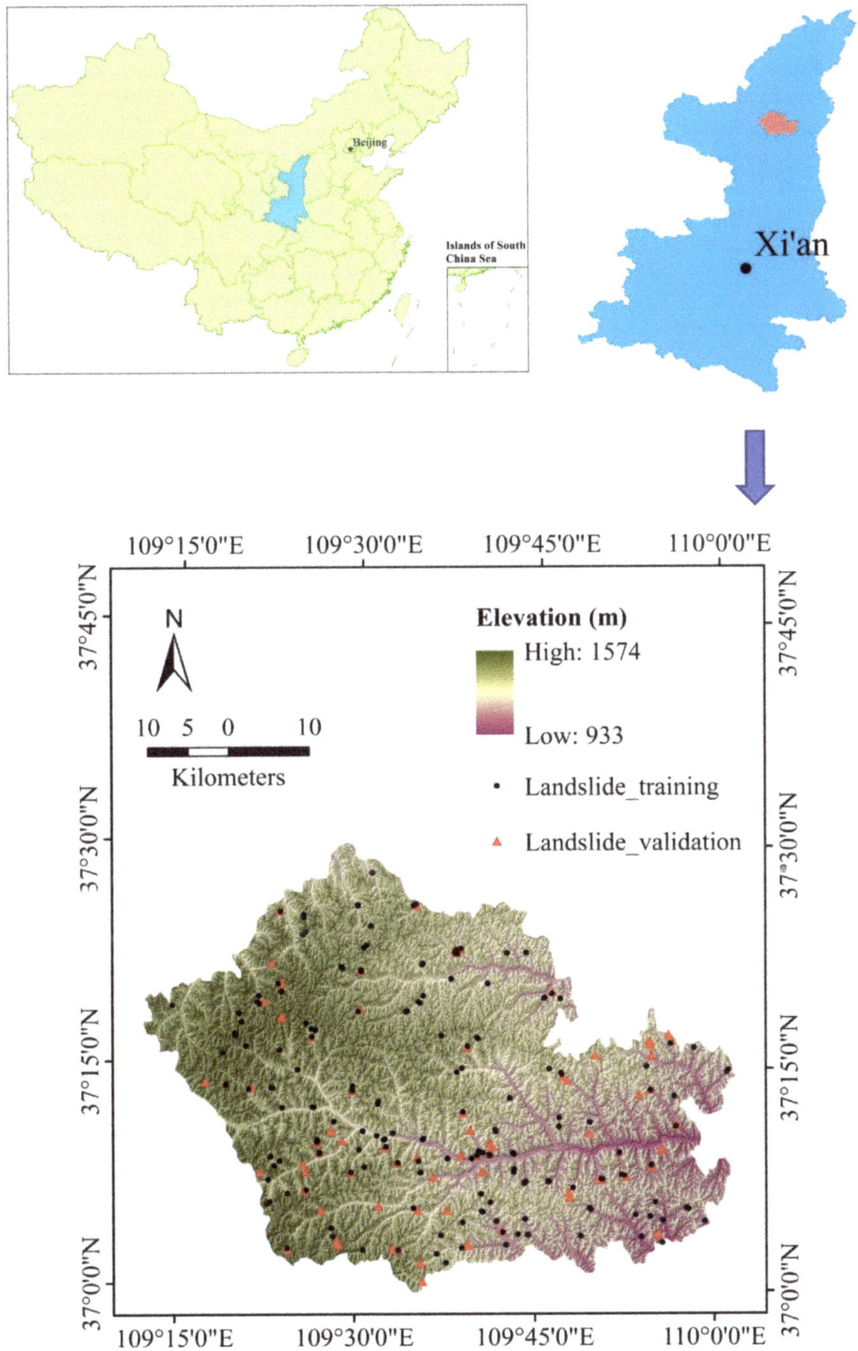

Figure 1. Study area.

After compiling the landslide inventory, it is necessary to choose the landslide conditioning factors to create the landslide susceptibility map [65]. The selection principle is to consider the mechanism and geo-environmental characteristics of landslide occurrence in the study area. Generally, the landslide conditioning factors used to evaluate landslide susceptibility include three categories: topographic factors, geological factors, and environmental factors. In this paper, 14 landslide conditioning factors were selected and transformed into the same resolution (30 × 30 m), including elevation, slope, aspect, plan curvature, profile curvature, sediment transport index (STI), stream power index (SPI), topographic wetness index (TWI), the normalized difference vegetation index (NDVI), land use, lithology, soil, distance to roads, and distance to rivers (Table 1, Figure 2).

Table 1. Source and scale of conditioning factors.

Factors	Data Source	Format Resolution/Scale
Elevation, slope, aspect, plan curvature, profile curvature, sediment transport index (STI), steam power index (SPI), topographic wetness index (TWI), distance to roads, distance to rivers	ASTER GDEM	Raster, 30 m
Normalized difference vegetation index (NDVI)	Landsat 8 operational land imager	Raster, 30 m
Lithology	Geological maps	Polygon, 1:200,000
Land use	Land use/land cover maps	Polygon, 1:100,000
Soil	Soil type maps	Polygon, 1:1,000,000

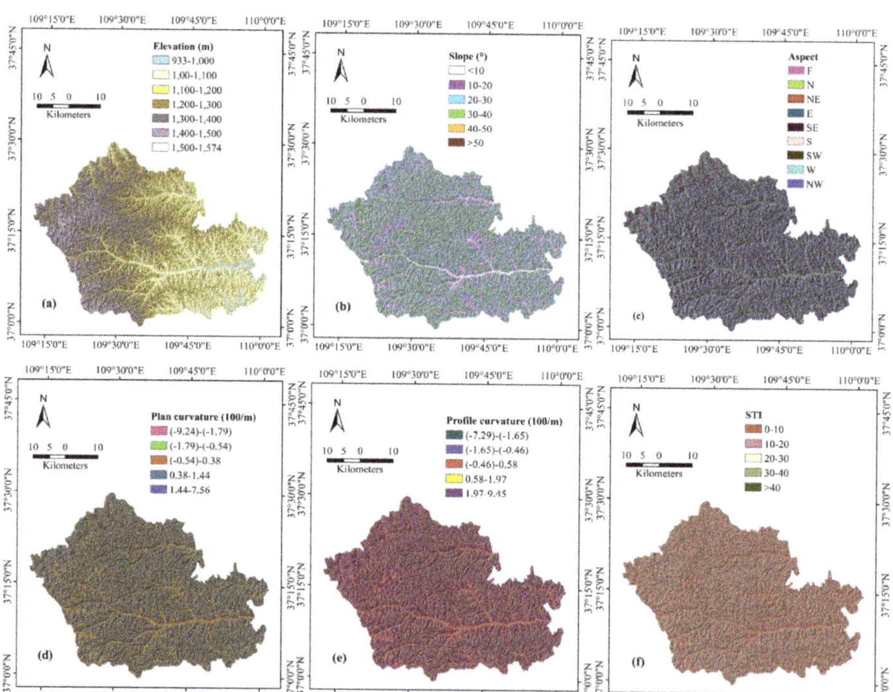

Figure 2. *Cont.*

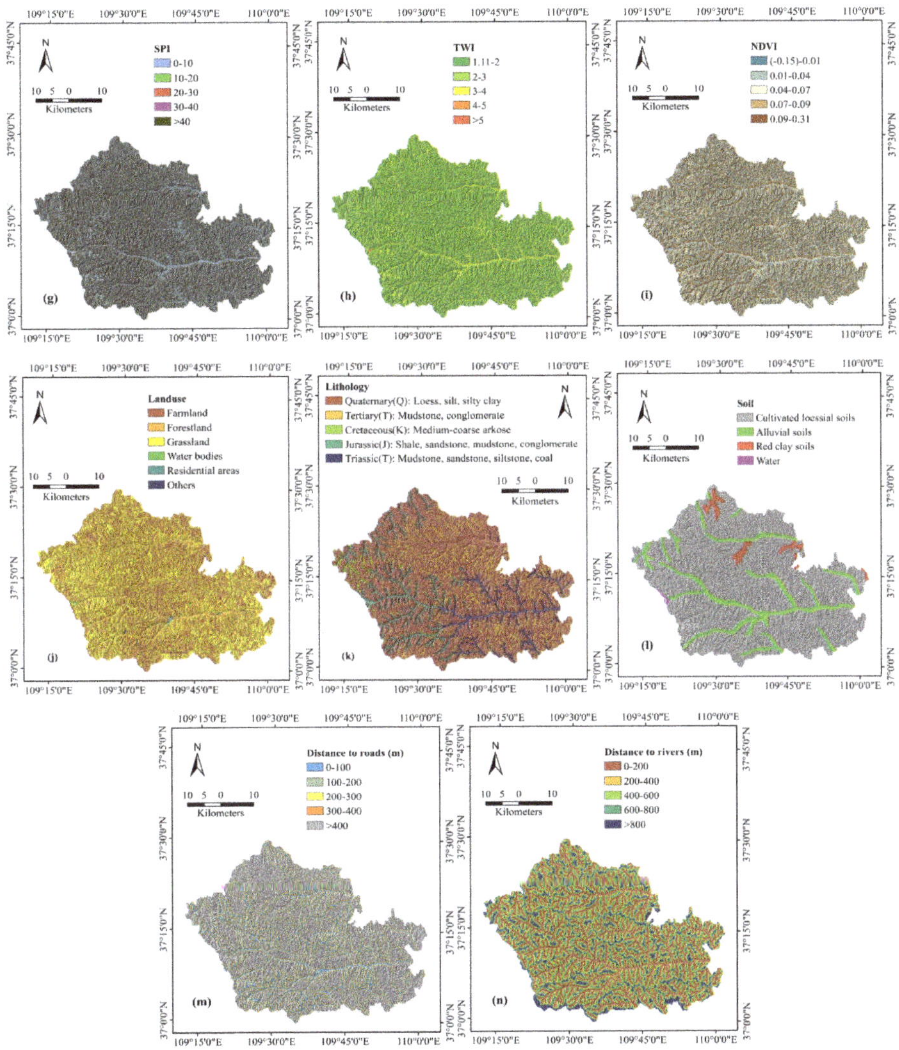

Figure 2. Thematic maps: (**a**) elevation, (**b**) slope, (**c**) aspect, (**d**) plan curvature, (**e**) profile curvature, (**f**) STI, (**g**) SPI, (**h**) TWI, (**i**) NDVI, (**j**) land use, (**k**) lithology, (**l**) soil, (**m**) distance to roads, (**n**) distance to rivers.

3. Modeling Approach

The chapter included the illumination of five models, namely certainty factors, functional trees, bagging, rotation forest and dagging. The certainty factors model was used to express the correlation between landslide and conditioning factors, the functional trees model was used as a base classifier, the bagging, rotation forest, and dagging were used as ensemble algorithms.

3.1. Certainty Factors

The certainty factor (CF) belongs to a probability function, which was first proposed in 1990 [66] and modified subsequently [67]. The certainty factor can be expressed as [68]:

$$CF = \begin{cases} \frac{PPa-PPs}{PPa(1-PPs)} & if \quad PPa \geq PPs \\ \frac{PPa-PPs}{PPs(1-PPa)} & if \quad PPa < PPs \end{cases} \tag{1}$$

where, PPa is the conditional probability of landslide in class a in study area A, PPs is the prior probability of the total number of landslides in study area A.

The range of CF is −1 to 1, the positive value indicates that the degree of certainty of landslide occurrence increases, while the negative value indicates that the degree of certainty of landslide occurrence decreases [69–71].

3.2. Functional Trees

Functional trees (FT) are a combination of a discriminant function and multivariable decision tree through constructive induction [72]. Functional trees use logistic regression functions to calculate the splitting of internal nodes (called oblique splitting) and estimation of leaves [73–75]. FT learns the classification tree based on the attributes of leaf nodes, decision nodes or nodes and leaves [38,76]. The decision nodes are built while the trees are growing, while the functional leaves construct when the trees are pruning [76]. Functional trees have the following three usage types: (1) the full functional tree using a regression model for internal nodes and leaves; (2) function tree internal-only uses the regression model for internal nodes; (3) functional tree leaves only use the regression model for leaves [75,76].

In the leaf logic regression function, the logic enhancement (iteration are weighted) of the least-squares function is determined for each output consisting of two classes [77]. Among them, training datasets of D and n samples (A_i, B_i) with $A_i \in R^n$, $B_i \in \{1,0\}$ [76]. A_i is the input vector containing all landslide condition factors [75,76]; whereas $P(A)$ is the probability prediction value of landslide occurrence; B_i is the coefficient of the i component of the input vector A_i. The posterior probability $P(A)$ of the left ventricle is calculated as follows [78]:

$$f_{B_i}(A) = \sum_{i=1}^{14} B_i A_i + B_0 \tag{2}$$

$$P(A) = \frac{e^{2f_{B_i}(A)}}{1 + e^{2f_{B_i}(A)}} \tag{3}$$

3.3. Bagging

Bagging is based on the concepts of bootstrapping and aggregating, which is used to obtain a more robust and accurate landslide model. Bagging is one of the most popular integration algorithms [79]. The process of a bagging algorithm includes:

Firstly, the bootstrap samples $S(x_i, y_i)$ are randomly resampled from a training set (x_i, y_i), forming a set of training subsets, where, $x_i \in R$, $y_i \in$ (landslide, non-landslide) [80]. Then, several models based on a classifier are constructed according to each subset, $C_i(x)$ is a classifier constructed from each guiding sample. All models based on classifier (L_i) are aggregated to generate the final model (L'), where, $L_1, L_2, \ldots, L_n$ generates a combined classifier (L'). L' predicts the class label of a given instance x by calculating the votes using the following equation [81]:

$$L'(x) = \arg\max_{y \in Y} \sum_{i=1}^{t} L(C_i(x) = y) \tag{4}$$

3.4. Rotation Forest

Rotation forest (RF) is a popular aggregation technique proposed by Rodriguez et al. [82]. RF is an effective technique for improving weak classifiers [83]. It uses principal component analysis (PCA), a multivariate technique used for analyzing large multivariate datasets, to reduce its dimensions [84]. In this method, features are extracted from the learning (training) dataset and a base classifier is used to generate learning sub training dataset [82].

For the use of RF: randomly divide the training dataset into D subsets, where D is the parameter of the algorithm, and construct the rotated sparse matrix by performing feature extraction for each subset. The classifier is based on the feature of a repeated matrix projection, and the result is obtained by combining the output of multiple classifiers [84]. RF can be used with any basic classifier, and the feature extraction of each classifier retains all the features that promote variability [84].

In the RF algorithm, $x = (x_1, x_2, \cdots, x_n)$ is the training sample set, and Y is the corresponding class label, that is used to consider landslides and non-landslides; $D_1, D_2, \cdots, D_L$ are the classifier in the set frame; and P is the set of landslide condition factors. The coefficients of the rotation matrix ${R_i}^a$ are obtained by transformation and base classifier. Obtain ${R_i}^a$ by rearranging R_i matrix [84]:

$$R_i = \begin{bmatrix} {b_{i,1}}^{(1)}, \cdots, {b_{i,1}}^{(M_1)} & 0 & \cdots & 0 \\ 0 & {b_{i,2}}^{(1)}, \cdots, {b_{i,2}}^{(M_2)} & \cdots & 0 \\ \vdots & \vdots & \ddots & \vdots \\ 0 & 0 & \cdots & {b_{i,K}}^{(1)}, \cdots, {b_{i,K}}^{(M_K)} \end{bmatrix} \quad (5)$$

For each sub training dataset extracted by the rotation matrix ${R_i}^a$, average grouping method is adopted to obtain the coefficients of each class in a given test sample [85]:

$${\mu_j}^{(x)} = \frac{1}{L}\sum_{i=1}^{L} d_{ij}(x{R_i}^a),\ j = 1, \ldots, c. \quad (6)$$

where ${\mu_j}^{(x)}$ is the maximum confidence specified on the class, classifier probability allocation D_i, and the d_{ij} regression $d_{ij}(x{R_i}^a)$ [85].

3.5. Dagging

Dagging is a well-known resampling integration technique originally proposed by Ting and Witten to generate many disjoint hierarchical folds from a dataset, and each data partition can be sent separately to the basic classifier [86]. The final forecast is based on a majority vote [86]. The main principle is to use a majority vote to combine multiple classifiers to improve the prediction accuracy of the basic classifier [86].

For a given training dataset, which has n samples, the dagging algorithm constructs M datasets (M is a free parameter) from the original training dataset [87]. Each dataset contains n samples [87], and no two datasets have the same sample. A basic classifier is trained for each dataset to build a classification model [87]. Therefore, the M dataset can be summarized into M classification models [86,87].

4. Results

This section consists of the detailed description of the results of the present study, which includes the following four sections: (1) the correlation between landslide and conditioning factors, and then the CF values are used as input to weight the classes of conditioning factors; (2) selection of landslide conditioning factors that are positive to the modeling process; (3) application of four hybrid models and generate landslide susceptibility maps; and (4) validation and comparison of models using ROC and Chi-squared methods.

4.1. Correlation Analysis of Landslide and Conditioning Factors Using the CF Method

The landslide density at each class was calculated by combining each thematic map and landslide inventory map. Meanwhile, this paper summarizes the spatial relationship between the landslides and conditioning factors using the CF method (Table 2). According to the calculation results in Table 2, the highest CF value (0.661) is found in the elevation category of 1500–1574 m, which indicates that the probability of landslide is the highest. Among the six classes classified by the slope, 40°–50° (0.324) is the highest CF value of the six categories. As far as aspect is concerned, the CF values of slopes facing south (0.309) and southwest (0.242) are the largest. Among the five classes classified by plan curvature, the classes of (−9.24)–(−1.79) have the lowest CF value (−0.495), and the classes of 1.44–7.56 have the highest CF value (0.244). Among the five classes classified by profile curvature, the classes of (−1.65)–(−0.46) have the lowest CF value (−0.346), and the classes of 0.58–1.97 have the highest CF value (0.277). For STI, the frequency of landslide occurrence is the most relevant in 20–30 categories, with the largest CF value (0.220). In TWI, the CF value is the largest in the classes of 2–3 (0.164) and the smallest in the classes of >5 (−1). For NDVI, the lowest CF value (−0.326) was found in the classes of 0.01–0.04, and the highest CF value was found in the categories of 0.07–0.09 (0.223). In terms of land use, landslides mostly occur in residential areas (0.465). Among the five types of lithology, the groups 2 and 4 were relatively more sensitive to landslide occurrence, with CF values of 0.430 and 0.465, respectively. For soil, the majority of landslides occurred in red clay soils with a CF value of 0.712. It can be seen from a distance to roads that the closer the distance is, the more sensitive the landslide. CF value is the largest in the categories of 0–100 m (0.452). For distance to rivers, CF value is the largest in the categories of 0–200 m (0.585).

Table 2. Relationship between landslides and conditioning factors using the certainty factor (CF) method.

Conditioning Factors	Classes	Percentage of Landslide	Percentage of Domain	CF
Elevation (m)	933–1000	2.17	1.14	0.476
	1000–1100	22.28	13.38	0.400
	1100–1200	33.15	28.22	0.149
	1200–1300	29.35	31.06	−0.055
	1300–1400	9.24	20.45	−0.548
	1400–1500	3.26	5.57	−0.415
	1500–1574	0.54	0.18	0.661
Slope (°)	<10	0.00	10.44	−1.000
	10–20	30.43	26.09	0.143
	20–30	36.96	35.14	0.049
	30–40	26.09	23.90	0.084
	40–50	6.52	4.41	0.324
	>50	0.00	0.02	−1.000
Aspect (°)	F (−1)	0.00	0.05	−1.000
	N (0–22.5; 337.5–360)	7.61	9.25	−0.177
	NE (22.5–67.5)	6.52	13.16	−0.504
	E (67.5–112.5)	17.39	16.34	0.060
	SE (112.5–157.5)	13.59	11.26	0.171
	S (157.5–202.5)	14.67	10.14	0.309
	SW (202.5–247.5)	16.85	12.77	0.242
	W (247.5–292.5)	17.93	15.44	0.139
	NW (292.5–337.5)	5.43	11.59	−0.531
Plan curvature (100/m)	(−9.24)–(−1.79)	2.72	5.38	−0.495
	(−1.79)–(−0.54)	15.76	17.98	−0.124
	(−0.54)–0.38	45.11	42.08	0.067
	0.38–1.44	25.54	26.34	−0.030
	1.44–7.56	10.87	8.22	0.244
Profile curvature (100/m)	(−7.29)–(−1.65)	7.07	8.08	−0.126
	(−1.65)–(−0.46)	15.76	24.10	−0.346
	(−0.46)–0.58	41.85	39.33	0.060
	0.58–1.97	29.35	21.23	0.277
	1.97–9.45	5.98	7.26	−0.177
STI	0–10	44.02	48.27	−0.088
	10–20	33.70	30.96	0.081
	20–30	15.22	11.21	0.263
	30–40	3.26	4.22	−0.227
	>40	3.80	5.34	−0.287

Table 2. *Cont.*

Conditioning Factors	Classes	Percentage of Landslide	Percentage of Domain	CF
SPI	0–10	20.11	32.46	−0.381
	10–20	25.00	19.69	0.212
	20–30	17.39	13.57	0.220
	30–40	10.33	8.20	0.206
	>40	27.17	26.08	0.040
TWI	1.11–2	55.43	56.33	−0.016
	2–3	39.67	33.16	0.164
	3–4	3.80	7.36	−0.483
	4–5	1.09	2.83	−0.617
	>5	0.00	0.31	−1.000
NDVI	(−0.15)–0.01	14.13	13.96	0.012
	0.01–0.04	11.41	16.94	−0.326
	0.04–0.07	16.85	22.45	−0.250
	0.07–0.09	35.33	27.44	0.223
	0.09–0.31	22.28	19.20	0.138
Land use	Farmland	25.54	36.96	−0.309
	Forestland	20.11	18.93	0.059
	Grassland	53.80	43.70	0.188
	Water bodies	0.00	0.10	−1.000
	Residential areas	0.54	0.29	0.465
	Others	0.00	0.02	−1.000
Lithology	1	60.33	75.17	−0.197
	2	21.74	12.38	0.430
	3	0.54	0.94	−0.421
	4	12.50	6.69	0.465
	5	4.89	4.82	0.015
Soil	Cultivated loessial soils	76.63	85.66	−0.105
	Alluvial soils	15.22	11.83	0.223
	Red clay soils	8.15	2.35	0.712
	Water	0.00	0.15	−1.000
Distance to roads (m)	0–100	27.72	15.20	0.452
	100–200	17.39	11.42	0.344
	200–300	11.96	11.35	0.050
	300–400	6.52	8.93	−0.270
	>400	36.41	53.10	−0.314
Distance to rivers (m)	0–200	69.02	28.64	0.585
	200–400	14.67	25.39	−0.422
	400–600	9.24	22.38	−0.587
	600–800	3.26	15.61	−0.791
	>800	3.80	7.98	−0.523

4.2. Selection of Landslide Conditioning Factors

In order to ensure the accuracy of landslide prediction results, it is necessary to remove unimportant or unrelated factors [88,89]. In this study, the Pearson correlation method [90,91] with 10-fold cross-validation was used as an effective feature selection method for evaluating the predictive ability of conditioning factors. The distance to rivers, slope, and lithology has the highest predictive abilities (Table 3). Since a no conditioning factor has a null predictive value, all are included in this analysis.

4.3. Application of Landslide Susceptibility Models

In this study, the training data and CF values were used to construct four models, namely the functional trees (FT) model, bagging-functional trees (BFT) model, rotation forest-functional trees (RFFT) model, and dagging-functional trees (DFT) model, respectively. To get the best performance of the model, the iteration times of the FT model and the minimum number of instances considering the separation of nodes from the training dataset are optimized to 15 and 36, respectively. When building the BFT, RFFT, and DFT models, the two parameters mentioned above were fixed firstly. After completing the above work, the optimized models were applied to the whole research area to create landslide susceptibility maps. The calculated landslide sensitivity index (LSI) values can be interpreted as the probability in the range of 0 and 1, and all LSI values can be converted to ArcGIS to generate the final landslide susceptibility map.

Table 3. Correlation attribute of landslide conditioning factors.

Factors	Average Merit	Standard Deviation
Distance to rivers	0.382	±0.015
Slope	0.224	±0.008
Lithology	0.180	±0.012
Elevation	0.185	±0.018
Distance to roads	0.172	±0.015
TWI	0.171	±0.014
SPI	0.152	±0.015
Aspect	0.147	±0.011
Soil	0.141	±0.013
Profile curvature	0.136	±0.020
NDVI	0.105	±0.024
Landuse	0.097	±0.013
Plan curvature	0.038	±0.011
STI	0.041	±0.015

Four landslide susceptibility maps generated by FT, BFT, RFFT, and DFT models are shown in Figure 3a–d respectively. The landslide susceptibility maps were reclassified into five classes, namely very low, low, moderate, high, and very high using the natural break method [92]. The comparison of area sizes for each category of the four models is shown in Figure 4. For the FT model, the largest area is the very low class (27.92%), followed by high class (23.47%), very high class (20.21%), low class (17.55%), and the smallest area is the moderate class (10.86%). For the BFT model, the percentages of very low, low, moderate, high, and very high classes are 24.02%, 22.87%, 19.88%, 18.10%, and 15.12%, respectively. The results of landslide susceptibility zoning using the RFFT model show that these percentages are 37.62% (very low), 21.41% (low), 7.79% (moderate), 12.25% (high), and 20.93% (very high), respectively. For the DFT model, the percentages of very low, low, moderate, high, and very high classes are 19.70%, 30.59%, 23.72%, 16.50%, and 9.49%, respectively.

4.4. Model Performances and Comparisons

In this study, the landslide susceptibility models were evaluated by using the areas under the ROC curves (AUC), standard error, 95% confidence interval, and significance level *p*-value. The ROC curve can be used as a useful tool to indicate the quality of deterministic and probabilistic prediction system [93–95]. The sensitivity (true positive rate) is shown as y-axis and 1-specificity (false positive rate) as x-axis [94,96]. The AUC values are in the range of 0.5 to 1 [97], and the excellent attributes of the model increase with the AUC values [98].

Using the training dataset, the performance of the landslide susceptibility models was evaluated (Table 4). The BFT model has the highest AUC value (0.947), the lowest standard error (0.011), and the narrowest 95% confidence interval (0.925–0.969). It is followed by the RFFT model, the FT model, and the DFT model. For the validation data, the calculation results are shown in Table 5. The BFT model has the highest AUC value (0.804), the lowest standard error (0.035), and the narrowest 95% confidence interval (0.736–0.871). It comes before the DFT model, the FT model, and the RFFT model. These results show that all performance in the validation dataset is slightly worse than those of the training data. These results show that the BFT model is the best model among the four models, and the ensemble model is not necessarily superior to the single model.

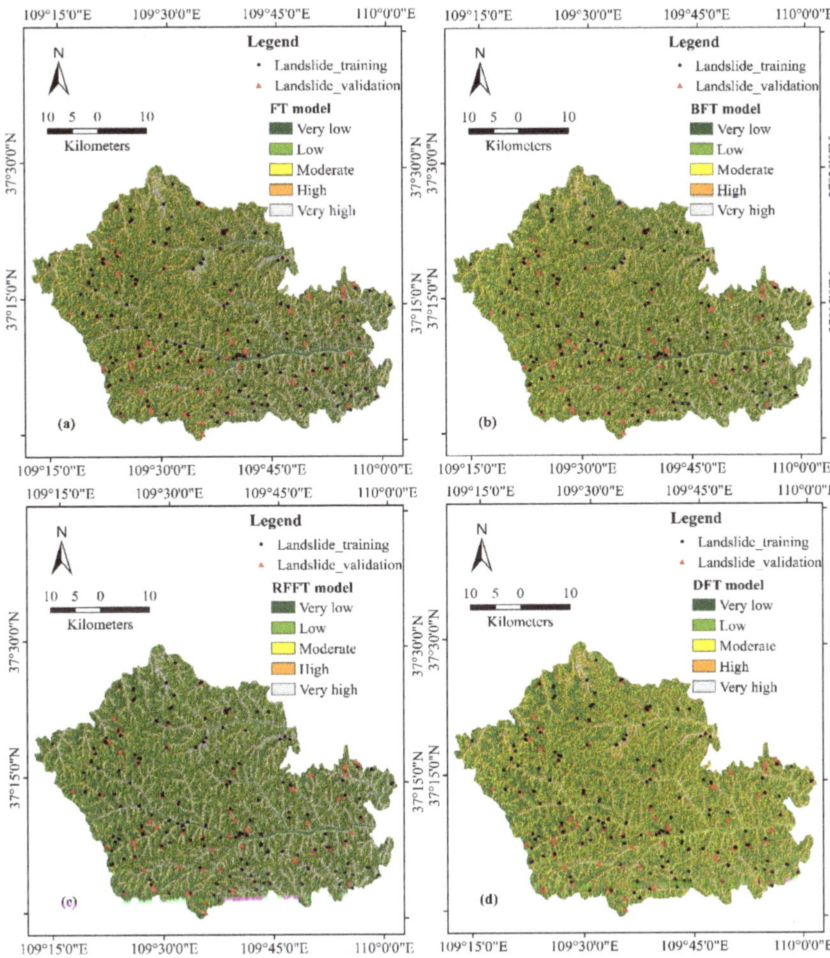

Figure 3. Landslide susceptibility maps: (**a**) functional trees (FT) model; (**b**) bagging-functional trees (BFT) model; (**c**) rotation forest-functional trees (RFFT) model; (**d**) dagging-functional trees (DFT) model.

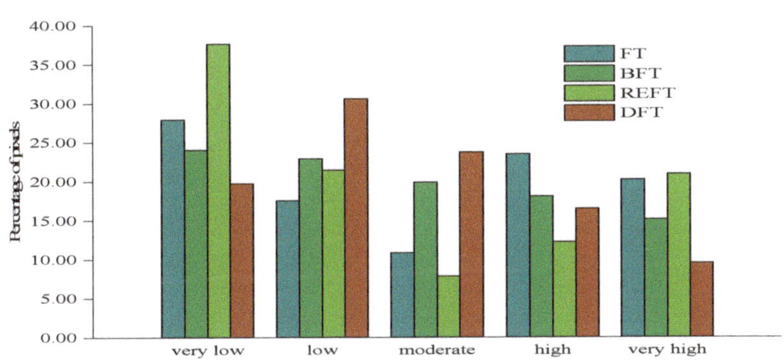

Figure 4. Percentages of landslide susceptibility classes.

Table 4. Parameters of receiver operating characteristic (ROC) curves with the training dataset.

Test Variables	FT Model	BFT Model	RFFT Model	DFT Model
ROC Curve Area	0.836	0.947	0.839	0.797
Standard Error	0.021	0.011	0.020	0.023
95% Confidence Interval	0.795–0.878	0.925–0.969	0.798–0.879	0.752–0.842
p-Value	<0.0001	<0.0001	<0.0001	<0.0001

Table 5. Parameters of ROC curves with the validation dataset.

Test Variables	FT Model	BFT Model	RFFT Model	DFT Model
ROC Curve Area	0.745	0.804	0.740	0.748
Standard Error	0.039	0.035	0.0394	0.039
95% Confidence Interval	0.668–0.822	0.736–0.871	0.663–0.817	0.672–0.824
p-Value	<0.0001	<0.0001	<0.0001	<0.0001

A Chi-squared test was used to analyze the significance of the four models (Table 6). It can be seen that only the comparison of FT and RFFT exhibits lower Chi-squared value (0.044) and higher p-value (0.834), which indicate no significant difference between the two models. The other five groups all present larger Chi-squared values and lower p-values. The significant differences between the models indicate that the differences between the models are good, which is more conducive to the modeling work and enables this study to obtain the susceptibility results smoothly.

Table 6. Pairwise comparison of four models.

Pair	FT vs. BFT	FT vs. RFFT	FT vs. DFT	BFT vs. RFFT	BFT vs. DFT	RFFT vs. DFT
Chi-squared	40.376	0.044	8.205	44.928	63.681	14.454
p-Value	<0.0001	0.834	0.004	<0.0001	<0.0001	0.000
Significance	Yes	No	Yes	Yes	Yes	Yes

5. Discussions

In this current study, the correlation analysis between conditioning factors and landslides was carried out by the CF method. The probability of landslide occurrence is in inverse correlation with elevation. This may be related to local rainfall and loess and may be related to human engineering activities. With the increase of slope angle, the degree of certainty of landslide occurrence decreases. This may be due to the larger slope angle, the less loose material or more weatherproof material. At the same time, it can be observed that most landslides occur on slopes facing south and southwest with the highest probability. This is mainly because more rain and sunshine are available to the south and landslides are prone to occur. The curvature of plan and profile shows anomalous results. The curvature of the plan (near zero) and convex plan (positive value) are highly sensitive. This anomaly may be related to the overweight effect [28,99,100]. In terms of land use, the probability of landslides in residential areas is the largest, which can explain the impact of human engineering activities on landslides. For the lithology, the second group (Tertiary (T): mudstone, conglomerate) and the fourth group (Triassic (T): mudstone, sandstone, songlomerate) are more sensitive to landslide occurrence. There is groundwater flow in the relatively fractured saturated sandstone and fractured conglomerates, resulting in additional load on the mudstone, resulting in landslides [28,101]. The linear characteristics of the road and river buffers are inversely correlated with landslide susceptibility in the distance. Such an important result has been repeated in many kinds of literature [6,102–104]. However, the remaining five variables make little contributions to the occurrence of landslides.

According to ROC curve analysis (Figures 5 and 6) and statistical index analysis (Tables 4 and 5), it can be concluded that the four machine learning methods selected in the training and testing data assemble a very small p-value and significant high performance in the 95% confidence interval. The BFT model has the highest AUC value (0.947), the lowest standard deviation (0.011), 95% confidence interval (0.925–0.969), and p-value (<0.0001). However, the DFT model has the worst results in this study area. The DFT model has the lowest AUC value (0.797), the highest standard deviation (0.023), 95% confidence interval (0.752–0.842), and p-value (<0.0001). There is no doubt that most ensemble models are superior to single models. However, there is still a phenomenon that the performance of hybrid machine learning methods is not always better than a single model. In order to find more optimal solutions, much more different set models should be applied to the research field.

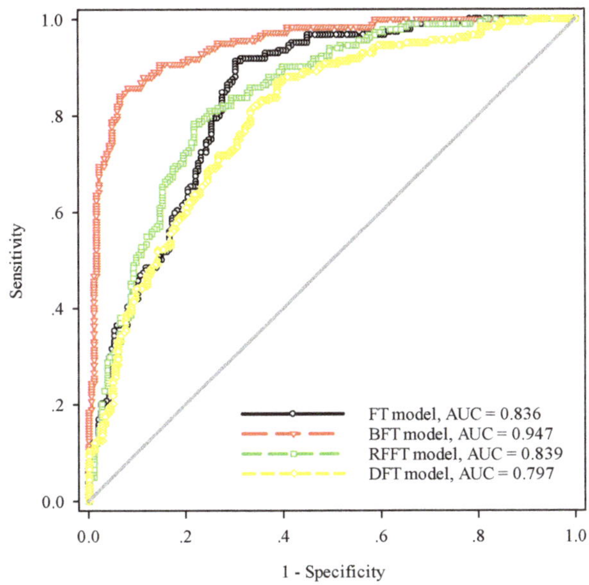

Figure 5. ROC curves of the models using the training dataset. AUC: area under the curve.

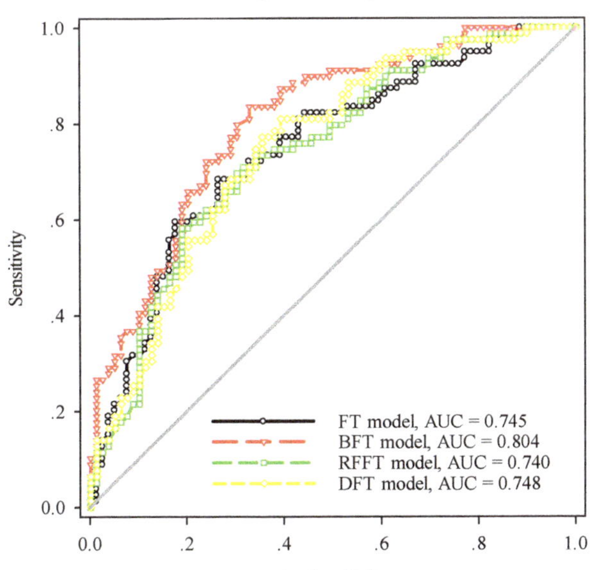

Figure 6. ROC curves of the models using the validation dataset.

According to the paired comparison of the performance of the models (Table 6), the Chi-squared test shows that the Chi-squared values are relatively large. Among them, the Chi-squared value of the FT and RFFT models is smaller, the *p*-value is larger, and the difference between these two models is not significant. The good results obtained from the other three groups can serve as a powerful basis for modeling in this study. At the same time, the BFT model is compared with the other three models in pairs, and the difference is significant. According to the evaluation results of various evaluation criteria, the performance of the BFT model is better than that of the RFFT model, FT model, and DFT model. As a final recommendation, the obtained results can be useful for policy planning and decision-making in areas prone to landslides. The proposed BFT model, based on performance and prediction accuracy, is suggested in the study area and other regions over the world where they have similar geo-environmental conditions with a logical caution.

6. Conclusions

This study applied functional tree-based ensemble techniques (FT model, BFT model, RFFT model, DFT model) for landslide susceptibility spatial modeling in Zichang County, China. Fourteen conditioning factors and the occurrence of landslides were used to analyze the correlation. Meanwhile, the ROC curve and statistical parameters were used to evaluate and compare the accuracy of the model results. The results showed that the prediction rate of the BFT model is the highest. Therefore, the BFT model is the best optimization ensemble model in this study, and it can be used as an advantageous and promising method for landslide susceptibility modeling. Finally, the landslide susceptibility map generated by this study can be used as an effective tool for future land planning and monitoring by government officials or research experts and scholars.

Author Contributions: X.Z. and W.C. contributed equally to the work. X.Z. and W.C. collected field data and conducted the landslide susceptibility mapping and analysis. X.Z. and W.C. wrote and revised the manuscript. W.C. provided critical comments in planning this paper and edited the manuscript. All the authors discussed the results and edited the manuscript. All authors have read and agreed to the published version of the manuscript.

Funding: This research was supported by the National Natural Science Foundation of China (Grant No. 41807192), Natural Science Basic Research Program of Shaanxi (Program No. 2019JLM-7, Program No. 2019JQ-094), China Postdoctoral Science Foundation (Grant No. 2018AT111084, 2017M61318), and project funded by Shaanxi Province Postdoctoral Science Foundation (Grant No. 2017BSHYDZZ07).

Conflicts of Interest: The authors declare no conflicts of interest.

References

1. Pham, B.T.; Tien Bui, D.; Prakash, I. Bagging based support vector machines for spatial prediction of landslides. *Environ. Earth Sci.* **2018**, *77*, 146. [CrossRef]
2. Chen, W.; Pourghasemi, H.R.; Naghibi, S.A. Prioritization of landslide conditioning factors and its spatial modeling in shangnan county, China using gis-based data mining algorithms. *Bull. Eng. Geol. Environ.* **2018**, *77*, 611–629. [CrossRef]
3. Guo, C.; Qin, Y.; Ma, D.; Xia, Y.; Chen, Y.; Si, Q.; Lu, L. Ionic composition, geological signature and environmental impacts of coalbed methane produced water in China. *Energy Sources Part A Recovery Util. Environ. Eff.* **2019**, 1–15. [CrossRef]
4. Corominas, J.; Westen, C.V.; Frattini, P.; Cascini, L.; Malet, J.P.; Fotopoulou, S.; Catani, F.; Eeckhaut, M.V.D.; Mavrouli, O.; Agliardi, F. Recommendations for the quantitative analysis of landslide risk. *Bull. Eng. Geol. Environ.* **2014**, *73*, 209–263. [CrossRef]
5. Zhang, T.Y.; Han, L.; Zhang, H.; Zhao, Y.H.; Li, X.A.; Zhao, L. Gis-based landslide susceptibility mapping using hybrid integration approaches of fractal dimension with index of entropy and support vector machine. *J. Mt. Sci.* **2019**, *16*, 1275–1288. [CrossRef]
6. Chen, W.; Pourghasemi, H.R.; Kornejady, A.; Zhang, N. Landslide spatial modeling: Introducing new ensembles of ann, maxent, and svm machine learning techniques. *Geoderma* **2017**, *305*, 314–327. [CrossRef]
7. He, Y.; Beighley, R.E. Gis-based regional landslide susceptibility mapping: A case study in Southern California. *Earth Surf. Process. Landf.* **2010**, *33*, 380–393. [CrossRef]

8. Hadmoko, D.S.; Lavigne, F.; Samodra, G. Application of a semiquantitative and gis-based statistical model to landslide susceptibility zonation in kayangan catchment, Java, Indonesia. *Nat. Hazards* **2017**, *87*, 437–468. [CrossRef]
9. Saha, A.K.; Gupta, R.P.; Arora, M.K. Gis-based landslide hazard zonation in the bhagirathi (ganga) valley, himalayas. *Int. J. Remote Sens.* **2002**, *23*, 357–369. [CrossRef]
10. Saha, A.K.; Gupta, R.P.; Sarkar, I.; Arora, M.K.; Csaplovics, E. An approach for gis-based statistical landslide susceptibility zonation—with a case study in the Himalayas. *Landslides* **2005**, *2*, 61–69. [CrossRef]
11. Mohammady, M.; Pourghasemi, H.R.; Pradhan, B. Landslide susceptibility mapping at golestan province, iran: A comparison between frequency ratio, dempster–shafer, and weights-of-evidence models. *J. Asian Earth Sci.* **2012**, *61*, 221–236. [CrossRef]
12. Milaghardan, A.H.; Delavar, M.; Chehreghan, A. Uncertainty in landslide occurrence prediction using dempster–shafer theory. *Model. Earth Syst. Environ.* **2016**, *2*, 188. [CrossRef]
13. Mezaal, M.; Pradhan, B.; Rizeei, H. Improving landslide detection from airborne laser scanning data using optimized dempster–shafer. *Remote Sens.* **2018**, *10*, 1029. [CrossRef]
14. Constantin, M.; Bednarik, M.; Jurchescu, M.C.; Vlaicu, M. Landslide susceptibility assessment using the bivariate statistical analysis and the index of entropy in the sibiciu basin (Romania). *Environ. Earth Sci.* **2011**, *63*, 397–406. [CrossRef]
15. Jaafari, A.; Najafi, A.; Pourghasemi, H.R.; Rezaeian, J.; Sattarian, A. Gis-based frequency ratio and index of entropy models for landslide susceptibility assessment in the caspian forest, northern iran. *Int. J. Environ. Sci. Technol.* **2014**, *11*, 909–926. [CrossRef]
16. Devkota, K.C.; Regmi, A.D.; Pourghasemi, H.R.; Yoshida, K.; Pradhan, B.; Ryu, I.C.; Dhital, M.R.; Althuwaynee, O.F. Landslide susceptibility mapping using certainty factor, index of entropy and logistic regression models in gis and their comparison at mugling–narayanghat road section in nepal himalaya. *Nat. Hazards* **2013**, *65*, 135–165. [CrossRef]
17. Bai, S.-B.; Wang, J.; Lü, G.-N.; Zhou, P.-G.; Hou, S.-S.; Xu, S.-N. Gis-based logistic regression for landslide susceptibility mapping of the zhongxian segment in the three gorges area, China. *Geomorphology* **2010**, *115*, 23–31. [CrossRef]
18. Chen, W.; Pourghasemi, H.R.; Zhao, Z. A gis-based comparative study of dempster-shafer, logistic regression and artificial neural network models for landslide susceptibility mapping. *Geocarto Int.* **2017**, *32*, 367–385. [CrossRef]
19. Chen, W.; Zhao, X.; Shahabi, H.; Shirzadi, A.; Khosravi, K.; Chai, H.; Zhang, S.; Zhang, L.; Ma, J.; Chen, Y.; et al. Spatial prediction of landslide susceptibility by combining evidential belief function, logistic regression and logistic model tree. *Geocarto Int.* **2019**, *34*, 1177–1201. [CrossRef]
20. Hong, H.; Chen, W.; Xu, C.; Youssef, A.M.; Pradhan, B.; Tien Bui, D. Rainfall-induced landslide susceptibility assessment at the chongren area (China) using frequency ratio, certainty factor, and index of entropy. *Geocarto Int.* **2017**, *32*, 16. [CrossRef]
21. Chen, W.; Li, W.; Chai, H.; Hou, E.; Li, X.; Xiao, D. Gis-based landslide susceptibility mapping using analytical hierarchy process (ahp) and certainty factor (cf) models for the baozhong region of baoji city, China. *Environ. Earth Sci.* **2016**, *75*, 63. [CrossRef]
22. Sujatha, E.R.; Rajamanickam, G.V.; Kumaravel, P. Landslide susceptibility analysis using probabilistic certainty factor approach: A case study on tevankarai stream watershed, India. *J. Earth Syst. Sci.* **2012**, *121*, 1337–1350. [CrossRef]
23. Chen, W.; Chai, H.; Sun, X.; Wang, Q.; Xiao, D.; Hong, H. A gis-based comparative study of frequency ratio, statistical index and weights-of-evidence models in landslide susceptibility mapping. *Arab. J. Geosci.* **2016**, *9*, 204. [CrossRef]
24. Hong, H.; Shahabi, H.; Shirzadi, A.; Chen, W.; Chapi, K.; Ahmad, B.B.; Roodposhti, M.S.; Hesar, A.Y.; Tian, Y.; Tien Bui, D. Landslide susceptibility assessment at the wuning area, China: A comparison between multi-criteria decision making, bivariate statistical and machine learning methods. *Nat. Hazards* **2019**, *96*, 173–212. [CrossRef]
25. Myronidis, D.; Papageorgiou, C.; Theophanous, S. Landslide susceptibility mapping based on landslide history and analytic hierarchy process (ahp). *Nat. Hazards* **2016**, *81*, 1–19. [CrossRef]

26. Gou, C.; Rui, L.; Feng, C. Gis-based landslide susceptibility mapping using analytical hierarchy process in Wenchuan. In Proceedings of the International Conference on Geoinformatics, Kaifeng, China, 20–22 June 2013.
27. Park, S.; Choi, C.; Kim, B.; Kim, J. Landslide susceptibility mapping using frequency ratio, analytic hierarchy process, logistic regression, and artificial neural network methods at the Inje area, Korea. *Environ. Earth Sci.* **2013**, *68*, 1443–1464. [CrossRef]
28. Chen, W.; Pourghasemi, H.R.; Panahi, M.; Kornejady, A.; Wang, J.; Xie, X.; Cao, S. Spatial prediction of landslide susceptibility using an adaptive neuro-fuzzy inference system combined with frequency ratio, generalized additive model, and support vector machine techniques. *Geomorphology* **2017**, *297*, 69–85. [CrossRef]
29. Hong, H.; Ilia, I.; Tsangaratos, P.; Wei, C.; Chong, X. A hybrid fuzzy weight of evidence method in landslide susceptibility analysis on the wuyuan area, China. *Geomorphology* **2017**, *290*, 1–16. [CrossRef]
30. Polykretis, C.; Chalkias, C. Comparison and evaluation of landslide susceptibility maps obtained from weight of evidence, logistic regression, and artificial neural network models. *Nat. Hazards* **2018**, *93*, 1–26. [CrossRef]
31. Dahal, R.K.; Hasegawa, S.; Nonomura, A.; Yamanaka, M.; Masuda, T.; Nishino, K. Gis-based weights-of-evidence modelling of rainfall-induced landslides in small catchments for landslide susceptibility mapping. *Environ. Geol.* **2008**, *54*, 311–324. [CrossRef]
32. Regmi; Deep, A.; Devkota; Chandra, K.; Yoshida; Kohki; Pradhan; Biswajeet; Pourghasemi; Reza, H. Application of frequency ratio, statistical index, and weights-of-evidence models and their comparison in landslide susceptibility mapping in central nepal himalaya. *Arab. J. Geosci.* **2014**, *7*, 725–742. [CrossRef]
33. Chen, W.; Li, W.; Hou, E.; Bai, H.; Chai, H.; Wang, D.; Cui, X.; Wang, Q. Application of frequency ratio, statistical index, and index of entropy models and their comparison in landslide susceptibility mapping for the baozhong region of baoji, China. *Arab. J. Geosci.* **2015**, *8*, 1829–1841. [CrossRef]
34. Samui, P. Multivariate adaptive regression spline (mars) for prediction of elastic modulus of jointed rock mass. *Geotech. Geol. Eng.* **2013**, *31*, 249–253. [CrossRef]
35. Chu, L.; Wang, L.J.; Jiang, J.; Liu, X.; Sawada, K.; Zhang, J. Comparison of landslide susceptibility maps using random forest and multivariate adaptive regression spline models in combination with catchment map units. *Geosci. J.* **2019**, *23*, 341–355. [CrossRef]
36. Alreja, J.; Parab, S.; Mathur, S.; Samui, P. Estimating hysteretic energy demand in steel moment resisting frames using multivariate adaptive regression spline and least square support vector machine. *Ain Shams Eng. J.* **2015**, *6*, 449–455. [CrossRef]
37. Gayen, A.; Saha, S. Application of weights-of-evidence (woe) and evidential belief function (ebf) models for the delineation of soil erosion vulnerable zones: A study on pathro river basin, jharkhand, India. *Model. Earth Syst. Environ.* **2017**, *3*, 1123–1139. [CrossRef]
38. Zhang, Z.; Fan, Y.; Han, C.; Wu, Y.; Tao, L.; Li, W.; Wang, Q.; Ping, L. Gis-based landslide susceptibility analysis using frequency ratio and evidential belief function models. *Environ. Earth Sci.* **2016**, *75*, 948. [CrossRef]
39. Althuwaynee, O.F.; Pradhan, B.; Lee, S. Application of an evidential belief function model in landslide susceptibility mapping. *Comput. Geosci.* **2012**, *44*, 120–135. [CrossRef]
40. Tien Bui, D.; Bui, Q.T.; Nguyen, Q.P.; Pradhan, B.; Nampak, H.; Trinh, P.T. A hybrid artificial intelligence approach using gis-based neural-fuzzy inference system and particle swarm optimization for forest fire susceptibility modeling at a tropical area. *Agric. For. Meteorol.* **2017**, *233*, 32–44. [CrossRef]
41. Chen, W.; Yan, X.; Zhao, Z.; Hong, H.; Tien Bui, D.; Pradhan, B. Spatial prediction of landslide susceptibility using data mining-based kernel logistic regression, naive bayes and rbfnetwork models for the long county area (China). *Bull. Eng. Geol. Environ.* **2019**, *78*, 247–266. [CrossRef]
42. Pham, B.T.; Tien Bui, D.; Pourghasemi, H.R.; Indra, P.; Dholakia, M. Landslide susceptibility assesssment in the uttarakhand area (India) using gis: A comparison study of prediction capability of naïve bayes, multilayer perceptron neural networks, and functional trees methods. *Theor. Appl. Climatol.* **2017**, *128*, 255–273. [CrossRef]
43. Tsangaratos, P.; Ilia, I. Comparison of a logistic regression and naïve bayes classifier in landslide susceptibility assessments: The influence of models complexity and training dataset size. *Catena* **2016**, *145*, 164–179. [CrossRef]

44. Krkač, M.; Špoljarić, D.; Bernat, S.; Arbanas, S.M.J.L. Method for prediction of landslide movements based on random forests. *Landslides* **2017**, *14*, 947–960. [CrossRef]
45. Paudel, U.; Oguchi, T.; Hayakawa, Y. Multi-resolution landslide susceptibility analysis using a dem and random forest. *Int. J. Geosci.* **2016**, *7*, 726. [CrossRef]
46. Chen, W.; Xie, X.; Wang, J.; Pradhan, B.; Hong, H.; Tien Bui, D.; Duan, Z.; Ma, J. A comparative study of logistic model tree, random forest, and classification and regression tree models for spatial prediction of landslide susceptibility. *Catena* **2017**, *151*, 147–160. [CrossRef]
47. Arnone, E.; Francipane, A.; Noto, L.V.; Scarbaci, A.; Loggia, G.L. Strategies investigation in using artificial neural network for landslide susceptibility mapping: Application to a sicilian catchment. *J. Hydroinform.* **2014**, *16*, 502. [CrossRef]
48. Tan, L.; Chen, G.; Zeng, R.Q.; Xiong, M.Q.; Meng, X.M. Application of artificial neural network in landslide susceptibility assessment. *J. Lanzhou Univ.* **2014**, *50*, 15–20.
49. Lee, S.; Ryu, J.H.; Min, K.; Won, J.S. Landslide susceptibility analysis using gis and artificial neural network. *Earth Surf. Process. Landf.* **2010**, *28*, 1361–1376. [CrossRef]
50. Chen, W.; Hong, H.; Panahi, M.; Shahabi, H.; Wang, Y.; Shirzadi, A.; Pirasteh, S.; Alesheikh, A.A.; Khosravi, K.; Panahi, S.; et al. Spatial prediction of landslide susceptibility using gis-based data mining techniques of anfis with whale optimization algorithm (woa) and grey wolf optimizer (gwo). *Appl. Sci.* **2019**, *9*, 3755. [CrossRef]
51. Hong, H.; Pradhan, B.; Xu, C.; Tien Bui, D. Spatial prediction of landslide hazard at the yihuang area (China) using two-class kernel logistic regression, alternating decision tree and support vector machines. *Catena* **2015**, *133*, 266–281. [CrossRef]
52. Chen, W.; Peng, J.; Hong, H.; Shahabi, H.; Pradhan, B.; Liu, J.; Zhu, A.X.; Pei, X.; Duan, Z. Landslide susceptibility modelling using gis-based machine learning techniques for Chongren County, Jiangxi Province, China. *Sci. Total Environ.* **2018**, *626*, 230. [CrossRef] [PubMed]
53. Chen, W.; Pourghasemi, H.R.; Naghibi, S.A. A comparative study of landslide susceptibility maps produced using support vector machine with different kernel functions and entropy data mining models in China. *Bull. Eng. Geol. Environ.* **2017**, *77*, 1–18. [CrossRef]
54. Pradhan, B. A comparative study on the predictive ability of the decision tree, support vector machine and neuro-fuzzy models in landslide susceptibility mapping using gis. *Comput. Geosci.* **2013**, *51*, 350–365. [CrossRef]
55. Lombardo, L.; Cama, M.; Conoscenti, C.; Marker, M.; Rotigliano, E. Binary logistic regression versus stochastic gradient boosted decision trees in assessing landslide susceptibility for multiple-occurring landslide events: Application to the 2009 storm event in messina (Sicily, Southern Italy). *Nat. Hazards* **2015**, *79*, 1621–1648. [CrossRef]
56. Tien Bui, D.; Pradhan, B.; Lofman, O.; Revhaug, I. Landslide susceptibility assessment in vietnam using support vector machines, decision tree, and naive bayes models. *Math. Probl. Eng.* **2012**, 974638. [CrossRef]
57. Ghorbanzadeh, O.; Rostamzadeh, H.; Blaschke, T.; Gholaminia, K.; Aryal, J. A new gis-based data mining technique using an adaptive neuro-fuzzy inference system (anfis) and k-fold cross-validation approach for land subsidence susceptibility mapping. *Nat. Hazards* **2018**, *94*, 497–517. [CrossRef]
58. Chen, W.; Shahabi, H.; Shirzadi, A.; Hong, H.; Akgun, A.; Tian, Y.; Liu, J.; Zhu, A.X.; Li, S. Novel hybrid artificial intelligence approach of bivariate statistical-methods-based kernel logistic regression classifier for landslide susceptibility modeling. *Bull. Eng. Geol. Environ.* **2019**, *78*, 4397–4419. [CrossRef]
59. Aghdam, I.N.; Pradhan, B. Landslide susceptibility mapping using an ensemble statistical index (wi) and adaptive neuro-fuzzy inference system (anfis) model at alborz mountains (iran). *Environ. Earth Sci.* **2016**, *75*, 553. [CrossRef]
60. Polykretis, C.; Chalkias, C.; Ferentinou, M. Adaptive neuro-fuzzy inference system (anfis) modeling for landslide susceptibility assessment in a mediterranean hilly area. *Bull. Eng. Geol. Environ.* **2019**, *78*, 1173–1187. [CrossRef]
61. Lee, S.; Ryu, J.H.; Won, J.S.; Park, H.J. Determination and application of the weights for landslide susceptibility mapping using an artificial neural network. *Eng. Geol.* **2004**, *71*, 289–302. [CrossRef]
62. Tien Bui, D.; Pradhan, B.; Revhaug, I.; Nguyen, D.B.; Pham, H.V.; Bui, Q.N. A novel hybrid evidential belief function-based fuzzy logic model in spatial prediction of rainfall-induced shallow landslides in the Lang son city area (Vietnam). *Geomat. Nat. Hazards Risk* **2015**, *6*, 243–271.

63. Zhou, C.; Yin, K.; Cao, Y.; Ahmed, B.; Li, Y.; Catani, F.; Pourghasemi, H.R. Landslide susceptibility modeling applying machine learning methods: A case study from longju in the three gorges reservoir area, China. *Comput. Geosci.* **2018**, *112*, 23–37. [CrossRef]
64. Chen, W.; Shahabi, H.; Zhang, S.; Khosravi, K.; Shirzadi, A.; Chapi, K.; Pham, B.; Zhang, T.; Zhang, L.; Chai, H. Landslide susceptibility modeling based on gis and novel bagging-based kernel logistic regression. *Appl. Sci.* **2018**, *8*, 2540. [CrossRef]
65. Süzen, M.L.; Kaya, B.Ş. Evaluation of environmental parameters in logistic regression models for landslide susceptibility mapping. *Int. J. Digit. Earth* **2012**, *5*, 338–355. [CrossRef]
66. Shortliffe, E.H. A model of inexact reasoning in medicine. *Math. Biosci.* **1975**, *23*, 351–379. [CrossRef]
67. Heckerman, D. Probabilistic interpretations for MYCIN's certainty factors. In *Machine Intelligence and Pattern Recognition*; Kanal, L.N., Lemmer, J.F., Eds.; Elsevier: Amsterdam, The Netherlands, 1986; Volume 4, pp. 167–196.
68. Lan, H.X.; Zhou, C.H.; Wang, L.J.; Zhang, H.Y.; Li, R.H. Landslide hazard spatial analysis and prediction using gis in the Xiaojiang watershed, Yunnan, China. *Eng. Geol.* **2004**, *76*, 109–128. [CrossRef]
69. Dou, J.; Tien Bui, D.; Yunus, A.P.; Jia, K.; Song, X.; Revhaug, I.; Xia, H.; Zhu, Z. Optimization of causative factors for landslide susceptibility evaluation using remote sensing and GIS data in parts of Niigata, Japan. *PLoS ONE* **2015**, *10*, e0133262. [CrossRef]
70. Dou, J.; Oguchi, T.; Hayakawa, Y.S.; Uchiyama, S.; Saito, H.; Paudel, U. Gis-based landslide susceptibility mapping using a certainty factor model and its validation in the chuetsu area, central Japan. *Landslide Sci. A Safer Geoenviron.* **2014**, *2*, 419–424.
71. Pourghasemi, H.R.; Pradhan, B.; Gokceoglu, C.; Mohammadi, M.; Moradi, H.R. Application of weights-of-evidence and certainty factor models and their comparison in landslide susceptibility mapping at Haraz watershed, Iran. *Arab. J. Geosci.* **2013**, *6*, 2351–2365. [CrossRef]
72. Gama, J. Functional trees. *Mach. Learn.* **2004**, *55*, 219–250. [CrossRef]
73. Kavzoglu, T.; Sahin, E.K.; Colkesen, I. Landslide susceptibility mapping using gis-based multi-criteria decision analysis, support vector machines, and logistic regression. *Landslides* **2014**, *11*, 425–439. [CrossRef]
74. Witten, I.H.; Frank, E.; Hall, M.A. *Data Mining: Practical Machine Learning Tools and Techniques*, 3rd ed.; Morgan Kaufmann Publishers Inc.: San Francisco, CA, USA, 2011; Volume 31, pp. 76–77.
75. Tien Bui, D.; Ho, T.-C.; Pradhan, B.; Pham, B.-T.; Nhu, V.-H.; Revhaug, I. Gis-based modeling of rainfall-induced landslides using data mining-based functional trees classifier with adaboost, bagging, and multiboost ensemble frameworks. *Environ. Earth Sci.* **2016**, *75*, 1–22. [CrossRef]
76. Pham, B.T.; Prakash, I. A novel hybrid intelligent approach of random subspace ensemble and reduced error pruning trees for landslide susceptibility modeling: A case study at Mu Cang Chai District, Yen Bai Province, Vietnam. In Proceedings of the International Conference on Geo-Spatial Technologies & Earth Resources, Hanoi, Vietnam, 5–6 October 2017.
77. Doetsch, P.; Buck, C.; Golik, P.; Hoppe, N.; Kramp, M.; Laudenberg, J.; Steingrube, P.; Forster, J.; Mauser, A. Logistic model trees with AUC split criterion for the KDD cup 2009 small challenge. In Proceedings of the International Conference on KDD-Cup 2009 Competition, New York, NY, USA, 28 June 2009; Volume 7, pp. 77–88.
78. Landwehr, N.; Hall, M.; Frank, E. Logistic model trees. *Mach. Learn.* **2005**, *59*, 161–205. [CrossRef]
79. Chen, W.; Pradhan, B.; Li, S.; Shahabi, H.; Rizeei, H.M.; Hou, E.; Wang, S. Novel hybrid integration approach of bagging-based fisher's linear discriminant function for groundwater potential analysis. *Nat. Resour. Res.* **2019**, *28*, 1239–1258. [CrossRef]
80. Breiman, L. Bagging predictors. *Mach. Learn.* **1996**, *24*, 123–140. [CrossRef]
81. Bauer, E.; Kohavi, R. An empirical comparison of voting classification algorithms: Bagging, boosting, and variants. *Mach. Learn.* **1999**, *36*, 105–139. [CrossRef]
82. Rodriguez, J.J.; Kuncheva, L.I.; Alonso, C.J. Rotation forest: A new classifier ensemble method. *IEEE Trans. Pattern Anal. Mach. Intell.* **2006**, *28*, 1619–1630. [CrossRef]
83. Ozcift, A. Svm feature selection based rotation forest ensemble classifiers to improve computer-aided diagnosis of Parkinson disease. *J. Med. Syst.* **2012**, *36*, 2141–2147. [CrossRef]
84. Hong, H.; Liu, J.; Tien Bui, D.; Pradhan, B.; Acharya, T.D.; Pham, B.T.; Zhu, A.-X.; Chen, W.; Ahmad, B.B. Landslide susceptibility mapping using j48 decision tree with adaboost, bagging and rotation forest ensembles in the Guangchang area (China). *Catena* **2018**, *163*, 399–413. [CrossRef]

85. Pham, B.T.; Prakash, I.; Singh, S.K.; Shirzadi, A.; Shahabi, H.; Tien Bui, D. Landslide susceptibility modeling using reduced error pruning trees and different ensemble techniques: Hybrid machine learning approaches. *Catena* **2019**, *175*, 203–218. [CrossRef]
86. Kai, M.T.; Witten, I.H. Stacking bagged and dagged models. In Proceedings of the Fourteenth International Conference on Machine Learning, Nashville, TN, USA, 8–12 July 1997.
87. Li, J.; Huang, T. Predicting and analyzing early wake-up associated gene expressions by integrating gwas and eqtl studies. *Biochim. Biophys. Acta-Mol. Basis Dis.* **2018**, *1864*, 2241–2246. [CrossRef] [PubMed]
88. Chen, W.; Shirzadi, A.; Shahabi, H.; Ahmad, B.B.; Zhang, S.; Hong, H.; Zhang, N. A novel hybrid artificial intelligence approach based on the rotation forest ensemble and naïve bayes tree classifiers for a landslide susceptibility assessment in Langao County, China. *Geomatics. Nat. Hazards Risk* **2017**, *8*, 1955–1977. [CrossRef]
89. Chen, W.; Xie, X.; Peng, J.; Wang, J.; Duan, Z.; Hong, H. Gis-based landslide susceptibility modelling: A comparative assessment of kernel logistic regression, naïve-bayes tree, and alternating decision tree models. Geomatics. *Nat. Hazards Risk* **2017**, *8*, 950–973. [CrossRef]
90. Guyon, I.; Elisseeff, A. An introduction to variable and feature selection. *J. Mach. Learn. Res.* **2003**, *3*, 1157–1182.
91. Chen, W.; Hong, H.; Li, S.; Shahabi, H.; Wang, Y.; Wang, X.; Ahmad, B.B. Flood susceptibility modelling using novel hybrid approach of reduced-error pruning trees with bagging and random subspace ensembles. *J. Hydrol.* **2019**, *575*, 864–873. [CrossRef]
92. He, Q.; Shahabi, H.; Shirzadi, A.; Li, S.; Chen, W.; Wang, N.; Chai, H.; Bian, H.; Ma, J.; Chen, Y.; et al. Landslide spatial modelling using novel bivariate statistical based naïve bayes, rbf classifier, and rbf network machine learning algorithms. *Sci. Total Environ.* **2019**, *663*, 1–15. [CrossRef]
93. Youssef, A.M.; Pradhan, B.; Jebur, M.N.; El-Harbi, H.M. Landslide susceptibility mapping using ensemble bivariate and multivariate statistical models in Fayfa area, Saudi Arabia. *Environ. Earth Sci.* **2015**, *73*, 3745–3761. [CrossRef]
94. Chen, W.; Panahi, M.; Pourghasemi, H.R. Performance evaluation of gis-based new ensemble data mining techniques of adaptive neuro-fuzzy inference system (anfis) with genetic algorithm (ga), differential evolution (de), and particle swarm optimization (pso) for landslide spatial modelling. *Catena* **2017**, *157*, 310–324. [CrossRef]
95. Chen, W.; Tsangaratos, P.; Ilia, I.; Duan, Z.; Chen, X. Groundwater spring potential mapping using population-based evolutionary algorithms and data mining methods. *Sci. Total Environ.* **2019**, *684*, 31–49. [CrossRef]
96. Akgun, A. A comparison of landslide susceptibility maps produced by logistic regression, multi-criteria decision, and likelihood ratio methods: A case study at İzmir, Turkey. *Landslides* **2012**, *9*, 93–106. [CrossRef]
97. Chen, W.; Panahi, M.; Khosravi, K.; Pourghasemi, H.R.; Rezaie, F.; Parvinnezhad, D. Spatial prediction of groundwater potentiality using anfis ensembled with teaching-learning-based and biogeography-based optimization. *J. Hydrol.* **2019**, *572*, 435–448. [CrossRef]
98. Chen, W.; Li, Y.; Xue, W.; Shahabi, H.; Li, S.; Hong, H.; Wang, X.; Bian, H.; Zhang, S.; Pradhan, B.; et al. Modeling flood susceptibility using data-driven approaches of naïve bayes tree, alternating decision tree, and random forest methods. *Sci. Total Environ.* **2020**, *701*, 134979. [CrossRef] [PubMed]
99. Kornejady, A.; Ownegh, M.; Rahmati, O.; Bahremand, A. Landslide susceptibility assessment using three bivariate models considering the new topo-hydrological factor: Hand. *Geocarto Int.* **2018**, *33*, 1155–1185. [CrossRef]
100. Pourghasemi, H.R.; Yousefi, S.; Kornejady, A.; Cerdà, A. Performance assessment of individual and ensemble data-mining techniques for gully erosion modeling. *Sci. Total Environ.* **2017**, *609*, 764–775. [CrossRef]
101. Kornejady, A.; Ownegh, M.; Bahremand, A. Landslide susceptibility assessment using maximum entropy model with two different data sampling methods. *Catena* **2017**, *152*, 144–162. [CrossRef]
102. Hong, H.; Pourghasemi, H.R.; Pourtaghi, Z.S. Landslide susceptibility assessment in Lianhua county (China): A comparison between a random forest data mining technique and bivariate and multivariate statistical models. *Geomorphology* **2016**, *259*, 105–118. [CrossRef]

103. Chen, W.; Li, H.; Hou, E.; Wang, S.; Wang, G.; Panahi, M.; Li, T.; Peng, T.; Guo, C.; Niu, C.; et al. Gis-based groundwater potential analysis using novel ensemble weights-of-evidence with logistic regression and functional tree models. *Sci. Total Environ.* **2018**, *634*, 853–867. [CrossRef]
104. Chen, W.; Sun, Z.; Han, J. Landslide susceptibility modeling using integrated ensemble weights of evidence with logistic regression and random forest models. *Appl. Sci.* **2019**, *9*, 171. [CrossRef]

© 2019 by the authors. Licensee MDPI, Basel, Switzerland. This article is an open access article distributed under the terms and conditions of the Creative Commons Attribution (CC BY) license (http://creativecommons.org/licenses/by/4.0/).

applied sciences

MDPI

Article

Spatiotemporal Dynamics and Obstacles of the Multi-Functionality of Land Use in Xiangxi, China

Hui Xiang [1,2], Qing-Yuan Yang [1,2,*], Kang-chuan Su [1,2] and Zhong-Xun Zhang [1,2]

1 College of Geographical Science, Southwest University, Chongqing 400715, China
2 Chongqing key Laboratory of Karst and Environment, Southwest University, Chongqing 400715, China
* Correspondence: yizyang@swu.edu.cn

Received: 28 June 2019; Accepted: 22 August 2019; Published: 4 September 2019

Abstract: The multi-functionality of land as the basis of land use and utilization is under increasing investigation. This study assesses the spatiotemporal dynamic multi-functionality of land use and analyzes the obstacle indicators in Xiangxi using two methods, i.e., the analytic hierarchy process (AHP) and the hierarchical weighting method (HWM). First, we found that the total function level of land use in Xiangxi was constantly optimized. Spatial heterogeneity was clearer. Land use had a trend toward diversification, with a focus on production or living function. The coordination of multifunctional land use has undergone certain changes. It was more apparent in the south than in central and northern regions. Second, we discovered that production function of land use in Xiangxi grew slowly and spatial differentiation was enhanced. The living function fluctuated with the trend of spatial equilibrium. Changes in ecological function and any spatial differentiation were not clear. Third, land use can be divided into living-ecological, production-living, and production obstacle types. Lastly, we state that, to narrow the gap between urban-rural areas and reduce the non-point pollution from agriculture in living-ecological barrier areas, we need to develop production and social public utilities in production-living barrier areas, and develop production and eliminate poverty in production barrier areas.

Keywords: land use; multi-functionality; production-living-ecology function; spatiotemporal dynamics; obstacle factors; Xiangxi

1. Introduction

Land is a complex system including topography, soil, hydrology, biology, climate, and other elements [1]. In addition to providing food, fresh water, and other material resources for humans, it is also essential for habitation, transport, leisure, and other activities [2]. Land use reflects the type and intensity of human activities, which directly affects the biodiversity of ecosystems [3], water security [4], and human health [5]. The changes of land use are dominated by human activities [6], which, in turn, affects people's livelihood and sustainable development of the economy. Several models have been used to assess land use change [7,8], but most of them have only focused on the economic or ecological functions of land [9]. Land resources are multifunctional, which is not only an important factor of the ecological environment, but also the main resources of human production and life. Land multi-function evaluation is a very effective method that takes economic, social, and ecological factors into consideration [10]. The concept of multi-functionality in land use originated from agriculture and refers to the ability of land to provide diversified products to meet various needs. This concept has gradually expanded into the non-agricultural sectors. In 2004, the SENSOR project, under the sixth framework of the European Union, considered that the function of land use refers to the various uses of land for the provision of multiple products and services [11], to meet the needs of humans for economic, environmental, cultural, and social services. The multi-functionality of land as the basis for land use and allocation is an important factor affecting regional land decision-making and spatial planning, and is related to the sustainable development of society and economy.

Research concerning the versatility of land use have become increasingly abundant. These studies have mainly focused on the types of land function, the multiple functions of agriculture, and the evolution of land function. Land multi-function includes regulation function, habitat function, production function, information function, and so on [12]. Scholars have generally discussed the relationship between agricultural versatility, agricultural policy [13], and sustainable development, believing that the main functions for agricultural land were food production and environmental sustainability [14], which had a positive impact on the life of the higher social capital class [15]. An index system for quantifying the versatility of agricultural land was constructed [16,17]. Based on results of this evaluation, the land functions were divided into different areas [18], and the zoning optimization strategy was proposed. Scholars have analyzed the rubber agroforestry system in Sumatra and China [19], concluding that the factors restricting multifunctional change are complex and diverse. A survey in Australia has found that ranches have changed from a single production function to a variety of functions [20], such as protecting biological communities and developing tourism. The study has discovered that family farms in Missouri provide different leisure services and play a diversified role in land use [21]. Since the 21st century, the land use of parks in Poland has undergone significant changes [22]. The reasons for the changes of forest landscape functions in Southwest Parks of Poland include population change, intensive agriculture, urbanization, and land-use policy change [23]. Evidence from Denmark shows that recreational hunting contributes to multi-functional maintenance and change of land use [24].

It can be seen from the above that scholars have carried out extensive studies concerning the types of functions, the current multi-functional use of agricultural land, and have made beneficial attempts to understand the spatio-temporal evolution of multi-functional land use. However, research concerning multi-functional dynamic changes to land use focus mainly on macro (national) and middle (provincial) scales, with insufficient research carried out on a micro (county) scale, and there are few discussions regarding the ethnic regions at the county-scale. Most research studies are taken from the perspective of the dimensions of time or space. Therefore, research using a combination of space and time with regard to land use is required. In addition, land use change has also brought a series of adverse effects [25,26], which restricts the sound development of the land systems. For measuring the function of land, it is also necessary to clarify factors that create obstacles for the versatile use of land, but few studies have conducted such analyses. Xiangxi is typically representative of less developed areas in China, and it is also a minority-concentrated area. Based on the land-use history of this area, we arrive at three questions.

(1). What changes have taken place with the land-use function of this region?
(2). What are the main factors restricting the land function there?
(3). What measures should be taken to mitigate the adverse effects?

To answer these questions, bases on the logic of "analysis of functions-diagnosis of barrier factors-policy recommendations," this study is conducted from the following aspects. First, we classified the land function types and constructed an evaluation index system. Second, we evaluated the temporal and spatial dynamics of land multifunction in the last five years. Third, we explored the obstacles in depth. Lastly, through comparative analyses between Xiangxi and other regions, we put forward the corresponding policy recommendations.

The aims of this study are: (1) to realize efficient positioning, quantitative expression [27], (2) identify obstacles to constrain land use efficiency in multifunctional land use, and (3) provide policy suggestions for the rational land use in minority areas.

2. Data and Methods

2.1. Research Area

Xiangxi (full name: Xiangxi tujia and miao Autonomous Prefecture) is located in the northwest part of the Hunan province (Figure 1). It is bordering the Hubei, Guizhou, and Chongqing provinces. It is the bridge connecting the central and western regions, and the link of communication between the Han and minority nationalities. Geographically, Xiangxi lies between 109°10′~110°22.5′ E and 27°44.5′~29°38′ N. The total area of the region is approximately 1.547 million hm^2, accounting for 7.3% of the Hunan province. It has seven counties and one city under its jurisdiction (Fenghuang county, Luxi county, Huayuan county, Guzhang county, Baojing county, Yongshun county, Longshan county, and Jishou city—a county-level city). Xiangxi is a typical winding territory of Wuling mountain, with high altitude terrain in the northwest, and low terrain in the southeast. Ethnic minorities live in compact communities in Xiangxi. In 2017, 80.73% of the total population of Xiangxi belonged to ethnic minorities, mainly Tujia and Miao.Xiangxi, with a relatively backward economy, which is one of the 14 areas included in the national pilot project for poverty alleviation and is currently in a critical period of new rural construction and modern industrial development. From 2013 to 2017, the economic and social development in Xiangxi increased rapidly, with an average annual GDP growth rate of 7.82%, a per capita GDP growth rate of about 7.26%, and an increase in the urbanization rate from 38.8% to 44.97%. The regional ecological environment is superior, with 70.24% forest coverage in 2017. The land use in Xiangxi is dominated by agriculture, with the area of land used for agriculture reaching 1.4118 million hm^2 (accounting for 91.26%), and an area of 0.0555 million hm^2 has been used for construction (accounting for 3.59%). Driven by economic interests, the land use structure has changed significantly, and the area of land used for construction has increased dramatically. For example, the total urban land in 2016 was 961.34 hm^2, which represented an annual increase of 15.8%. Therefore, the efficient use of land in Xiangxi currently faces two major problems. First, the agricultural land area is large but the production efficiency is low. Second, urban construction is limiting the space available for production and ecology, and the contrast between land supply and land demand is increasingly prominent. Therefore, it is necessary to improve the function of production in land use in Xiangxi, and to raise people's awareness that changes in the structure of land use lead to changes in land function.

2.2. Data Sources

The data used in this study concerning the current situation of land use comes from the file issued by the Xiangxi state bureau of land and resources "Suggestions for general land use planning (2006–2020)." The DEM digital elevation data comes from the geographical spatial data cloud (http://www.gscloud.cn/), and all other data (population, land area, land for traffic area, the output value of farming and animal husbandry, etc.) comes directly or indirectly from the Xiangxi statistical Yearbook (2013–2017) and the Hunan statistical Yearbook (2014–2018) (Xiangxi statistical Yearbook is based on the data of that year, and Hunan statistical Yearbook is based on the data of the previous year).

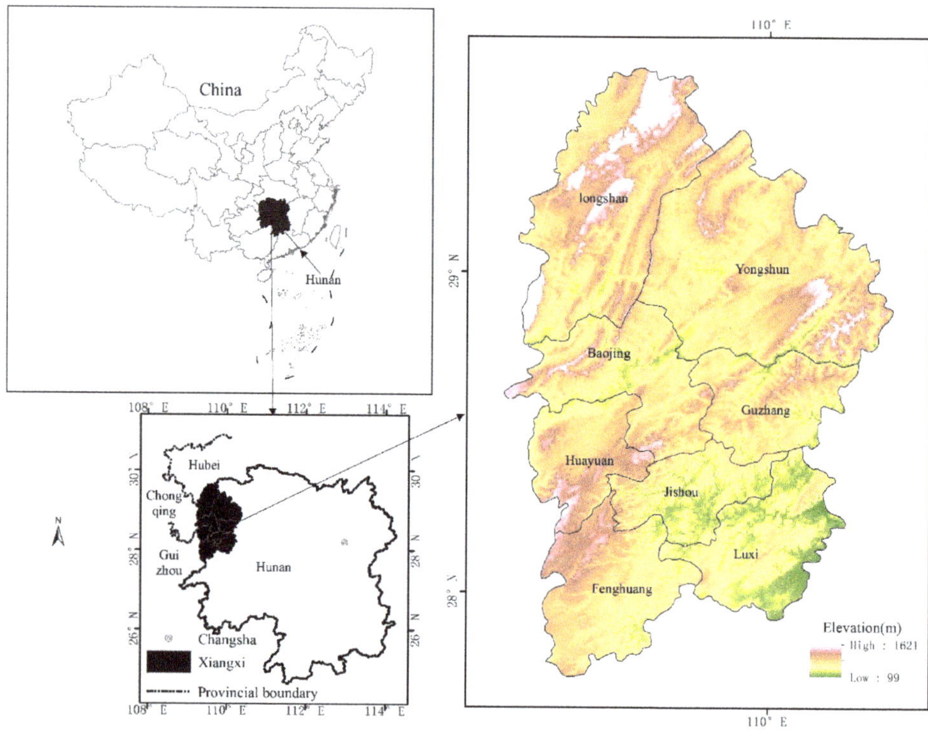

Figure 1. Map of the study area.

2.3. Research Methods

2.3.1. Classifying the Multifunctional Types of Land Use

The land system is composed of economic, social, and ecological subsystems [28], and the land is rendered whole with comprehensive functions via the organic coupling of each subsystem [29]. This study is based on the economic, social, and environmental dimensions of sustainable development, combined with the idea of a national "production-living-ecology (PLE)," based on the optimization of land use, by subdividing the total functions (TF) for which land is used as PLE sub-functions (Table 1): production functions (referred to as PF), living functions (LF), and ecological functions (EF). Production functions are based on the key needs for human survival. Land provides agricultural products to guarantee the capacity of non-agricultural economic output and transportation. These functions are, therefore, measured as the three aspects of agricultural production, economic growth, and transportation security [30]. Living functions are the ability of the land to meet the needs for human development [31,32], which are mainly reflected by four aspects: employment support, social security, cultural leisure, and the residential home. Ecological functions are related to the high-quality production and living needs of human beings. This function is evaluated from the three dimensions of maintaining ecological balance, providing resources, and keeping the environment clean. This study, therefore, divides the function types of land use into three levels from top to bottom: total function, sub-function, and single function. The number of functions at each level is 1, 3, and 10, respectively.

Table 1. Data sources.

Data Types	Data Names	Units	Sources
General Situation of Xiangxi	Agricultural land area, construction land area	hm^2	Land Bureau of Xiangxi
	DEM elevation data	m	http://www.gscloud.cn
	Population composition, GDP, Urbanization rate, Forest Coverage Rate	%	Statistical Yearbook of Xiangxi
The 23 evaluation indexes of land use function	X1, X2 X3, X6, X7, X10, X11, X13, X18, X19,X20, X21	yuan/person %	Statistical Yearbook of Xiangxi or Hunan
	X5, X12, X15	m^2/person	
	X4	10^4 yuan/km^2	
	X8, X14	yuan	
	X9	person	
	X16	t/person	
	X17	$km^2/10^4$ yuan	
	X22, X23	t/hm^2	

Note: The meaning of X1–X23 is detailed in Table 2.

2.3.2. The Establishment of the Evaluation Index System

The selection of indicators for evaluating multifunctional land use follows the basic principles. (1) Indicators should take the regional development situation into account. For example, the output from agriculture and animal husbandry in Xiangxi accounts for more than 95% (the data of 2017) of the regional output of agriculture, forestry, animal husbandry, and fisheries. The two indexes of per capita agricultural output value and per capita animal husbandry output value were, therefore, selected, according to the needs of regional production and development. (2) Indicators should be chosen that can be quantified and easily obtained. Most indicators selected come either directly or indirectly from the public websites of government departments, which can be directly accessed. (3) The indicators were independent and complementary (Sun et al., 2017). For example, the employment security function for rural and urban land was, respectively, represented by the number of rural employees and the average annual wage of urban employees. (4) Using direct effective indicators. Specific indexes were selected that could minimize the total quantity of indexes, which directly reflects the functional level of land. From the above principles, a total of 23 typical sensitive indicators representing land use function in Xiangxi were selected (Table 2).

2.3.3. Determination of the Index Weight

The analytic hierarchy process (AHP) was used to calculate the weight of each factor. Through modeling and quantitative analysis, the AHP simplifies complex problems and is widely used in the field of land evaluation.

The software of Yaahp (full name is yet another AHP) was invented by Zhang Jianhua. It can be download from this website: http://www.yaahp.cn. It is simple and efficient, and widely used to determine weights [33,34].

First, Yaahp v.10.3 software is used to build a four-level hierarchical structure model, which is a function-subfunction-single function-index layer.

Second, the evaluation factors are scored. The judgment matrix A_{mk} was constructed. Experts were invited to evaluate the relative importance of two factors (m, k) at the same level. The evaluation results were divided into five levels: absolutely important, very important, relatively important, slightly important, and equally important. They were assigned 1, 3, 5, 7, and 9 points, respectively. If the evaluation results landed in the middle, the median value was taken.

Table 2. Assessment indicator system of multi-functionality of land use in Xiangxi.

Total Function	Sub-Function and Weight	Single Function and Weight	Index Layer and Weight	Indicator Description and Attributes
Land function	Production functions (PF), 0.3451	Agricultural and animal husbandry production function (PF1), 0.1149	Per capita agricultural output (X1), 0.0856	Output value of agriculture/Regional population +
			Per capita output value of animal husbandry (X2), 0.0293	Animal husbandry output value/Regional population+
		Economic growth function (PF2), 0.1382	Proportion of secondary and tertiary industries (X3), 0.0493	Output value of the second and third industries/GDP +
			The density of economic (X4), 0.0889	GDP/Gross land area +
		Traffic support function (PF3), 0.0920	Per capita road area (X5), 0.0500	Directly come from the statistical Yearbook +
			The proportion of passenger transport turnover (X6), 0.0200	Directly from the statistical Yearbook +
			Freight turnover as a proportion of the state (X7), 0.0220	Regional freight turnover/Statewide passenger turnover +
	Living functions (LF), 0.3359	Employment support function (LF1), 0.0850	The average annual salary of urban employees (X8), 0.0425	Directly from the statistical Yearbook +
			Rural agricultural personnel (X9), 0.0425	Directly from the statistical Yearbook +
		Social security function (LF2), 0.0851	Urban-rural income balance index (X10), 0.0465	Per capita disposable income ratio of urban and rural residents +
			Incidence of poverty (X11), 0.0386	Directly from the statistical Yearbook -
		Cultural and leisure function (LF3), 0.0792	Per capita park green space (X12), 0.0264	Directly from the statistical Yearbook +
			The proportion of residents' expenditure on culture, education, and entertainment (X13), 0.0528	Residents' expenditure on culture, education, and entertainment/residents' total expenditure +
		Residential home function (LF4), 0.0866	Rural per capita housing value (X14), 0.0288	Directly from the statistical Yearbook +
			Per capita housing area (X15), 0.0578	Directly from the statistical Yearbook +
	Ecological function (EF), 0.3190	Resource supply function (EF1), 0.0878	Urban water supplies per capita (X16), 0.0439	Total urban water supply/urban population +
			Per capita land area (X17), 0.0439	Land area/area population +
		Ecological maintenance function (EF2), 0.1068	Green coverage of the built-up area (X18), 0.0551	Directly from the statistical Yearbook +
			Forest coverage (X19), 0.0517	Directly from the statistical Yearbook +
		Environmental purification function (EF3), 0.1244	Urban sewage treatment rate (X20), 0.0622	Directly from the statistical Yearbook +
			Centralized disposal rate of rural garbage (X21), 0.0388	Directly from the statistical Yearbook +
			Amount of fertilizer applied per unit cultivated land (X22), 0.0078	Fertilizer application/cultivated area -
			Pesticide application amount per unit of cultivated land (X23), 0.0156	Pesticide application/cultivated area -

The third step involved consistent checking and weight calculation. To test the rationality of the pairwise judgment matrix, consistency testing is required. If the test value is ≤0.1, it shows that the matrix evaluation is reasonable. However, the results have to be revised. Running software found that the evaluation results passed the consistency test.

Lastly, the weight of all factors was calculated (details in Table 2).

2.3.4. Measuring the Versatility of Land Use

The first step was to standardize the data. According to the land use function evaluation index system, a sample matrix X of m evaluation indexes in n areas from 2013 to 2017 was constructed, where $X = (X_{ij})_{n\times m}$, $i = 1,2,\ldots N$, $j = 1, 2,\ldots M$. In order to make an evaluation of the quantitative comparison of different attributes and dimension indexes, the optimal value of each index (the maximum value of the positive index and the minimum value of the negative index) during the study period was selected as the reference value X_o in order to conduct a dimensionless quantization of X. The calculation formula used was as follows.

$$Y_{ij} = \begin{cases} X_{ij}/X_{oj} \text{ the positive index} \\ X_{oj}/X_{ij} \text{ the negative index} \end{cases} \quad (1)$$

where Y_{ij} was the standardized value of the j evaluation index in area i, X_{ij} was the original value of the j valuation index in area i, and X_{0j} was the optimal value of the j evaluation index. Evaluation samples Y were obtained after standard treatment, $Y = (Y_{ij})_{n\times m}$, $Y \in (0\ 1]$.

The second step was to calculate a value for the land use function. Value (F) represents the value for the level of land use with regard to functions. The larger F is, the better the level of land use is, and vice versa. According to the standardized values of the evaluation samples and the weight of factors at all levels, the land single functional value ($F1$), sub-functional value ($F2$), and total functional value ($F3$) were calculated. The formula for calculating the functional values is shown below.

$$F1 = \sum W_j Y_{ij} \; F2 = \sum W_i F1 \; F3 = \sum W_n F2 \quad (2)$$

where W_j was the weight of evaluation index, W_i was the weight of a single function ($F2$), and Wn was the weight of a sub-function ($F3$).

The third step was to calculate the degree of dynamic change in land use and the degree of functional advantage of the land. The degree of dynamic change in land function (d) refers to the degree of change in land function levels within a certain period. The overall degree of dynamic change in land function was, therefore, calculated for the study period of five years. The calculation used was as follows.

$$d = \frac{F_{t+4} - F_t}{F_t} \times 100\% \quad (3)$$

where t represents the year, F_t represents the land function value of the year t, $d < 0$ indicates that the land function remains unchanged, $d >$ indicates that the land function is enhanced, and $d < 0$ indicates that the land function is degraded.

The dynamic dominance of land function reflected the differences in land use, calculated with:

$$s = \frac{|d|_{max}}{\sum |d|} \quad (4)$$

where $|d|_{max}$ was the maximum absolute value of the degree of dynamic change in land function, and $\Sigma|d|$ was the sum of the absolute value of the dynamic changes of land function. The larger S is, the more diverse the land function change is, which means that the land use tends to be simplified. Smaller values for S point to diversification.

The fourth step was to measure the multifunctional coordination degree of land use. The difference in the standard deviation of the functional value of land use (σ) reflects the coordination of the functional level of the land. The calculation formula was as follows.

$$\sigma = \sqrt{\frac{\sum (F - \overline{F})^2}{N}} \tag{5}$$

where $\overline{F}$ was the average value for land function and N was the function number. The value of σ is inversely related to the coordination degree of land function. The higher the value is, the lower the coordination degree of land function is, and vice versa.

2.3.5. Diagnosing the Obstacles to Land Use Function

The degree of land dysfunction is represented by the degree of the total function that has been hindered. The higher the degree of land dysfunction, the stronger its influence. The calculation formulas for the obstacle degree (Qi) of a single function and the obstacle degree (Ki) of a sub-function are shown below.

$$Q_i = \frac{(1 - Y_{ij})W_j}{\sum(1 - Y_{ij})W_j} \times 100\% \tag{6}$$

$$K_i = \sum Q_i \tag{7}$$

where $(1 - Y_{ij})$ represents the gap between the i index of land use function and the goal of function, which is the difference between the index standard value (Y_{ij}) and the optimal standard value (1).

At last, we placed the calculation results into the attributes of each research units, and created the column chart using the spatial analysis function of software ArcMap v.10.2 (ArcMap is one of the three user desktop components of ArcGIS, and it was developed by the Environmental System Research Institute in 1978. Its official website is http://argmaps.com). The length of the column represented the size of each data, and labeled the values to visualize the results of the research.

3. Results

3.1. The Spatio-Temporal Dynamic Evaluation of the Total Function of Land Use

We got the map of total function of land use change in Xiangxi over the past five years (Figure 2), to analyze the trends in spatio-temporal dynamic evolution.

(1) The total function level of land use was constantly optimized and regional differences were found to be greater. During the study period, the value of land function level in Xiangxi showed an upward trend (Figure 2a). The highest increase occurred in Jishou (24.05%), and the lowest (8.90%) increase occurred in Longshan, with the other areas falling in between (10%–20%). In 2013, Xiangxi Autonomous Prefecture committee put forward the developmental policy of "5-4-2" (i.e., the "five constructions," industry, infrastructure, new towns, ecological construction for civilization, "four Xiangxi"—green, civilized, an open and harmonious Xiangxi, "two take the lead," take the lead in development, take the lead out of poverty). Under the lead of an open strategy for Hunan province and land use activities in the area remained stable and improved. With these changes in the total function of land use, the differences in the land function level in Xiangxi were increasingly clear, and the standard deviation in the value for total land function between the counties increased from 0.0344 to 0.0515. On the whole, the land use function level in Xiangxi presented a spatial pattern of high levels in the north and south (Jishou, Luxi, Fenghuang, and Huayuan in the south, Longshan, and Yongshun in the north), and low levels in the middle (Guzhang and Baojing).

(2) The functions for which land was used tended toward diversification, focusing on those associated with production or living. During the study period, the function value for the change of land use in each region was characterized by rapid growth of the lower (production and living) and

slow change of the higher (ecological) functions, which reflects the trend in land use diversification. Production and living functions (Figure 2b,c) underwent the largest changes in value, with the highest degree of dynamic changes in production functions observed in Jishou (37.48%) and the fastest change in living functions in Guzhang (39.57%). Ecological functions changed slowly or even declined (dynamic change degree-3.09%~12.08%). Over the five years studied, the dominance of land production or living functions in eight areas was 40% to 65%. The west of the region (except Huayuan) was dominated by diversified land use that focused on production functions, which was strongest in Longshan (64.46%). The east and west of Huayuan were marked by diversified land use dominated by living functions, and the dynamic dominance of living function was found to be the highest in Guzhang (55.90%). This indicates that the current notion of land use in Xiangxi is in the promotion of economic growth and the development of social undertakings to improve the livelihood of its residents.

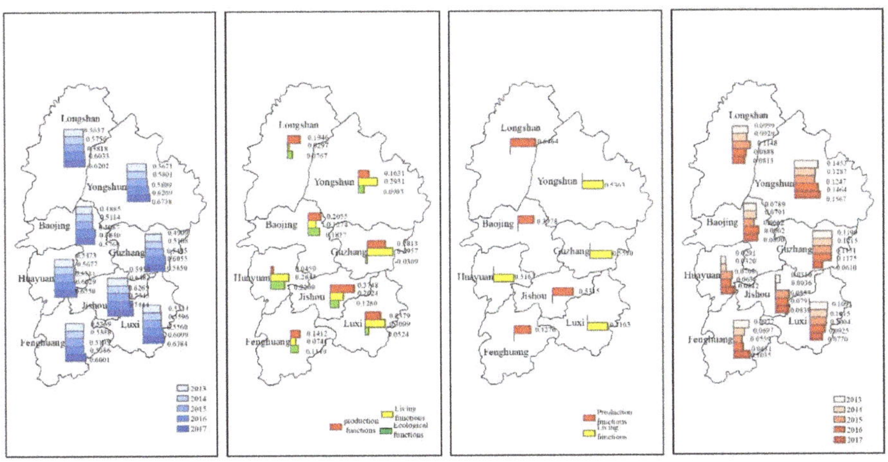

(a) (b) (c) (d)

Figure 2. Changes of land use total function from 2013 to 2017 in Xiangxi. (**a**) The value of the function. (**b**) The dynamic degree. (**c**) The dynamic dominance. (**d**) The standard deviation.

(3) The coordination of multi-functional land use was found to be changing slowly, with high coordination in the south and low in the central and northern regions. The standard deviation of land function in Jishou, Huayuan, Fenghuang, Baojing, and Yongshun underwent increasing fluctuations, which indicates a volatile decrease in functional coordination (Figure 2d). The standard deviation of land function in Luxi, Guzhang, and Longshan decreased, which reflects that the degree of functional coordination increased. However, the standard deviation for the land function in each area did not change significantly during the study period (variable rate −5.89% to 5.51%), which reflects the slow changes in the coordination of regional land use. The southern regions (Jishou, Huayuan, Fenghuang, and Yongshun) demonstrated a high degree of land use coordination, with the highest coordination seen in Jishou (the standard deviation was as low as 0.0036 in 2014), and the central (Baojing and Guzhang) and northern regions (Yongshun and Longshan) had a low degree of land use coordination, of which Yongshun was the lowest (with a standard deviation as high as 0.1567 in 2017). This was mainly because of the strengthening of the approach to development within Jishou (the core of Xiangxi) and the policy guidance and technical support of the local government, which promotes coordinated development in economy, society, and environment. However, the northern areas are far away from Jishou, and the effect of such policy changes reaching this area is limited. In addition, the industrial structure is unbalanced, with traditional industries, agriculture, and animal husbandry accounting for a large proportion. However, the degree of coordination in land resource utilization is low.

3.2. The Spatio-Temporal Dynamic Analysis of the Sub-Function of Land Use

The sub-function of the spatial and temporal evolution of land use during the study period was analyzed from three aspects: the value of functions, the dynamic degree of the functions, and the dynamic dominance of functions, using the data processed from the evaluation samples (Figure 3).

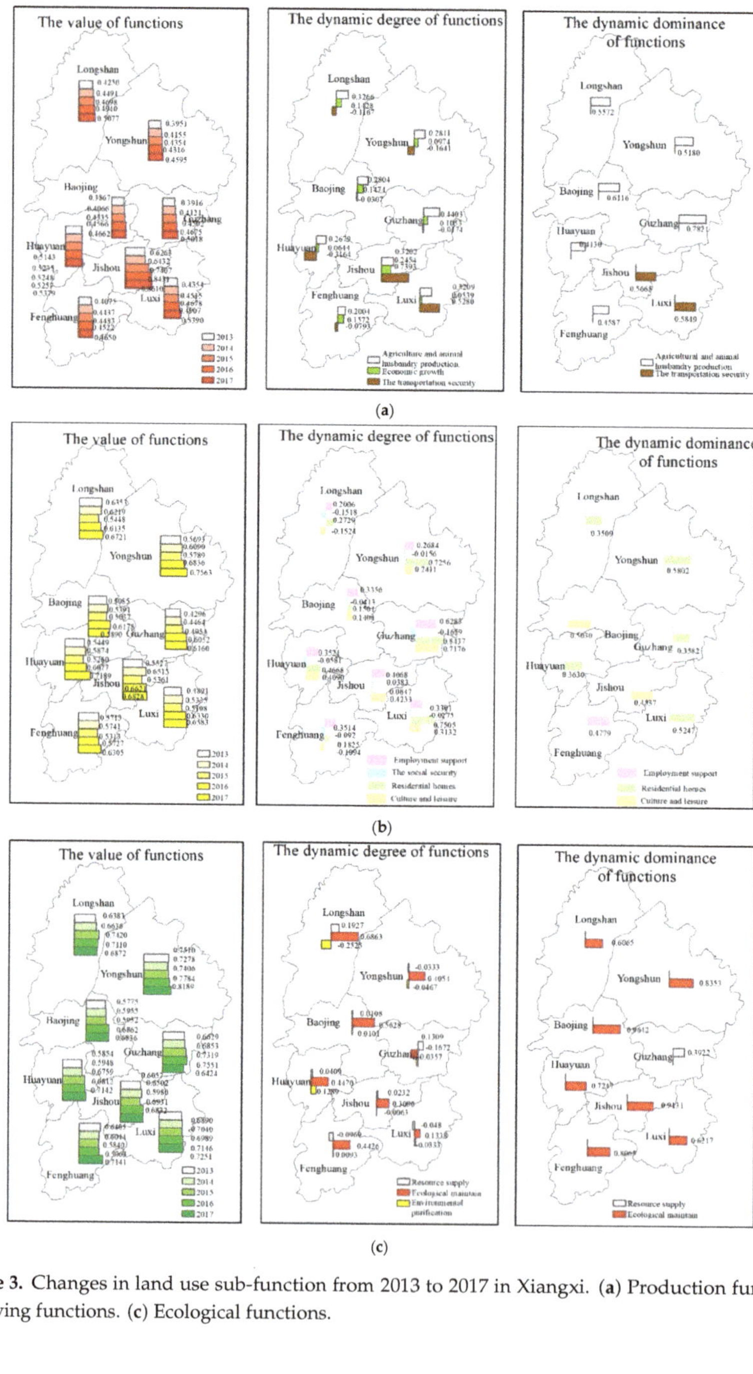

Figure 3. Changes in land use sub-function from 2013 to 2017 in Xiangxi. (**a**) Production functions. (**b**) Living functions. (**c**) Ecological functions.

3.2.1. The Production Function

Production grew slowly, with unbalanced development over the whole region. The production functions of the region were generally not high and all increased slowly, with a faster growth rate in the south and the middle than in the north (Figure 3a). Jishou demonstrates the best production function and the fastest development, because it is the political, economic, and cultural center in that region, with relatively superior production conditions and rapid development in transportation and economy. The production function values of Baojing, Guzhang, Yongshun, Longshan, Huayuan, and Fenghuang were always at a low level and have changed slowly, which is due to the fact that these areas have been dominated by traditional agriculture and animal husbandry production, which is a decline in transportation and slow economic growth. The function of security from transportation was clearly differentiated over the region, but the spatial differentiation of agricultural and animal husbandry production and economic growth was not so clear. In 2012, the Aizhai bridge was opened to traffic and the Chongqing-Hunan highway was completed, which strengthened the economic connection and material exchange between Jishou, Chongqing, and Changsha. This promoted development in the flow of people and cargo, which rendered Jishou the regional transportation hub. Luxi is the only area through which this route passes, and the proportion of freight volume moving through the state has increased significantly (from 18.02% to 35.98% within five years). Therefore, Jishou and Luxi have both undergone a relatively rapid increase in traffic function. The expressways connecting other areas in the prefecture were developed later, such as that of LongYong and YongJi, which were completed in 2016 and 2017, respectively, and this has limited the external exchanges between these regions to a certain extent. Within the 5 years studied, the per capita agricultural output value and per capita output value from animal husbandry both increased. The agricultural and animal husbandry production of Guzhang underwent the largest increase rate (28.79%). The economic density of all areas increased, and the proportion of the secondary and tertiary industries has also risen in most counties (Huayuan and Luxi declined slightly). However, the overall economic growth of such industries within Xiangxi was slow during the study period. Jishou has always had a high proportion of secondary and tertiary industries (with 10.58% higher than the average level of the whole state in 2017), and the economic density has been growing rapidly (with a dynamic degree of 43.13%), so the dynamic degree of economic growth was the highest (24.54%).The production function of the whole prefecture focuses on agriculture and animal husbandry. Except for Jishou and Luxi with clear benefits from the development of transportation, the production functions of agriculture, and animal husbandry have developed rapidly in other areas.

3.2.2. The Living Function

The living function fluctuated and tended toward a balance. The values for the living function of the land in the eight areas of Xiangxi have fluctuated, but have increased overall (Figure 3b). The living function of Yongshun has always been at the forefront (reaching the highest value of 0.7563 in 2017). Longshan and Jishou also have certain advantages in this area. The value for the living function in Guzhang was always low, but growth has been significant and the gap with other regions has narrowed. The level of employment support, cultural leisure, and residential homes have all improved to a certain degree. However, the urban-rural income balance index and the incidence of poverty in all areas of Xiangxi (except Jishou) have also increased to some extent. The wealth gap between urban and rural areas has widened, and the social security function has deteriorated. With the intensification of the urbanization process, the imbalance between urban and rural development has become increasingly prominent, and the incidence of rural poverty has increased. In recent years, local governments have encouraged the development of commercial housing, along with the renovation of dilapidated houses in rural areas and shanty towns in urban areas. The living conditions and living environment of residents have constantly improved. During the study period, the growth rate per capita in the area of Xiangxi was 13.77% to 112.61%, with the highest in Guzhang. With increasing communication with the outside world, job opportunities provided by tea production and tourism in Xiangxi have increased, and the land employment guarantee function is, therefore, constantly optimized. The growth rate in the southern and central regions was faster than that of the north, which can be associated with the development of the regional production function. The improvement to production and the optimization of the industrial structure have enhanced the ability for increased employment opportunities. The functional level of cultural and leisure in regions other than Longshan and Fenghuang has been improved to a certain extent. The per capita green park space in Guzhang has increased 2.23 times, which is a key factor in the rapid growth of the cultural and leisure functions. The center of the living function was mainly residential homes (Longshan, Yongshun, Huayuan, Guzhang, and Luxi), which was followed by the cultural and leisure functions (Jishou, Baojing), and the employment support function (Fenghuang).

3.2.3. The Ecological Function

The ecological function has changed slowly, and regional differences in this functional level have narrowed. The level of ecological function has slightly decreased in Guzhang, whereas it has increased in all other areas, albeit not significantly (Figure 3c). The value for the ecological function in land use over all eight areas of Xiangxi was always high, and was closely related to the significant supply of resources (due to the sparse population) and ecological maintenance (from high green coverage and forest coverage). The ecological advantage of land use in Yongshun was relatively clear because the three associated functions of resource supply, ecological maintenance, and environmental purity in Yongshun were all excellent. (In 2017, the three functions of Yongshun in all areas were respectively ranked 2, 2, and 1). Fenghuang was found to require improvement, since the level for ecological function in the area during 2015 and 2016 was the lowest. The standard deviation of the ecological function for each area decreased as a whole from 0.054 (2013) to 0.048 (2017), which reflects that the regional ecological function tended to be balanced. The three functions of ecological function (resource supply, ecological maintenance, and environmental purification) were clearly differentiated during the study period. With the promotion of ecological environmental construction by a local government, the rate of increase in green coverage and forest in all regions has been greatly improved, which means that the growth of ecological maintenance function is the clearest. The growth rate was 13.38% to 68.63% (except for Guzhang). In some counties (Fenghuang, Luxi, Yongshun), the resource supply capacity decreased, which reflects the lack of regional reserve resources and restricts the consequences of regional development. With the increased input of agricultural chemicals and the high intensity use of chemical fertilizers (taking Huayuan as an example, the index increased by 170.83% in the study period), the ecological purification capacity decreased, which posed a significant threat to the ecological environment of the regional land resources. In addition to the high dynamic dominance of resource supply in Guzhang, the ecological function of land use in the other seven areas focused on ecological maintenance.

3.3. *The Analysis of Obstacle Factors*

According to Equations (5) and (6), the degree of obstacles affecting both single functions and sub-functions was calculated, and the obstacle factors for the sub-functions and the total function were, respectively, obtained (Table 3). The obstacle factors of a sub-function were obtained in the following way: the obstacle degree of 10 single functions was ranked from high to low, with 1–3 as obstacle factors, 4–7 as intermediate factors, and the rest as dominant factors. The obstacle factors for total function were obtained according to the average value of the three sub-functions (production, living, and ecological functions). Those higher than the average value were counted as obstacle factors.

Table 3. The main obstacles of land use function in Xiangxi.

	Year	Jishou	Luxi	Fenghuang	Huayuan	Baojing	Guzhang	Yongshun	Longshan
The obstacles of total function	2013	LF	PF, LF	PF	PF	PF	PF, LF	PF	PF
	2014	PF	PF, LF	PF	PF	PF	PF, LF	PF	PF
	2015	LF, EF	PF, LF	PF	PF, LF	PF	PF, LF	PF	PF, LF
	2016	LF, EF	PF	PF	PF	PF	PF	PF	PF
	2017	LF, EF	PF	PF	PF	PF	PF	PF	PF
The obstacles of sub-function	2013	PF1, LF1, EF3	PF1, PF2, PF3	PF1, PF2, PF3	PF1, PF2, PF3	PF2, PF3, EF1	PF2, PF3, LF1	PF2, PF3, LF3	PF2, PF3, EF2
	2014	PF1, PF2, EF3	PF1, PF2, PF3	PF1, PF2, PF3	PF1, PF2, PF3	PF2, PF3, LF3	PF2, PF3, LF3	PF2, PF3, LF3	PF2, PF3, LF3
	2015	PF1, LF3, EF3	PF2, PF3, LF3	PF2, PF3, EF3	PF1, PF2, PF3	PF2, PF3, LF3	PF2, PF3, LF3	PF2, PF3, LF3	PF2, PF3, LF3
	2016	PF1, LF2, EF3	PF2, PF3, LF1	PF2, PF3, EF3	PF2, PF3, EF1	PF2, PF3, EF3	PF2, PF3, LF1	PF2, PF3, LF3	PF2, PF3, LF3
	2017	PF1, LF2, EF3	PF2, PF3, LF1	PF2, PF3, EF3	PF1, PF2, PF3	PF2, PF3, LF3	PF2, PF3, EF2	PF2, PF3, LF2	PF2, PF3, EF3

Note: PF is production function, LF is living function, and EF is ecological function. PF1 is the function for agriculture and animal husbandry production, PF2 is the function for economic growth, and PF3 is the function for transportation security. LF1 serves as employment support function, LF2 as the social security function, LF3 as cultural and leisure function, LF4 as the residential home function, EF1 as the resource supply function, EF2 as the ecological maintenance function, and EF3 as the environmental purification function.

According to the frequency of total dysfunction factors in the past five years, this study divided the land use function of the eight areas of Xiangxi into three types: living-ecological obstacle, production-living obstacle, and production obstacle.

3.3.1. The Living-Ecological Type of Obstacle

Jishou is representative of the living-ecological obstacle type. Except for 2014, the degree of obstacles against production in Jishou was slightly higher than average (35%). In other years, the obstacle degree of production and living dysfunction (or one of the two) was notably higher, with significant room for improvement. From the perspective of this sub-function, Jishou suffers from a high degree of obstacles to the agricultural and animal husbandry production function (PF1), social security function (LF2), and environmental purification function (EF3). Statistics show that, in the recent five years, the per capita agricultural output value and per capita animal husbandry output value in Jishou were both in the middle and lower reaches. In 2013, the two indicators were respectively 2334.70 yuan/person and 609.58 yuan/person, which are both lower than the average value of the whole state. The urban-rural income balance index in Jishou was low, in the range of 31.29% to 34.86%, which was lower than the statewide rankings. Meanwhile, within five years, the amount of chemical fertilizers and pesticides applied per unit of cultivated land in Jishou was two to four times higher than that of other areas, which results in greater pressure on the ecological environment of the land, and the capacity for environmental purity was, therefore, weak.

3.3.2. The Production-Living Type of Obstacle

Luxi and Guzhang were both undergoing obstacles to production and living. The main factors restricting the land versatility of the two areas were the economic growth function (PF2), the transportation security function (PF3), the employment support function (LF1), and the cultural leisure function (LF3). According to statistical data, in 2017, the economic density of Luxi and Guzhang accounted for 25.60% and 13.86% of Jishou, respectively, and the low output value per unit of land was one of the main factors restricting land use in the two areas. In 2017, passenger transport turnover in Luxi and Guzhang accounted for 7.58% and 4.98% of the total, respectively, and the backward transportation facilities also limited the development of the region. The average salary of workers in both areas was not high (all were less than 88% of Jishou in 2017), which was less attractive for the labor force. In addition, farmers have to leave the area to work and do business frequently, and there are few rural employees (Guzhang retains less than one-third of the rural workers of Yongshun in 2017). The per capita green park space and the proportion of residents' cultural, educational, and entertainment expenditure in the two areas are less than the average level of the whole state, and there remains a significant room for improvement in land use for the cultural and leisure function (LF3).

3.3.3. The Production Type of Obstacle

The geographical location of Xiangxi is in a remote region that restricts the efficient development and utilization of the land. Fenghuang, Huaguan, Baojing, Yongshun, and Longshan all suffer from production obstacles. The agriculture and animal husbandry production function (PF1), economic growth function (PF2), transportation support function (PF3), and environmental purification function (EF3) are all hindered by significant obstacles. The agricultural and animal husbandry production of Baojing, Yongshun, and Longshan have certain advantages, especially the agricultural output per capita and animal husbandry output per capita of Longshan and Yongshun in the last five years ranking in the top 3. However, the development of regional agricultural production is extensive, and is dominated by traditional agriculture and animal husbandry. The the level of productivity is low. The Wuling mountain area is high terrain, covering a large area with few people. It is limited by the influence of the surrounding economic influence, which limits the development of the economy and transport in the five areas. For example, in 2017, the sum of economic density of the five areas was 1.26 times that of Jishou. The sum of road area per capita was 1.04 times that of Jishou, and the sum of freight turnover was only 0.31 times that of Jishou. Although the land use intensity of Xiangxi is low, the ecological function has certain advantages. However, in the last five years, the large input of agricultural chemicals in these five areas (in 2017, the average amount of fertilizer applied in the five areas increased by 1.52 times that of 2013) led to an enhancement of human disturbances to the land and an increase in the environmental purification dysfunction degree of the land.

As a whole, the obstacle factors for land use in Xiangxi are part of the dynamic change in spatial and temporal dimensions. Although there are spatial and temporal differences in the types and degrees of obstacle factors, the main type of obstacles against efficient land use in Xiangxi are due to production, agriculture and animal husbandry (PF1), economic growth (PF2), transportation (PF3), and environmental purity (EF3), which are still the main factors restricting the overall land use function in Xiangxi.

4. Discussion

This study analyzed the multi-functional spatial and temporal evolution in the pattern of land use in eight areas of Xiangxi during the period of 2013 to 2017. This is based on an evaluation index system using production, living, and ecology as the main obstacle factors, which is divided into three obstacle types. To restrain the obstacles and promote sustainable land use in Xiangxi, it is necessary to learn from the experiences of other regions.

4.1. Comparison in Land Use Management

An assessment of 150 agricultural grasslands in Germany found that land intensive use increased feed production, but also led to loss of biodiversity and changes of land functions [35]. To improve the social and environmental functions of land as well as measurements of land management, irrigation and fertilization must be improved. Soil erosion, soil pollution, and soil degradation were serious problems in the corn belt of the United States, and agricultural production was seriously threatened. To coordinate the contradiction between grain production and environmental protection, and to achieve multi-functional agriculture, it was necessary to integrate market, technology, and policy measures [36]. From 1990 to 2010, the total land function of Guangzhou increased [37], but agricultural production and resource supply capacity became the main factors restricting its development. Consequently, protecting farmland and improving resource utilization efficiency were the main measures to restrain functional impairment.

From the above cases, we can see that multi-functional land use is an inevitable trend. It is beneficial for the long-term use of land resources to formulate differentiated measures according to regional situations.

4.2. Policy Suggestions for the Land Use in Xiangxi

Based on the principle of a specific policy enforced in a certain city, we proposed the following suggestions.

(1) Jishou, which is a city with obstacles to both living and ecology, aims to narrow the urban-rural gap and reduce agricultural non-point source pollution. We will coordinate the construction of an urban and rural land use policy system, and establish a mobile market for urban and rural land elements. We will increase investment and policy for the support of infrastructure in rural areas, where the level of education, medical care, and social security will be enhanced. This will ensure that urban and rural residents both enjoy equal opportunities from development and receive equal protection for their rights. We will strictly protect basic farmland, delimit areas where basic farmland is concentrated, and transform medium-and low-yielding farmland. We will expand the planting of flue-cured tobacco, Chinese herbal medicines, fruits, and other cash crops, develop green and organic agriculture, and reduce pollution from non-point agricultural sources.

(2) To overcome the production-living barriers of Luxi and Guzhang, we should promote the development of production and public utilities. We will improve the distribution of residential areas and guide rural settlements to gather in towns and cities. We will make reasonable plans for the construction of transportation, ensure development of important transport routes such as the Zhang-Ji-Huai corridor, and promote the all-around development of highways, railways, and air transportation, etc. Natural resources and national cultural endowments will be instrumental in developing regional tourism. Featured agricultural products (bacon, alpine vegetables, etc.) will be intensively produced, to extend the industrial chain and expand related industries, so as to promote economic growth and improve the income level of employees. We will improve the land supply system, the planning and decision-making system, and optimize the structure of construction land. Land for cultural and recreational facilities in cities and towns shall be guaranteed. The renovation of old cities will be strengthened to create a beautiful urban environment. We will attract social funds, to increase the input for the livelihood of residents, and improve the functions of the central urban areas.

(3) The top priority for other counties with production barriers is given toward developing production and to quickly eliminating poverty. The mode of production in which smallholders operate should be changed by promoting the consolidation of farmland to achieve scale operation. We will develop modern urban agriculture and tourism agriculture, perfecting the agricultural industry system to improve the diversification of agricultural functions. The measures concerning transportation and economic development have been mentioned in the production-living obstacle countermeasures, and will not be repeated.

4.3. Limitation and Future Research

The research of land functions in Xiangxi is of great importance for optimizing land-use patterns, adjusting land use structure, and promoting coordinated development of regional economy and ecology. However, due to the difficulty of obtaining data, we did not select a countryside or a town as the case for analysis. No information of a specific case was obtained. Simultaneously, the research period is relatively short: only five years. Thus, there was a lack of a long-term historical evolution process analysis. Therefore, research of typical cases and long-term scales should be further studied.

5. Conclusions

The main conclusions are as shown below.

(1). With the continuous optimization of the overall functional level of land use in Xiangxi, the spatial heterogeneity was increasingly clear. The coordination of land use has changed to a certain extent, and the overall situation was better in the south than in the middle or the north.

(2). The production function of land use in Xiangxi was observed to be slowly increasing with more rapid growth in the southern and central regions than in the northern regions. The spatial differentiation between these areas was clear.

(3). Obstacles against efficient land use in Xiangxi can be divided into three types: the living-ecological obstacle (Jishou), the production-living obstacle (Luxi, Guzhang), and the production obstacle (Fenghuang, Huayuan, Baojing, Yongshun, and Longshan).

(4). To promote the multi-functional use of land, we should formulate differentiated policies for different regions. In the future, we can choose a smaller area or a longer time scale for further research.

Author Contributions: Conceptualization, Q.-Y.Y. and H.X. Methodology, Q.-Y.Y. Investigation, H.X. Writing—original draft preparation, H.X. Writing—review and editing, K.-c.S. and Z.-X.Z.

Funding: The Key Project of Chongqing Key Research Base of Humanities and Social Sciences, grant number 14SKB014, funded this research.

Acknowledgments: The authors thank the X.X. Statistical Bureau for providing the data used in this research work. The authors thank the anonymous reviewer and W.Y.-Teacher of Southwest University, for their valuable comments.

Conflicts of Interest: The authors declare no conflict of interest.

References

1. Verburg, P.H.; Van De Steeg, J.; Veldkamp, A.; Willemen, L. From land cover change to land function dynamics: A major challengeto improve land characterization. *J. Environ. Manag.* **2009**, *90*, 1327–1335. [CrossRef] [PubMed]
2. Steffen, W.; Crutzen, P.J.; McNeill, J.R. The anthropocene: Are humans now overwhelming the great forces of nature. *AMBIO* **2007**, *36*, 614–621. [CrossRef]
3. Scholes, R.J.; Biggs, R. A biodiversity intactness index. *Nature* **2005**, *434*, 45–49. [CrossRef] [PubMed]
4. Vörösmarty, C.J.; McIntyre, P.B.; Gessner, M.O.; Dudgeon, D.; Prusevich, A.; Green, P.; Glidden, S.; Bunn, S.E.; Sullivan, C.A.; Liermann, C.R.; et al. Global threats to human water security and river biodiversity. *Nature* **2010**, *467*, 555–561. [CrossRef] [PubMed]
5. Patz, J.A.; Daszak, P.; Tabor, G.M.; Aguirre, A.A.; Pearl, M.; Epstein, J.; Wolfe, N.D.; Kilpatrick, A.M.; Foufopoulos, J.; Molyneux, D.; et al. Unhealthy Landscapes: Policy recommendations on land use change and infectious disease emergence. *Environ. Health Perspect.* **2004**, *112*, 1092–1098. [CrossRef] [PubMed]
6. Ellis, E.C. Ecology in an anthropogenic biosphere. *Ecol. Monogr.* **2015**, *85*, 287–331. [CrossRef]
7. Matthews, R.B.; Gilbert, N.G.; Roach, A.; Polhill, J.G.; Gotts, N.M. Agent-based land-use models: A review of applications. *Landsc. Ecol.* **2007**, *22*, 1447–1459. [CrossRef]
8. Heisterman, M.; Christoph, M.; Ronneberger, K. Land in sight? Achievements, deficits and potentials of continental to global scale land-use modeling. *Agric. Ecosyst. Environ.* **2006**, *114*, 141–158. [CrossRef]
9. Wiggering, H.; Dalchow, C.; Glemnitz, M.; Helming, K.; Müller, K.; Schultz, A.; Stachow, U.; Zander, P. Indicators for multifunctional land use—Linking socio-economic requirements with landscape potentials. *Ecol. Indic.* **2006**, *6*, 238–249. [CrossRef]
10. Pérez-Soba, M.; Petit, S.; Jones, L.; Bertrand, N.; Briquel, V.; Omodei-Zorini, L.; Contini, C.; Helming, K.; Farrington, J.H.; Mossello, M.T.; et al. *Land Use Functions—A Multifunctionality Approach to Assess the Impact of Land Use Changes on Land Use Sustainability*; Springer: Berlin/Heidelberg, Germany, 2008; pp. 375–404.
11. SchoBer, B.; Helming, K.; Wiggering, H. Assessing land use change impacts: A comparison of the SENSOR land use function approach with other frameworks. *J. Land Use Sci.* **2010**, *5*, 159–178.
12. De, G.R. Function-analysis and valuation as a tool to assess land use conflicts in planning for sustainable, multifunctional landscapes. *Landsc. Urban Plan.* **2006**, *75*, 175–186.
13. David, A. Multifunctionality, Agricultural Policy, and Environmental Policy. *Agric. Resour. Econ. Rev.* **2004**, *33*, 8–17.

14. Aubry, C.; Ramamonjisoa, J.; Dabat, M.H.; Rakotoarisoa, J.; Rakotondraibe, J.; Rabeharisoa, L. Urban agriculture and land use in cities: An approach with the multi-functionality and sustainability concepts in the case of Antananarivo (Madagascar). *Land Use Policy* **2012**, *29*, 429–439. [CrossRef]
15. Kiminami, L.; Kiminami, A.; Furuzawa, S. Impacts of multi-functionality of urban agriculture on the CCs in Japan. *Asia Pac. J. Reg. Sci.* **2018**, *2*, 507–527. [CrossRef]
16. Tipraqsa, P.; Craswell, E.T.; Noble, A.D.; Schmidt-Vogt, D. Resource integration for multiple benefits: Multifunctionality of integrated farming systems in Northeast Thailand. *Agric. Syst.* **2007**, *94*, 694–703. [CrossRef]
17. Mittenzwei, K.; Fjellstad, W.; Dramstad, W.; Flaten, O.; Gjertsen, A.K.; Loureiro, M.; Prestegard, S.S. Opportunities and limitations in assessing the multifunctionality of agriculture within the CAPRI model. *Ecol. Indic.* **2007**, *7*, 827–838. [CrossRef]
18. Liu, C.; Xu, Y.; Huang, A.; Liu, Y.; Wang, H.; Lu, L.; Sun, P.; Zheng, W. Spatial identification of land use multifunctionality at grid scale in farming-pastoral area: A case study of Zhangjiakou City, China. *Habitat Int.* **2018**, *76*, 48–61. [CrossRef]
19. Van Noordwijk, M.; Tata, H.L.; Xu, J.; Dewi, S.; Minang, P.A. *Segregate or Integrate for Multifunctionality and Sustained Change through Rubber-Based Agroforestry in Indonesia and China*; Springer: Dordrecht, The Netherlands, 2012; pp. 69–104.
20. Hughes, M.; Jones, R. From productivism to multi-functionality in the Gascoyne-Murchison Rangelands of Western Australia. *Rangel. J.* **2010**, *32*, 175–185. [CrossRef]
21. Barbieri, C.; Valdivia, C. Recreation and agroforestry: Examining new dimensions of multifunctionality in family farms. *J. Rural Stud.* **2010**, *26*, 465–473. [CrossRef]
22. Krajewski, P. Monitoring of landscape transformations within landscape parks in Poland in the 21st century. *Sustainability* **2019**, *11*, 2410. [CrossRef]
23. Krajewski, P.; Solecka, I.; Karol, M. Forest landscape change and preliminary study on its driving forces in Ślęża Landscape Park (Southwestern Poland) in 1883–2013. *Sustainability* **2018**, *10*, 4526. [CrossRef]
24. Lund, J.F.; Jensen, F.S. Is recreational hunting important for landscape multi-functionality? Evidence from Denmark. *Land Use Policy* **2017**, *61*, 389–397. [CrossRef]
25. Zhang, W.; Ricketts, T.H.; Kremen, C.; Carney, K.; Swinton, S.M. Ecosystem services and dis-services to agriculture. *Ecol. Econ.* **2007**, *64*, 253–260. [CrossRef]
26. Hanson, J.D.; Liebig, M.A.; Merrill, S.D.; Tanaka, D.L.; Krupinsky, J.M. Dynamic cropping systems: Increasing adaptability amid an uncertain future. *Agron. J.* **2007**, *99*, 939–943. [CrossRef]
27. Andersen, P.S.; Vejre, H.; Dalgaard, T.; Brandt, J. An indicator-based method for quantifying farm multifunctionality. *Ecol. Indic.* **2013**, *25*, 166–179. [CrossRef]
28. Gaodi, X.; Lin, Z.; Caixia, Z.; Xiangzheng, D.; Koenig, H.J.; Tscherning, K.; Helming, K. Assessing the multi-functionalities of land use in China. *J. Resour. Ecol.* **2010**, *1*, 311–318.
29. Jordan, N.; Warner, K.D. Enhancing the multifunctionality of US agriculture. *Bioscience* **2010**, *60*, 60–66. [CrossRef]
30. Zhang, Y.; Yan, J.; Jiang, P.; Yang, N. Normal cloud model based evaluation of land resources ecological security in Hubei province. *Trans. Chin. Soc. Agric. Eng.* **2013**, *29*, 252–258.
31. König, H.J.; Podhora, A.; Helming, K.; Zhen, L.; Wang, C.; Wübbeke, J.; Baumeister, T.; Du, B.; Yan, H. Confronting in-ternational research topics with stakeholders on multifunc-tional land use: The case of Inner Mongolia, China. *iForest* **2014**, *7*, 403–413. [CrossRef]
32. Paracchini, M.L.; Pacini, C.; Jones, M.L.; Pérez-Soba, M. An aggre-gation framework to link indicators associated with multi-functional land use to the stakeholder evaluation of policyoptions. *Ecol. Indic.* **2011**, *11*, 71–80. [CrossRef]
33. Tong, O.; Shao, S.; Zhang, Y.; Chen, Y.; Liu, S.L.; Zhang, S.S. An AHP-based water-conservation and waste-reduction indicator system for cleaner production of textile-printing industry in China and technique integration. *Clean Technol. Environ. Policy.* **2012**, *14*, 857–868. [CrossRef]
34. Li, S.H. Research on evaluation system of new energy vehicles based on AHP. *Appl. Mech. Mater.* **2013**, *260*, 321–325. [CrossRef]
35. Allan, E.; Manning, P.; Alt, F.; Binkenstein, J.; Blaser, S.; Blüthgen, N.; Böhm, S.; Grassein, F.; Hölzel, N.; Klaus, V.H.; et al. Land use intensification alters ecosystem multifunctionality via loss of biodiversity and changes to functional composition. *Ecol. Lett.* **2015**, *18*, 834–843. [CrossRef]

36. Jordan, N.; Warner, K.D. *Towards Multifunctional Agricultural Landscapes for the Upper Midwest Region of the USA*; Wiley: New York, NY, USA, 2013; pp. 139–156.
37. Wang, F.; Dong, Y.X. Dynamic evaluation of land use functions based on grey relation projection method and diagnosis of its obstacle indicators: A case study of Guangzhou city. *J. Nat. Resour.* **2015**, *30*, 1698–1713.

© 2019 by the authors. Licensee MDPI, Basel, Switzerland. This article is an open access article distributed under the terms and conditions of the Creative Commons Attribution (CC BY) license (http://creativecommons.org/licenses/by/4.0/).

applied sciences

MDPI

Article

Analyzing Spatial Community Pattern of Network Traffic Flow and Its Variations across Time Based on Taxi GPS Trajectories

Wenhao Yu [1,2], Menglin Guan [1,*] and Zhanlong Chen [1]

1 School of Geography and Information Engineering, China University of Geosciences, Wuhan 430074, China; yuwh@cug.edu.cn (W.Y.); chenzl@cug.edu.cn (Z.C.)

2 State Key Laboratory of Resources and Environmental Information System, Beijing 100101, China

* Correspondence: gmlcug@gmail.com; Tel.: +86-27-67883728

Received: 10 April 2019; Accepted: 14 May 2019; Published: 18 May 2019

Abstract: The transport system is a critical component of the urban environment in terms of its connectivity, aggregation, and dynamic functions. The transport system can be considered a complex system due to the massive traffic flows generated by the spatial interactions between land uses. Benefiting from the recent development of location-aware sensing technologies, large volumes of traffic flow data (e.g., taxi trajectory data) have been increasingly collected in spatial databases, which provides new opportunities to interpret transport systems in cities. This paper aims to analyze network traffic flow from the perspective of the properties of spatial connectivity, spatial aggregation, and spatial dynamics. To this end, we propose a three level framework to mine intra-city vehicle trajectory data. More specifically, the first step was to construct the network traffic flow, with nodes and edges representing the partitioned regions and associated traffic flows, respectively. We then detected community structures of network traffic flow based on their structural and traffic volume properties. Finally, we analyzed the variations of those communities across time for the dynamic transport system. Through experiments in Beijing city, we found that the method is effective in interpreting the mechanisms of urban space, and can provide references for administrative divisions.

Keywords: traffic flows; taxi trajectory; float cars; spatial community; transport system

1. Introduction

Transport systems are of great importance to urban environments for their connectivity, aggregation, and dynamic functions. Land uses are connected by the transport network to improve the accessibility of human activities. Considering the spatial heterogeneity of traffic flows, multiple land uses are also attracted by each other, showing an agglomeration (or aggregation) pattern in the space. In addition, such characteristics of a transport system depend largely on the temporal dimension. Therefore, mining the connectivity, aggregation, and dynamic patterns of transport flows can be helpful for revealing traffic structures and the associated mechanisms of socioeconomic phenomena, e.g., logistics, neighborhood, living habitats, and urban function zones [1–3].

In reality, the regionalization of urban areas is often non-adjacent. For example, working areas and residential areas belong to the same group in terms of their functions, while in the physical space, they are often distant from each other. Therefore, only using geometric indicators such as geographic distance to measure the connectivity of land uses of interest is limited, and the potential solution could come from the function space of transport. Instead of the static condition of geometric space, a transport system implies the real interactions between land uses across space and time. For example, in the morning, the interaction between residential areas and working areas is much intense, while at lunch time, the interaction between working areas and catering service areas is more intense. In addition, the interactions of land uses in terms of traffic flows can reveal the traffic conditions on different routes in the urban space. It is believed that different routes serve different roles in the daily transportation. For example, the routes connecting residential areas and working areas are likely to be chosen by commuters in the morning. Besides the main roads, some minor roads, for example detours, could also be favored by residents. However, how to extract the functional associations between land uses remains a challenging task.

Considering the importance of transport systems, many studies have been done to discover the hidden regularities in urban transportation. However, due to limited data sources, it is rather difficult to identify the changes of traffic flows across space and time [1,4]. In addition, China's transport infrastructures are developing very fast, and the associated transport systems are becoming more complex and dynamic. It is necessary to study the transport system in a more effective and timely way. Benefiting from the recent development of location-aware sensing technologies, there is an unprecedented opportunity for us to obtain big traffic data with agent trajectory information [5,6]. For example, nowadays, most of vehicles are equipped with a global positioning system (GPS), which records the locations and other semantic properties (e.g., speed and direction) of agents. Analyzing these data within the context of a transport system could provide a detailed view of the traffic flow, and then reveal typical spatial interaction patterns in the urban space, e.g., popular routes for driving, associations among land uses, and the spatial structures of urban space.

For example, Ahas et al. [7] used mobile phone positioning data to explore the movement patterns of suburban commuters in Tallinn, Estonia. They found that there is a remarkable temporal rhythm to respondents' locations. Based on mobile phone data, Sevtsuk et al. [8] also discovered that there is significant temporal regularity in human mobility. Other data sources can be also used to analyze traffic flows, e.g., location data of buses [9], smart card transaction data of subways [10], and taxi trajectory data [11,12].

Compared to other modes of transport, the taxi trajectory has no limitations of a fixed line, and thus is more flexibly able to reflect real traffic flows in an urban environment. There are also many relevant studies analyzing taxi trajectory data under different application contexts. For example, Zheng et al. [13] analyzed taxi trajectory data to construct interaction relationships between local regions, and then applied the result to assist in city planning. Guo et al. [14] and Yuan et al. [15] tried to extract the operation status of a traffic system from taxi trajectory data. To identify the city structure, Zhou et al. [16] proposed a field-based data clustering analysis method to detect the changing patterns of constant hotspot areas and inconstant hotspot areas. Since the pick-up and drop-off points of taxi trajectories often imply facilities of interest, Yue et al. [17] used taxi trajectory data to discover the attractive areas that people often visit, e.g., hot shopping and leisure land uses or living and working areas. Recently, Liu et al. [18] proposed an approach to identify traffic congestion regions and their spatiotemporal distributions from taxi trajectory data. Liu et al. [19] viewed the trips of taxis as a displacement in the random walk model, and found that the distribution of directions of taxi trajectories in Shanghai shows a characteristic northeast east–southwest west dominant direction. In addition, they implemented the Monte Carlo simulation and found that geographical heterogeneity leads to a faster observed decay of trips, and the distance decay effect makes the spatial distribution of trips more concentrated in the urban area. Liu et al. [20] proposed the use of spatially embedded networks and network analysis techniques to model intra-city spatial interactions. Zhou et al. [21] allocated Origin/Destination points to land use parcels for describing regional activities, and then combined a series of relevant indicators to explore the land use patterns of Wuhan city. Pan et al. [22] analyzed the characteristics of the pick-up/drop-off points extracted from taxi GPS trajectories, and, based on these features of pick-up/drop-off points, extracted the regular patterns that correspond to the land-use classes within different regions in Hangzhou city. In addition to these aspects, taxi trajectory data can be also used for human mobility pattern mining [5,23,24] and environmental pollution analysis [25,26].

Most previous studies have focused on the connectivity and aggregation of land uses, and few studies have been conducted to comprehensively understand the spatio-temporal community structures of a transport system. With continuous transportation development, there is an increasingly urgent demand for analysis of the spatial structures of transport flows and their variations across time. In general, a traffic flow can be considered a link that connects two land uses, and thus, with multiple flows of this type, a transport system can be modeled as a weighted graph. In this regard, we can introduce community detection and graph matching methods to explore the complex organization of land uses in a transport flow system [27,28].

In this study, we propose a three-levels framework to mine intra-city vehicle trajectory data and detect the spatio-temporal relationships between land uses in the traffic flow system. Therefore, the main contributions of this study are the following:

- Modeling the connectivity structure of traffic flows. We first constructed a spatially embedded network to model the connectivity of land uses, in which the node represents the partitioned region, the edge represents the linkage between adjacent nodes, and the weight of the edge depends on the volume of traffic flows between the corresponding nodes.
- Extracting the aggregation patterns of land uses (i.e., nodes). Based on the network traffic flow, we then employed a community detection technique (i.e., K-Medoids clustering) to classify all the nodes. Instead of the simple geographic distance, our community detection method takes into account the real traffic volume and graph structure properties. In this way, the land uses that have a strong relationship could be aggregated in the same group.
- Analyzing the dynamic patterns of transportation communities. Since the transport system is a highly dynamic system, we propose a graph matching method to detect the change of network traffic flows across time. In this way, we can not only identify the structure of traffic flows across space, but also its variation across time.

The rest of this paper is organized as follows. Section 2 introduces the related definitions, and how to construct the network traffic flows and to generate the communities. To explore the variation of communities across time, this section then introduces an indicator to measure the similarity of two communities. Section 3 describes our extensive experiments based on the taxi trajectory data in Beijing, showing the potential of the proposed approach for transport system analysis and urban applications. Finally, Section 4 concludes the paper.

2. Community Detection across Space and Its Variation across Time

Our method improves the traditional K-Medoids method based on traffic flow volume and network structure properties. Firstly, we partition the study region into equally sized square cells, and then model each cell as a node and the connectivity between each pair of cells as a link. Based on the network traffic flow, we propose that the similarity of nodes can be calculated based on the attraction degree and structure similarity. In this way, community clustering can be implemented. Finally, we use the graph matching technique to calculate the similarity between the community structures within different time periods. Therefore, our method can be considered a spatio-temporal analysis tool.

2.1. Network Construction and Its Variation with Different Cell Sizes

2.1.1. Network Construction

Due to signal loss or degradation, taxi trajectories are usually recorded with spatial uncertainty. Even if a set of trajectory flows comes from (or drives to) the same regions, the recorded trajectory points are unlikely to share the same coordinates. Therefore, in order to extract the collective regularities from these massive trajectory points, our method proposes that the study region be partitioned into equally sized square cells, each of which represents a place in the urban space. In this way, each trajectory point could be assigned to its nearest cell, and a traffic flow consisting of multiple trajectory points could be represented as a set of partitioned cells. More specifically, by specifying the cell size k, the whole region would be transformed into a grid of size k × k. There are also other ways to partition the region, e.g., traffic analysis zones (TAZs). However, TAZs are constrained to a fixed scale, and our method can be used for multi-scale spatial analysis with different cell sizes (please see Section 3). In general, for fine scale applications, a small cell can be used, while for coarse scale applications, a large cell would be used. With the tessellation of space, we can then define the nodes and edges of network traffic flow as follows.

Definition 1. *(Node): Assuming the range of Euclidean coordinates of a cell C is* $\{[x_{min}, x_{max}], [y_{min}, y_{max}]\}$*, this cell can be modeled as a node only when there is at least one trajectory point (with coordinate* (x, y)*) falling into the cell. Specifically, this constraint can be formalized as follows.*

$$\begin{aligned} x_{min} &< x \leq x_{max} \\ y_{min} &< y \leq y_{max} \end{aligned} \tag{1}$$

Definition 2. *(Edge): Assuming there are n trajectory flows between nodes* C_1 *and* C_2*, the connectivity between* C_1 *and* C_2 *can be modeled as an edge with weight n. As presented in Figure 1, the weighted edge is used here to represent the flow transitions between cells (i.e., sub-regions).*

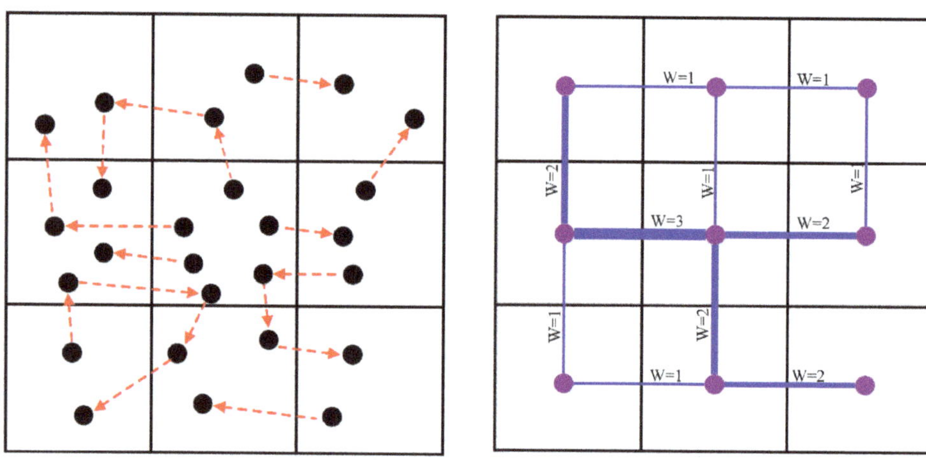

(a) (b)

Figure 1. Example of the traffic flow and network: (**a**) taxi trajectory data, (**b**) flow network with edge weight equal to corresponding probabilities of movement between sub-regions.

The movement of vehicles implies the complex interaction between land uses, and connects distant regions into an integrated system. Since the basic characteristic of this system is connectivity, we propose the construction of a spatially embedded network consisting of nodes (Definition 1) and edges (Definition 2) to represent traffic flow. Based on such a network, we can then employ graph analysis techniques (see Sections 2.2 and 2.3) to discover the hidden regularities in a transport system.

2.1.2. Variation of Network with Different Cell Sizes

Dividing the whole study area with different cell sizes would make the distribution patterns of trajectory points different, and would also lead to different results in the detection of communities. Figure 2 shows the effect of different cell sizes on the construction of network flows. Although Figure 2a,c have the same trajectory flow, they are divided by different cell sizes. As a result, the constructed networks have different granularities (Figure 2b,c). Therefore, with different cell sizes, we can observe the variation of the network from different scales of space.

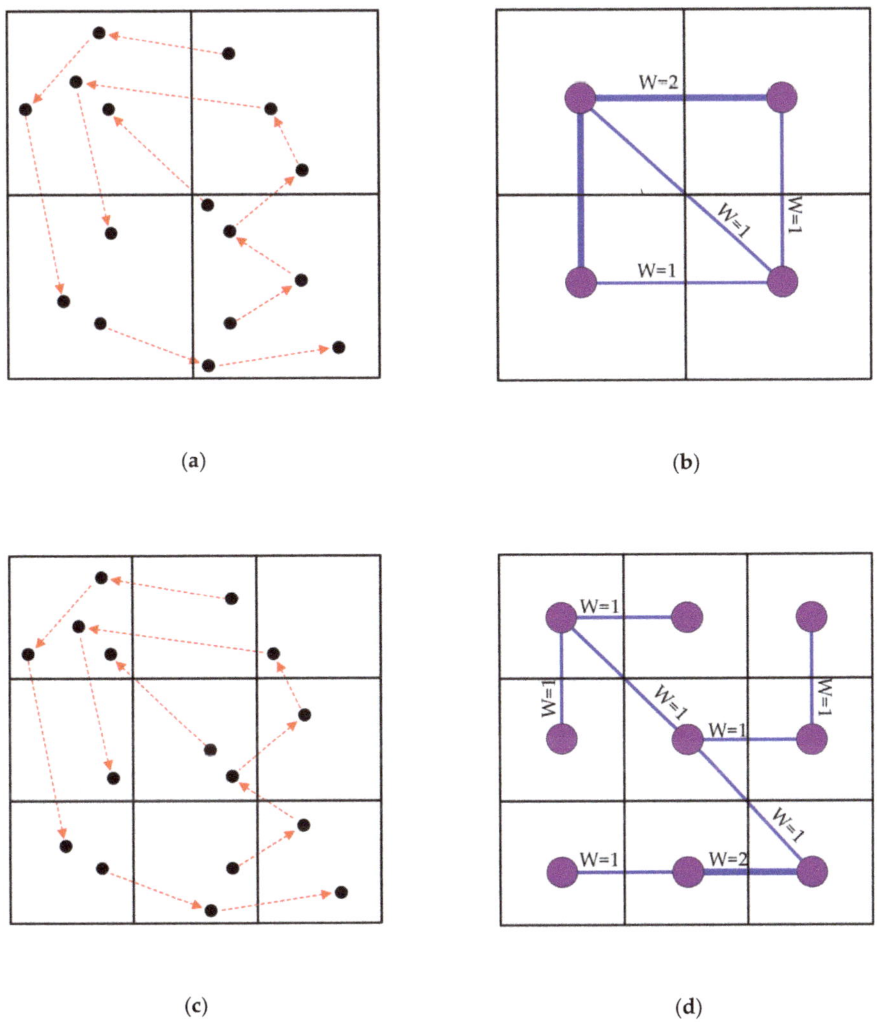

Figure 2. Variation of network with different cell sizes. (**a**,**c**) represent the variation of tessellation using different cell sizes, and (**b**,**d**) are the corresponding networks.

2.2. Community Detection across Space

2.2.1. Similarity of Nodes

Besides the connectivity property, a transport system has a spatial heterogeneity in urban space. In other words, some land uses are more attractive to each other in terms of transportation, and in the function space they form aggregation patterns, i.e., community. Such community could imply a popular route at a specific time, or an agglomeration of living areas and work areas. Generally, the more intense the traffic flow interaction between land uses is, the higher probability that the land uses have to be grouped together. Since the transport system is modeled as a spatially embedded network, we can then use the community detection method to extract the clustering patterns of land uses. It should be noted that, compared to the classic graph measures, the concept of community in our context has its own characteristics. Specifically, community detection in a network traffic flow should not only take into account the graph structure factor, but should also consider the traffic flow volumes among different regions. Therefore, we propose that the measures of attraction degree and structure similarity be integrated to group the nodes of network traffic flow.

Before defining the attraction degree, we first introduce the related concept, as follows.

Definition 3. *(Attraction factor): In a network traffic flow, the traffic volume characteristics between a pair of directly connected nodes reveal their closeness relationship, and we term this connection the attracting factor. For the directly connected nodes* N_1 *and* N_2*, their degrees* d_{N_1} *and* d_{N_2} *are defined as the number of their connected edges, respectively. Assuming the edge between* N_1 *and* N_2 *is* $E_{(N_1,N_2)}$ *with associated weight* $W_{(N_1,N_2)}$*, the attracting factor* $f_{(N_1,N_2)}$ *between* N_1 *and* N_2 *is calculated as follows:*

$$f_{(N_1,N_2)} = \ln\left(1 + \frac{d_{N_1}}{\sum_{j=1}^{d_{N_1}} W_{1j}} * W_{(N_1,N_2)}\right) \tag{2}$$

In Figure 3, there are six nodes, including N_1, N_2, N_3, N_4, N_5 and N_6, with their associated edges. In this community, $d_{N_1} = 3$, $W_{(N_1,N_5)} = 1$, $\sum_{j=1}^{d_{N_1}} W_{1j} = 3+2+1 = 6$, $f_{(N_1,N_5)} = \ln\left(1 + \frac{3}{6} * 1\right) = 0.4055$, $d_{N_5} = 3$, $W_{(N_5,N_1)} = 1$, $\sum_{j=1}^{d_{N_5}} W_{1j} = 3+1+1 = 5$, $f_{(N_5,N_1)} = \ln\left(1 + \frac{3}{5} * 1\right) = 0.4700$.

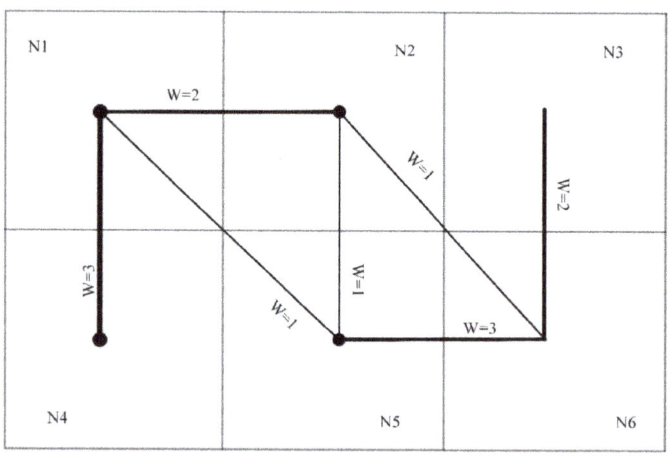

Figure 3. A sample of a community.

Definition 4. *(Attraction degree): Assuming nodes* N_1 *and* N_2 *are directly connected, their attracting degree can be measured as following.*

$$Attr_{(N_1,N_2)} = \frac{f_{(N_1,N_2)} + f_{(N_2,N_1)}}{2} \tag{3}$$

In Figure 3, the attraction degree between N_1 and N_5 is $Attr_{(N_1,N_5)} = \frac{f_{(N_1,N_5)}+f_{(N_5,N_1)}}{2} = \frac{0.4055+0.4700}{2} = 0.43775$.

Definitions 3 and 4 model the force of attraction between any two directly connected nodes (i.e., cells). However, in reality, a node is attracted not only by its directly connected nodes, but also by its indirectly connected nodes. Therefore, we propose the extension of Definition 4 to take into account the attraction degrees from both the directly connected nodes and the indirectly connected nodes. Assuming nodes N_i and N_j are indirectly connected by the path PT (N_i, N_{i+1}, N_{i+2}, … , N_j), their attractive degree can be calculated as the product of the attractive degrees of all the pairs of directly connected nodes in PT, i.e., $f_{(N_k,N_{k+1})}$ $(i \le k < j)$. The detail is as follows:

$$Attr_{(N_i,N_j)} = \prod_{k=i}^{k=j-1} f_{(N_k,N_{k+1})} \tag{4}$$

It should be noted that there might be multiple paths between nodes N_i and N_j, and thus there could be more than one attraction degree value for our analysis. In this regard, we choose the largest-weight path between N_i and N_j to calculate the attraction degree. In general, the nodes directly or indirectly connected by a larger-weight path have a stronger mutual relationship. Compared to the indirectly connected nodes, the directly connected nodes have a higher probability to be attracted by each other.

In Figure 3, there are three paths from N_1 to N_6, including path 1 (PT1: $N_1 \rightarrow N_2 \rightarrow N_6$), path 2 (PT2: $N_1 \rightarrow N_5 \rightarrow N_6$), and path 3 (PT3: $N_1 \rightarrow N_2 \rightarrow N_5 \rightarrow N_6$). The weight of PT1 is W_{PT1} = 2 + 1 = 3, the weight of PT2 is W_{PT2} = 1 + 3 = 4, and the weight of PT3 is W_{PT3} = 2 + 1 + 3 = 6. The largest-weight path, between N_1 and N_6, is PT3. Thus, according to Equation (2) and Equation (3), $Attr_{(N_1,N_2)} = 0.8047$, $Attr_{(N_2,N_5)} = 0.5148$, $Attr_{(N_5,N_6)} = 0.9730$, and thus $Attr_{(N_1,N_6)} = 0.8047 \times 0.5148 \times 0.9730 = 0.4031$ (Equation (4)).

Besides the strength of connectivity between nodes, the local structure of a graph is also critical to cluster nodes. More specifically, for any pair of nodes connected by a path, the greater proportion their path weight has in the total weight of their neighbors, the more similar the two nodes are. In a local structure, the nodes with a relatively stronger linkage tend to be grouped together. As a comparison, two nodes with a large connection could also be separated into different groups if one of them were to have another, stronger linkage to other nodes. To this end, we introduce structure similarity into our method, as follows.

Our structure similarity indicator is inspired by the Jaccard similarity coefficient, which has been widely applied to describe the relevance among objects. Assuming X and Y are two sets, the Jaccard similarity coefficient is defined as follows:

$$sim_{(X,Y)} = \frac{|X \cap Y|}{|X \cup Y|} \tag{5}$$

In addition, in graph theory, it is believed that the critical structural factors of a graph are the links that have relatively larger weight [29]. In this regard, we define structure similarity based on local edges and associated weights, as presented in Definition 5.

Definition 5. *(Structure similarity): For two directly connected nodes* N_1 *and* N_2*, their structure similarity is as follows:*

$$sim_{(N_1,N_2)} = \frac{W_{(N_1,N_2)}}{\sum_{c=1}^{d_{N_1}} W_{(N_1,N_{1c})} + \sum_{c=1}^{d_{N_2}} W_{(N_2,N_{2c})} - W_{(N_1,N_2)}} \tag{6}$$

where $W_{(N_1,N_{1c})}$ *is the weight of the edge connecting node* N_1 *and its neighbor* N_{1c}*, and* $W_{(N_2,N_{2c})}$ *is the weight of the edge connecting node* N_2 *and its neighbor* N_{2c}*. Equation (6) can be only used to measure the structure similarity of directly connected nodes, and in order to analyze the relationships between indirectly connected nodes, we extend Equation (6), as follows:*

$$sim_{(N_i,N_j)} = \prod_{k=i}^{k=j-1} sim_{(N_k,N_{k+1})} \tag{7}$$

where N_k and N_{k+1} are two directly connected nodes on the path connecting N_i and N_j.

In addition, we choose the largest-weight path between N_i and N_j to calculate the structure similarity. Generally, the directly connected nodes have a larger structure similarity than the indirectly connected nodes do.

In Figure 3, $\sum_{j=1}^{d_{N_1}} W_{1j} = 6$, $\sum_{j=1}^{d_{N_5}} W_{5j} = 5$, $W_{(N_1,N_5)} = 1$, $sim_{(N_1,N_5)} = \frac{1}{6+5-1} = 0.1$. The largest-weight path between N_1 and N_6 is $N_1 \to N_2 \to N_5 \to N_6$, and, thus, $sim_{(N_1,N_2)} = 0.2$, $sim_{(N_2,N_5)} = 0.1111, sim_{(N_5,N_6)} = 0.2727$. Therefore, according to Equation (7), $sim_{(N_1,N_6)} = 0.2 * 0.1111 * 0.2727 = 0.0061$.

Finally, since both the attraction degree (Equation (4)) and the structure similarity (Equation (7)) have been normalized, we can integrate them into a single measure, as follows:

$$fsim_{(N_i,N_j)} = sim_{(N_i,N_j)} + Attr_{(N_i,N_j)} \tag{8}$$

2.2.2. Algorithm

Based on the integrated similarity measure, we then calculated the final distance for each pair of directly or indirectly connected nodes (i.e., cells) in the network traffic flow.

$$fdis_{(N_i,N_j)} = \frac{1}{fsim_{(N_i,N_j)}} \tag{9}$$

In the process of detecting community, the dissimilarity index for each pair of nodes is adopted, with which one can measure the extent of proximity between the nodes of a network and signify to what extent two nodes would 'like' to be in the same community [30]. This proximity reflects the connectivity property of nodes in a diffusion process. The final minimization problem under this distance can also be solved by a k-means algorithm [31].

For our community detection algorithm, we adopted the K-Medoids algorithm, which belongs to the family of k-means clustering. More specifically, we first calculate the distances between all the pairs of nodes, and then select k nodes (i.e., initial k medoids) which have the largest distance to each other. Secondly, we assign each node (except the nodes that have already been labeled) to its nearest cluster according to the distances measured on the network traffic flow. This process is iteratively conducted until the medoids do not change or the number of iterations is equal to the threshold. In addition, in the end of each iteration, the node that has the minimum sum of distances within the cluster is selected as the medoid. Our community detection algorithm is as follows (Algorithm 1):

Algorithm 1. Community Clustering

Input: A spatially embedded network consisting of nodes and edges with weight, the number of communities k, the maximum number of iterations $MaxI$.
Output: A set of communities: $C = \{C_1, C_2, \ldots, C_k\}$
1. Initialization:
$fdis_{(N_i,N_j)} = 0$, $iteration$=0, ClusterCentriod [] C=null;
2. Node distance calculating: //calculate the distance between each pair of nodes.
  for each pair of nodes N_i and N_j ($i \neq j$)
    $fsim_{(N_i,N_j)} = sim_{(N_i,N_j)} + Attr_{(N_i,N_j)}$;
    $fdis_{(N_i,N_j)} = \frac{1}{fsim_{(N_i,N_j)}}$;
  end for
  for each pair of nodes N_i and N_j ($i = j$)
    $fdis_{(N_i,N_j)} = 1$;
  end for
3. Community detection based on the K-Medoids framework:
  Select k nodes that have the largest distance to each other as initial k medoids, i.e., $\{C_1, C_2, \ldots, C_k\}$;
  Assign each node to the closest medoid;
  While (the medoids do not change or $iteration \leq MaxI$)
    for each node N_i
      Assign N_i to the closest medoid C_m with min $\{fdis_{(N_i,C_m)}\}$;
    end for
    for each cluster C_j ($j \leq k$)
      Update the medoid of each community by detecting the node that has the minimum sum of distances within the cluster;
    end for
    iteration++;
  end while
Return the structure consisting of k communities: $C = \{C_1, C_2, \ldots, C_k\}$.

2.3. Variation of Community across Time

Besides the spatial heterogeneity, a transport system also has the dynamic property, and, in order to analyze such variation of a transport system across time, we propose a graph structure matching measurement (GSMM) between two network traffic flows sharing the node set. Specifically, as presented in previous sections, a network traffic flow in a specific time slice can be divided into several communities. In other words, the variation of a transport system across time can be represented as the change of the corresponding community structures. Hence, the GSMM measures the degree of matching between two community structures, i.e., two node sets.

Definition 6. *(Similarity of two node sets): Let S_1 and S_2 be two node sets, the similarity between S_1 and S_2 is defined as follows:*

$$Ssim_{(S_1,S_2)} = \frac{2|S_1 \cap S_2|}{|S_1| + |S_2|} \quad (10)$$

where $|S|$ is the number of the nodes of set S and $|S_1 \cap S_2|$ is the number of the nodes that S_1 and S_2 share. For example, if $S_1 = S_2$, $Ssim_{(S_1,S_2)} = 1$; if $S_1 \cap S_2 = \Phi$, $Ssim_{(S_1,S_2)} = 0$.

Equation (10) only measures the similarity between two node sets (i.e., two communities), each of which plays a different role in the corresponding graph structure. Specifically, for a graph, some communities are more important than others, and, in order to measure the global similarity between two graphs (i.e., two sets of communities), we propose calculation of the sum of the weighted similarity between two sets of communities. In this process, we define the weight of a community as its contribution rate in the corresponding graph. In general, the more nodes the community has, the larger contribution rate it has for the whole system. Assuming the graph C consists of the communities $\{C_1, C_2, \ldots, C_k\}$ and $S_1, S_2, \ldots, S_k$ are the node sets of $C_1, C_2, \ldots, C_k$, respectively, the contribution rate of C_i $(i = 1, 2, \ldots, k)$ is defined as follows:

$$comR_{C_i} = \frac{S_i}{S} \quad (11)$$

Based on the contribution rate of community, we then calculate the weighted similarity between two network traffic flows as follows.

Definition 7. *(Similarity of spatially embedded networks): Let C_p and C_q be two spatially embedded networks with k-size community structures $\{C_1^p, C_2^p, \ldots, C_k^p\}$ and $\{C_1^q, C_2^q, \ldots, C_k^q\}$, respectively. $comR_1^p, comR_2^p, \ldots, comR_k^p$ are the contribution rates of the communities $C_1^p, C_2^p, \ldots, C_k^p$, respectively, and $comR_1^q, comR_2^q, \ldots, comR_k^q$ are the contribution rates of the communities $C_1^q, C_2^q, \ldots, C_k^q$, respectively. $S_1^p, S_2^p, \ldots, S_k^p$ are the node sets of the communities $C_1^p, C_2^p, \ldots, C_k^p$, respectively, and $S_1^q, S_2^q, \ldots, S_k^q$ are the node sets of the communities $C_1^q, C_2^q, \ldots, C_k^q$, respectively. Then, the similarity between the two graphs C_p and C_q is defined as:*

$$FsimC = \max_{i_1, i_2 \ldots i_k} \frac{2 * PsimC_{(i_1, i_2 \ldots i_k)}}{\sum_{j=1}^{k}\left(comR_j^p + comR_j^q\right)} \quad (12)$$

where $PsimC_{(i_1, i_2, \ldots, i_k)}$ is calculated as follows:

$$\begin{aligned} PsimC_{(i_1, i_2 \ldots i_k)} = {} & \tfrac{1}{2} * \left(comR_1^p + comR_{i1}^q\right) * Ssim_{(S_1^p, S_{i1}^q)} + \tfrac{1}{2} * \left(comR_2^p + comR_{i2}^q\right) * Ssim_{(S_2^p, S_{i2}^q)} \\ & + \ldots + \tfrac{1}{2} * \left(comR_k^p + comR_{ik}^q\right) * Ssim_{(S_k^p, S_{ik}^q)} \end{aligned} \quad (13)$$

where $(i_1, i_2 \ldots i_k)$ is a full permutation of the set $I = \{1, 2 \ldots k\}$.

The community structure is a partition of all the land uses (i.e., nodes) of the network traffic flow for a specific time slice. Hence, the variation of networks across time can be analyzed by measuring the similarity between the corresponding community structures. The better the matching between two community structures, the more similar the corresponding traffic conditions in different time slices.

3. Data Sets and Settings

We conducted a series of experiments to explore the transport system of Beijing city using taxi trajectory points. As the capital of China, Beijing is the national political, economic, and administrative center. By the end of 2016, the number of taxis in Beijing had reached 71,600, with a permanent population of 21.729 million. More than 55% of residents take taxis every week [32]. In this paper, we chose the central zone of the city (i.e., the range of latitude 39° 49′ 41″–39° 59′ 17″ N and longitude 116° 15′ 47″–116° 29′ 09″ E) as the study region (Figure 4). The research area is within the third ring area of Beijing. The total number of valid records was more than 20,000,000, which covers a time period of 24 h (from 0 a.m. to 12 p.m.) (Figure 4). Because of signal loss or degradation, geospatial locations of trajectory may have been recorded with spatial and temporal uncertainties. Considering these factors, our data were preprocessed by the provider to indicate whether the record was valid or invalid with respect to the GPS signal. In our research, only the valid records were used. Nevertheless, there were some small deviations in position information relative to the actual position. In this respect, we adopted the technique of region tessellation, which is used to model collective behaviors between regions. The internal variation of the region is not considered. Hence, it can handle small deviations of location information and can be used to reveal the collective travel patterns between regions. In order to have a macroscopic understanding of people's travel patterns, we extracted the taxi pick-up and drop-off points (Figure 5). It can be observed that the hotspots of pick-up and drop-off points are distributed within the second ring roads of Beijing.

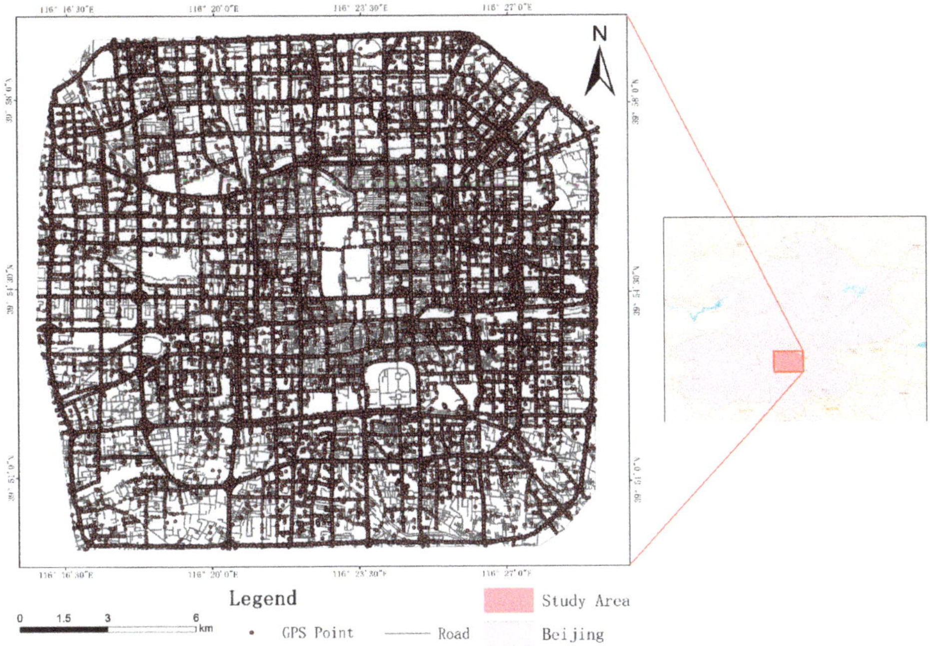

Figure 4. Study area: the core area of Beijing.

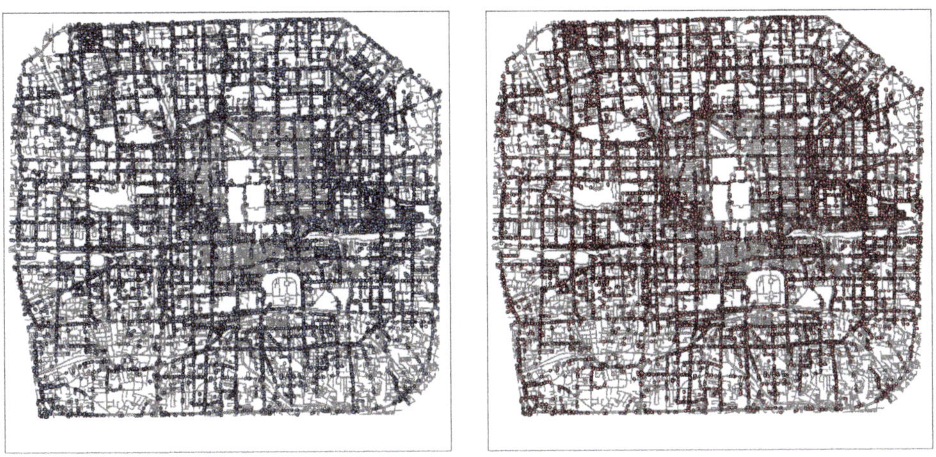

(**a**) Passengers pick-up location (**b**) Passengers drop-off location

Figure 5. Passengers pick-up/drop-off points.

4. Experiment and Result

4.1. Result and Analysis

We first partitioned the study area into a set of cells with size 1 km × 1 km. This scale was determined on the basis of relevant studies suggesting that the cell size is fine enough to depict urban structure [33]. We then used these cells to construct spatially embedded networks to analyze the interactions between land uses (i.e., sub-regions). Moreover, we obtained the results in different time periods to discover the dynamic patterns of the transport system.

In the spatially embedded network, the nodes represent the regions of the city and the edges represent the traffic linkage between different regions. Furthermore, the intensity of the connections between different regions varies with time. In order to clearly reveal the travel patterns, we constructed networks for typical time periods (Figure 6), i.e., morning rush hour, noon rush hour, evening rush hour, and midnight.

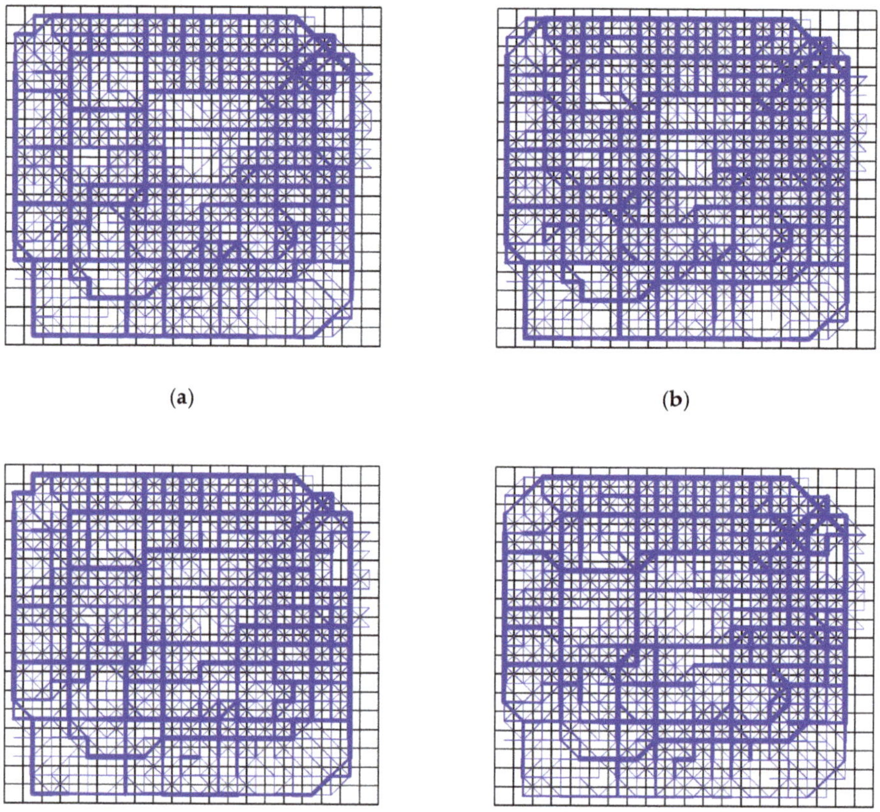

(a) (b) (c) (d)

Figure 6. Spatially embedded networks for traffic flows in different time periods with cell size k = 1000 m: (**a**) morning rush hour; (**b**) noon rush hour; (**c**) evening rush hour; (**d**) midnight.

During different time periods of the day, the traffic flows showed different spatial connectivity patterns to meet the varying travel demands of people. In the morning rush hour, the interactions between residential areas, working areas, and schools were more intense than those between the other areas. Later, entering the period of the noon rush hour, the traffic flow volume and the associated connectivity patterns became more significant in the central areas and main roads of the city. Most of the central areas belong to the working zone and commercial zone, and thus the strong connectivity indicates the frequent interactions between working and lunch activities. In addition, although the volume of traffic flows in the evening rush hour was less than that in morning rush hour, their network structures were similar. In Beijing, in order to avoid traffic congestions, many people choose to get off work and go home or go to recreational areas after 19:00. This may be a reason that the traffic volume in evening rush hour was less than that in morning rush hour. Additionally, the routes that people choose to go to work and get off work are similar, and thus the networks in the morning rush hour and evening rush hour had a similar structure. Furthermore, the networks at midnight had a multi-center structure, which depended on the hubs of recreational areas and commercial areas. In general, except the morning rush hour, the other time periods depended more on the eastern areas than the western areas.

Among these hours, the noon rush hour had the largest volume of traffic flows. From the morning rush hour to the noon rush hour, the volume of trajectory flow increased substantially and some new links emerged in local regions. This implies that the interactions between land uses in the noon rush hour are more intense than in the morning rush hour. From the noon rush hour to the evening rush hour, the interactions decreased not only through the main roads but also across the western areas. The obvious feature of interactions at midnight is that there were significant connections to or from recreational land (i.e., the eastern area).

Besides the connectivity property, the spatially embedded network can also imply the aggregation patterns of land uses in the functional space. The land uses that have a strong connectivity relationship in the network traffic flow tend to be grouped together (Figure 7), and, in this way, we can explore the city structure and transport system using the resulting clustering patterns of land uses.

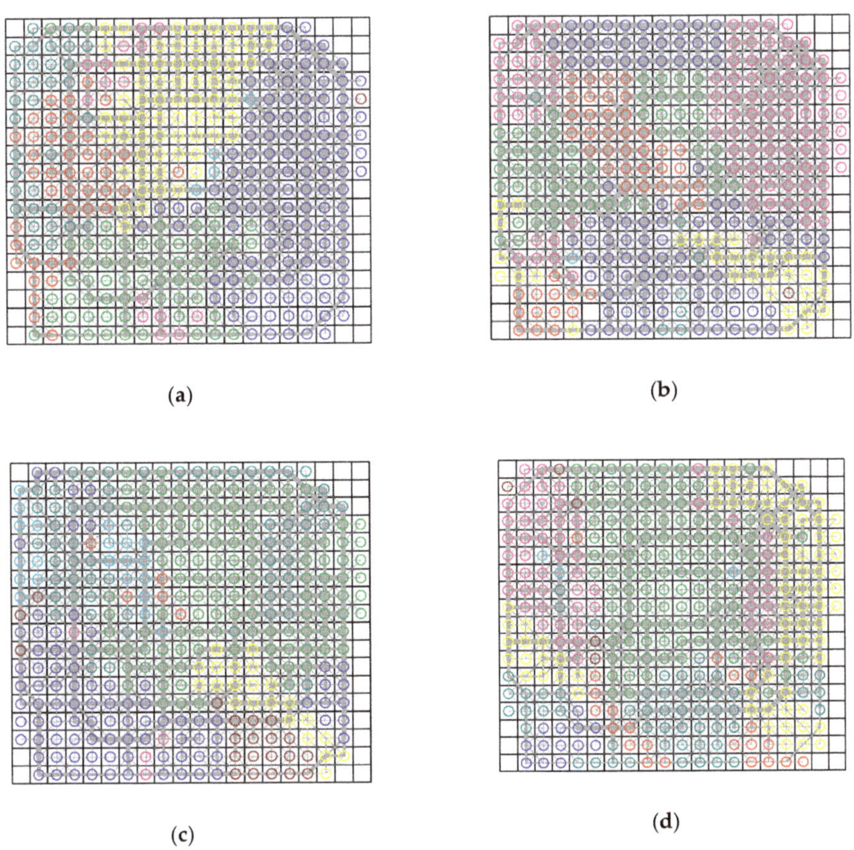

Figure 7. Communities in different time periods with cell size k = 1000 m: (**a**) morning rush hour; (**b**) noon rush hour; (**c**) evening rush hour; (**d**) midnight.

As presented in Figure 7, we classified the land uses into eight communities and there were some cells with no nodes. This is because there were no trajectory flows traversing across these cells. In addition, the land uses in the same community do not have to be contiguous in space. The reason is that our study aimed to cluster land uses from the perspective of their transport function, and the final similarity of the nodes was decided by both their attractive degree and structure similarity, which were calculated based on the spatially embedded network. Some neighboring land uses could have a low similarity due to their weak connectivity in the spatially embedded network, and some distant land uses could be grouped together if they are connected by a route with large traffic flows.

More specifically, as presented in Figure 7a, the study region in the morning rush hour was partitioned into five main communities, in each of which the land uses had a strong transport connection. Such relations also imply the actual aggregation of human activities and urban functions, e.g., the business district in the eastern part, the residential area in the southern part, and the universities in the northern area. In addition, there were four main communities in the noon rush hour. The most significant feature in this time period was that many communities were non-contiguous, or spanned multiple regions. For example, the community labeled by blue dots spanned the northern area (i.e., universities and high-technology regions) and the southern area (i.e., business districts and railway stations). In this time period, the traffic connection between distant land uses became stronger and cross-regional human activities were more frequent. In the period of the evening rush hour, the city was partitioned into two main communities which correspond to interactions among the residential areas, business districts, and universities. The small part (pink dots) corresponded to the connection between residential areas and train station areas. At midnight, there were four main communities, in which the large volume of traffic flows was directed for entertainment (e.g., bar). For example, the central Hohai entertainment area (green dots) attracted most of the neighboring land uses.

We then used the GSMM method to quantify the similarity between community structures in different time periods. In such a way, we were able to find out the degree to which the transport system changed across time. As presented in Figure 8, the GSMM measure values were calculated at the macro level rather than the micro level. First, it can be observed that community structures in successive time periods usually had a high similarity. For example, the community similarities in the successive time slices of [4:00–8:00], [11:00–14:00], and [15:00–18:00] were higher than those in non-successive time slices. Secondly, most of the community structures in the rush hours (e.g., the similarity between [7:00–8:00] and [8:00–9:00]) had a high similarity. Hence, the distributions of traffic flows are so regular in these periods that urban planners could estimate the associated travel behavior patterns. Note that the community structures between [12:00–14:00] and [4:00–7:00] were similar at the macro scale. Considering the routines and habits of residents, there are relatively few traffic flows in these periods and the arterial roads provide the main functions of transportation in the city. The travel origins and destinations are concentrated in a few business districts and railway stations. Hence, the traffic conditions in these time periods showed a similar characteristic. In addition, it can be observed that the community structure in [21:00–22:00] was very different from most of the structures in the other time periods. The reason may be that there are many different activities happening (e.g., working and entertainment) in [21:00–22:00], and thus the connectivity between local regions is much more complex than those in other time periods. Furthermore, the community structure in the evening rush hour of [18:00–19:00] was also very different from most of the other structures. This could be because in this time period the travel activities become increasingly active, and most of the residents in Beijing choose to travel along different routes. The land uses were also aggregated into different communities in this period compared to those in the successive time periods.

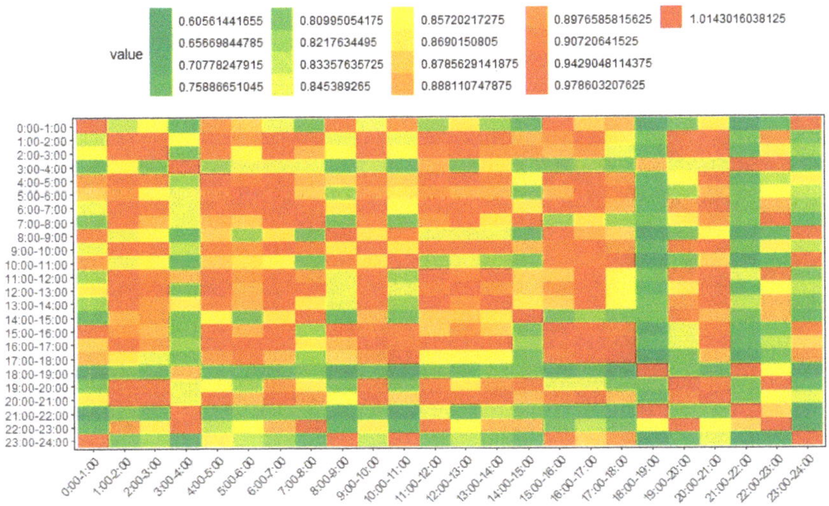

Figure 8. Similarity of the community structures in different time periods with 1000 m cell size.

4.2. Variation in Different Spatial Scales

Our proposed method is adaptive to applications with different spatial scales. Hence, the next experiment refined the tessellation of space using a 500 m grid and analyzed the associated spatial community patterns across space and time.

As the size of cells becomes smaller, there are more cells with no nodes, which correspond to the buildings (e.g., Imperial Palace) or lakes (e.g., Beihai Park) (Figure 9). We can easily observe the distributions of these land uses from the spatially embedded networks. In addition, compared to networks with 1000 m cell size, networks with 500 m cell size can present more detailed structures of street network infrastructure and associated traffic flows in the city. With the refined tessellation, we could observe more detailed interactions from the results. In the morning rush hour, strong connectivity existed mainly among the regions of residential areas, business districts, and high-tech areas. Entering the noon rush hour, the ring-like structure of the transport system became most significant. In addition, in the periods of the evening rush hour and midnight, the traffic flows were concentrated in the eastern and northern parts, which are the business cores of the city. Therefore, the transport system of Beijing depends largely on the loop lines, with a significant temporal pattern.

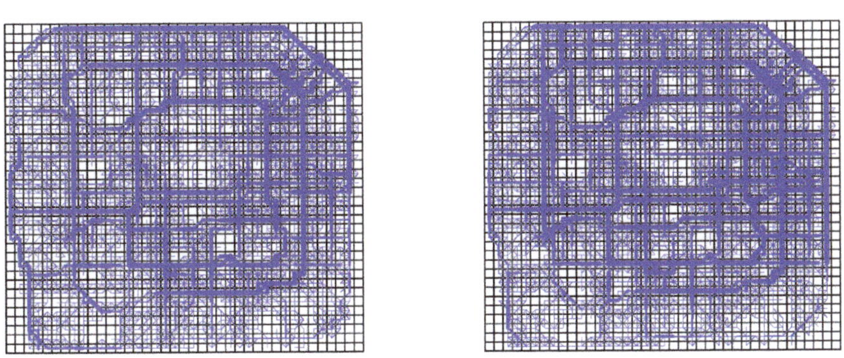

(**a**) (**b**)

Figure 9. *Cont.*

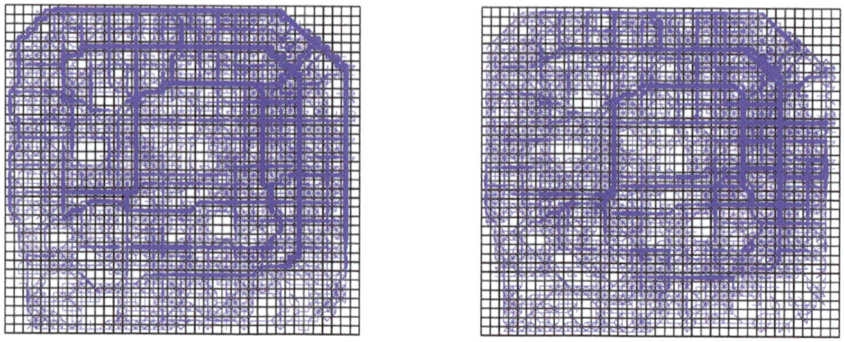

(c) (d)

Figure 9. Spatially embedded networks for traffic flows in different time periods with cell size k = 500 m: (**a**) morning rush hour; (**b**) noon rush hour; (**c**) evening rush hour; (**d**) midnight.

In order to compare the results under the two scales, we also classified the land uses into eight communities. As presented in Figure 10a, the study region was partitioned into three main communities in the morning rush hour. The eastern region was divided into two parts, and the community represented by green dots implies the aggregation of residential area (i.e., western part) and commercial business districts (i.e., eastern part). Another part in the eastern region was merged with the northern region, and the resulting community implies the aggregation of residential areas, universities, and high-technology regions. Later, entering the noon rush hour and the evening rush hour, the study region was partitioned into three communities and four communities, respectively. In addition, the aggregation of regions in both of the two time periods seems to be more significant than those in the corresponding time periods with cell size 1000 m. At midnight, the study area was partitioned into three communities. Compared to the result with cell size 1000 m, the interaction between land uses was weakened at midnight with cell size 500 m, and the region was divided into two main parts: one part was merged into the community of the business district (green dots), and the other one was merged into the community of business district and residential areas (blue dots). In general, the communities with cell size 500 m can reveal more detailed information of aggregation of land uses.

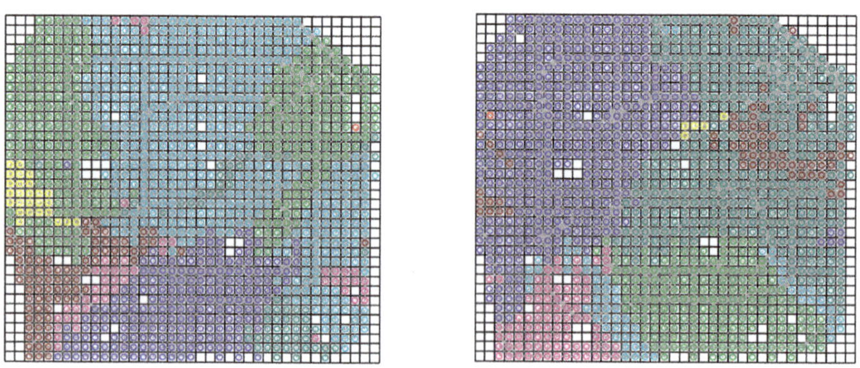

(a) (b)

Figure 10. *Cont.*

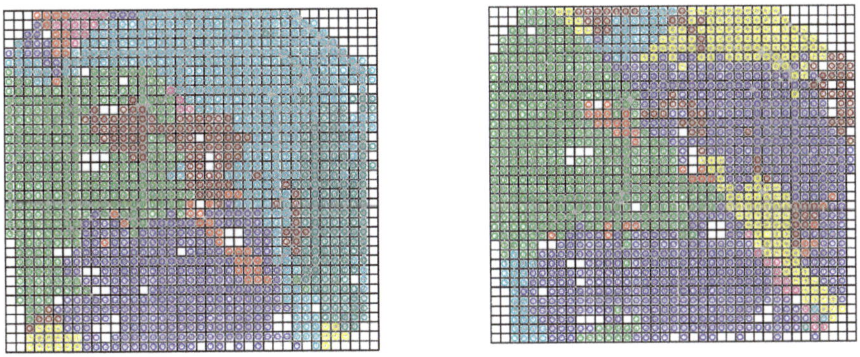

(c) (d)

Figure 10. Communities in different time periods with cell size k = 500 m: (**a**) morning rush hour; (**b**) noon rush hour; (**c**) evening rush hour; (**d**) midnight.

Using the GSMM method, we also calculated the similarity matrix, as presented in Figure 11. It can be observed that in the rush hours, the measured values with 500 m cell size were similar to those with 1000 m cell size. In addition, during the non-rush hours, the measured values with 500 m cell size were higher than the corresponding values with 1000 m cell size. The reason could be that, with the refining of space tessellation, more cells had traffic flows and the number of the matching cells across time increased. Specifically, the small size grid can capture the local interactions between regions, which were more regular in the non-rush hours than in the rush hours (see Figure 9).

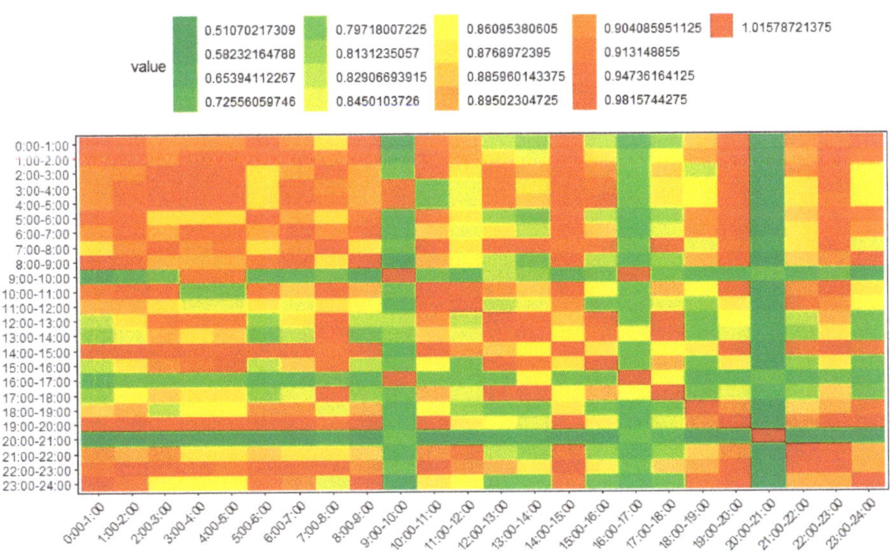

Figure 11. Similarity of the community structures in different time periods with 500 m cell size.

4.3. Algorithm Efficiency with Different Cell Sizes

The results above show the effect of different cell sizes on community detection. In order to further explore the algorithm efficiency with different cell sizes, we implemented the method with cell sizes of 500 m, 600 m, 700 m, 800 m, 900 m, and 1000 m.

With the increase of cell size, the total running time decreased (Figure 12a), and when the cell size changed from 500 m to 700 m, the algorithm efficiency increased sharply. When the cell size changed from 700 m to 1000 m, the running time remained stable. The reason could be that the number of nodes decreased with the increase of cell size, and thus the algorithm cost less time during the clustering of nodes. Furthermore, when the cell size was larger than 700 m, the number of iterations in k-means algorithm did not change much, and thus the running time of the algorithm remained stable. In addition, as presented in Figure 12b, the running time of the algorithm changed across the time periods. In general, the running time of the algorithm in the rush hours (e.g., 7:00–9:00, 11:00–13:00, 18:00–20:00, 22:00–23:00) was larger than those in the other hours. The reason could be that the flow structures are more complex in rush hours.

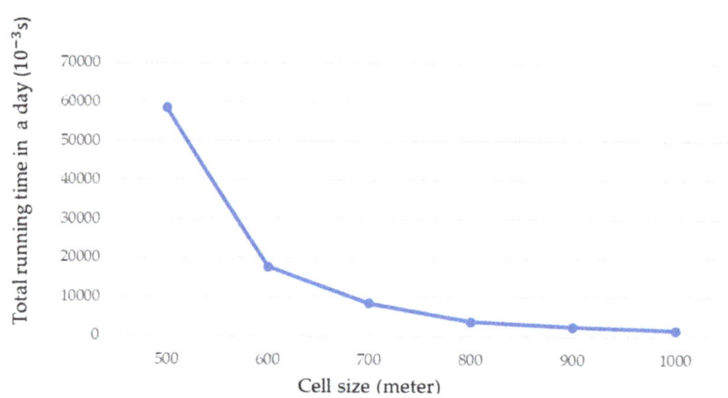

(a)

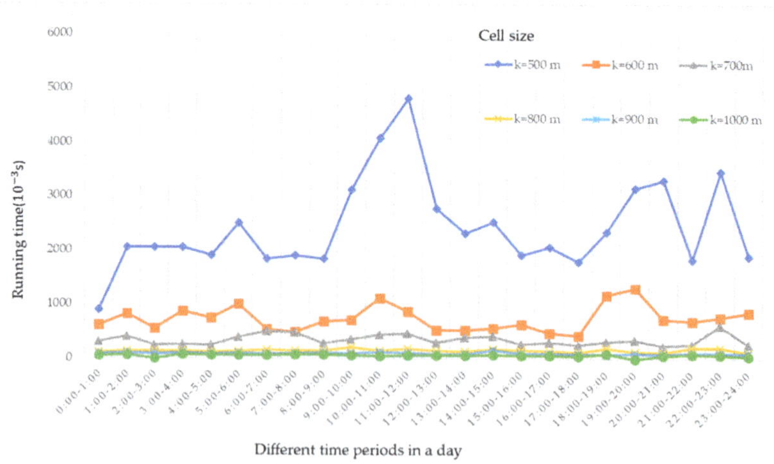

(b)

Figure 12. The algorithm efficiency with different cell sizes: (**a**) the total running time of the algorithm with different cell sizes, and (**b**) the running time of the algorithm in different time periods.

5. Conclusion and Future Directions

Based on the collective intra-city trips extracted from the emerging taxi GPS trajectory data, this paper explored network traffic flows towards a deep understanding of city structure. We introduced network science techniques (e.g., community detection) to reveal the regular patterns of traffic flows across space and time. More specifically, aiming at the connectivity, aggregation, and dynamic properties of transport system, we proposed a three level framework to explore the complex traffic network. It firstly partitions the study region and constructs a spatially embedded network for representing the connectivity relationships between local regions. In order to extract the aggregation patterns of land uses, the method then uses the community detection techniques based on the volume of traffic flows and structural properties of the network. Furthermore, our method employs a graph structure matching measure to uncover the regularities of the transport system across time. The proposed method is also adaptive to multi-scale applications in space and time.

Through the case study, we found that the interactions of land uses show different characteristics in different time periods, and the aggregation patterns of functional areas is dynamic across the time. This result is highly associated with the travel behaviors of residents in the city, and thus can be used further in social science research. In addition, the result can provide references for the dispatching of the traffic system. For example, we can plan for the prevention of traffic jams in regions which have intense interactions of traffic flows. Moreover, it can be used to assist urban structure analysis. As presented in our case study, Beijing has a polycentric form with significant loop structure.

In this paper, we took the taxi trajectory data into consideration because taxi accounts for a large proportion of public transport in Beijing city. Taxi drivers are very familiar with the city of concern, and thus there are increasingly more studies focusing on the use of taxi trajectory data for urban analysis [3,12,33]. In addition, since the taxi is a common mode of transport, our method could be adaptable to other cities. We would like to regard this research as a beginning of detecting spatial interaction communities based on vehicle datasets. With the rapid development of big data, more traffic data (e.g., bus trajectories, passenger car data, and biking trajectories) can be introduced into our framework for exploring city structures comprehensively. Further study can also use more methods (e.g., complex system) to understand the mechanisms of the traffic flow space. It would be interesting to analyze the associations between the physical space and virtual space of a city using social media data.

Author Contributions: Methodology Development & Implementation, Data Acquisition, and Analysis: W.Y.; M.G.; Z.C.; Writing & Revision: led by W.Y. with contributions from all other co-authors.

Funding: The project was supported by the National Natural Science Foundation of China (No. 41701440, 41871305), the Natural Science Foundation of Hubei Province (No. 2018CFB513), the Fundamental Research Funds for the Central Universities, China University of Geosciences (Wuhan) (No. CUG170640), and a grant from State Key Laboratory of Resources and Environmental Information System.

Conflicts of Interest: The authors declare no conflict of interest.

References

1. Zheng, L.; Xia, D.; Zhao, X.; Tan, L.Y.; Li, H. Spatial–temporal travel pattern mining using massive taxi trajectory data. *Phys. A Stat. Mech. Appl.* **2018**, *501*, 24–41. [CrossRef]
2. Zhu, X.; Guo, D. Mapping large spatial flow data with hierarchical clustering. *Trans. GIS* **2014**, *18*, 421–435. [CrossRef]
3. Yu, W. Discovering frequent movement paths from taxi trajectory data using spatially embedded networks and association rules. *IEEE Trans. Intell. Transp. Syst.* **2019**, *23*, 855–866. [CrossRef]
4. Yuan, Y.; Raubal, M.; Liu, Y. Correlating mobile phone usage and travel behavior—A case study of Harbin, China. *Comput. Environ. Urban Syst.* **2012**, *36*, 118–130. [CrossRef]
5. Jiang, B.; Yin, J.; Zhao, S. Characterizing the human mobility pattern in a large street network. *Phys. Rev. E* **2009**, *80*, 021136. [CrossRef] [PubMed]
6. Song, C.; Koren, T.; Wang, P.; Barabási, A.-L. Modelling the scaling properties of human mobility. *Nat. Phys.* **2010**, *6*, 818–823. [CrossRef]

7. Ahas, R.; Aasa, A.; Silm, S.; Tiru, M. Daily rhythms of suburban commuter's movements in the Tallinn metropolitan area: Case study with mobile positioning data. *Transp. Res. Part C* **2010**, *18*, 45–54. [CrossRef]
8. Sevtsuk, A.; Ratti, C. Does urban mobility have a daily routine? Learning from the aggregate data of mobile networks. *J. Urban Technol.* **2010**, *17*, 41–60. [CrossRef]
9. Jiang, S.; Guan, W.; Zhang, W.; Chen, X.; Yang, L. Human mobility in space from three modes of public transportation. *Phys. A Stat. Mech. Appl.* **2017**, *483*, 227–238. [CrossRef]
10. Wang, Z.; Hu, Y.; Zhu, P.; Qin, Y.; Jia, L. Ring aggregation pattern of metro passenger trips: A study using smart card data. *Phys. A Stat. Mech. Appl.* **2018**, *491*, 471–479. [CrossRef]
11. Wang, W.; Pan, L.; Yuan, N.; Zhang, S.; Liu, D. A comparative analysis of intra-city human mobility by taxi. *Phys. A Stat. Mech. Appl.* **2015**, *420*, 134–147. [CrossRef]
12. Cai, H.; Zhan, X.; Zhu, J.; Jia, X.; Chiu, A.S.F.; Xu, M. Understanding taxi travel patterns. *Phys. A Stat. Mech. Appl.* **2016**, *457*, 590–597. [CrossRef]
13. Zheng, Y.; Liu, Y.; Yuan, J.; Xie, X. Urban computing with taxicabs. In Proceedings of the 13th International Conference on Ubiquitous Computing, Beijing, China, 17–21 September 2011; ACM: New York, NY, USA, 2011; pp. 89–98.
14. Guo, D.; Zhu, X.; Jin, H.; Gao, P.; Andris, C. Discovering spatial patterns in origin-destination mobility data. *Trans. GIS* **2012**, *16*, 411–429. [CrossRef]
15. Yuan, J.; Zheng, Y.; Zhang, C.; Xie, W.; Xie, X.; Sun, G.; Huang, Y. T-drive: Driving directions based on taxi trajectories. *Int. Arch. Photogramm. Remote Sens. Spat. Inf. Sci.* **2016**, *16*, 1682–1750.
16. Zhou, Q.; Qin, K.; Chen, Y.X.; Li, Z.X. Hotspots detection from taxi trajectory data based on data field clustering. *Geogr. Geo-Inf. Sci.* **2016**, *32*, 51–56.
17. Yue, Y.; Zhuang, Y.; Li, Q.; Mao, Q. Mining time-dependent attractive areas and movement patterns from taxi trajectory data. In Proceedings of the International Conference on Geoinformatics, Fairfax, VA, USA, 12–14 August 2009; IEEE: Piscataway, NJ, USA, 2009.
18. Liu, C.; Qin, K.; Kang, C.G. Exploring time-dependent traffic congestion patterns from taxi trajectory data. In Proceedings of the 2015 2nd IEEE International Conference on Spatial Data Mining and Geographical Knowledge Services (ICSDM), Fuzhou, China, 8–10 July 2015; pp. 39–44.
19. Liu, Y.; Kang, C.; Gao, S.; Xiao, Y.; Tian, Y. Understanding intra-urban trip patterns from taxi trajectory data. *J. Geogr. Syst.* **2012**, *14*, 463–483. [CrossRef]
20. Liu, X.; Gong, L.; Gong, Y. Revealing travel patterns and city structure with taxi trip data. *J. Transp. Geogr.* **2015**, *43*, 78–90. [CrossRef]
21. Zhou, Y.; Fang, Z. Labeling residential community characteristics from collective activity patterns using taxi trip data. *ISPRS Int. Arch. Photogramm. Remote Sens. Spat. Inf. Sci.* **2017**, *XLII-2/W7*, 1481–1486. [CrossRef]
22. Pan, G.; Qi, G.; Wu, Z.; Zhang, D.; Li, S. Land-use classification using taxi gps traces. *IEEE Trans. Intell. Transp. Syst.* **2013**, *14*, 113–123. [CrossRef]
23. Gonzalez, M.C.; Hidalgo, C.A.; Barabasi, A.L. Understanding individual human mobility patterns. *Nature* **2008**, *453*, 779–782. [CrossRef]
24. Liu, L.; Andris, C.; Ratti, C. Uncovering cabdrivers' behavior patterns from their digital traces. *Comput. Environ. Urban Syst.* **2010**, *34*, 541–548. [CrossRef]
25. Prapat, P.; Nguyen, T.K.O. Assessment of potential long-range transport of particulate air pollution using trajectory modeling and monitoring data. *Atmos. Res.* **2007**, *85*, 3–17.
26. Dias, D.; Tchepel, O. Modelling of human exposure to air pollution in the urban environment: A GPS-based approach. *Environ. Sci. Pollut. Res.* **2014**, *21*, 3558–3571. [CrossRef]
27. Cho, E.; Myers, S.A.; Leskovec, J. Friendship and mobility: User movement in location-based social networks. In Proceedings of the 17th ACM SIGKDD International Conference on Knowledge Discovery and Data Mining, San Diego, CA, USA, 21–24 August 2011; ACM: New York, NY, USA, 2011; pp. 1082–1090.
28. Ratti, C.; Sobolevsky, S.; Calabrese, F.; Andris, C.; Reades, J.; Martino, M.; Claxton, R.; Strogatz, S.H. Redrawing the map of Great Britain from a network of human interactions. *PLoS ONE* **2010**, *5*, e14248. [CrossRef] [PubMed]
29. Kaufman, L.; Rousseeuw, P.J. Finding groups in data: an introduction to cluster analysis. *J. R. Stat. Soc. Ser. C* **1991**, *40*, 401–423.
30. Liu, J.; Liu, T. Detecting community structure in complex networks using simulated annealing with k-means algorithms. *Phys. A Stat. Mech. Appl.* **2010**, *389*, 2300–2309. [CrossRef]

31. Lafon, S.; Lee, A. Diffusion maps and coarse-graining: A unified framework for dimensionality reduction, graph partitioning, and data set parameterization. *IEEE Trans. Pattern Anal. Mach. Intell.* **2006**, *28*, 1393–1403. [CrossRef] [PubMed]
32. Wang, B. Research on Taxi Travel Demand Based on Beijing Passenger Hot Spot Area. Master's Thesis, Beijing Jiaotong University, Beijing, China, 2018.
33. Liu, Y.; Wang, F.; Xiao, Y.; Gao, S. Urban land uses and traffic 'source-sink areas': Evidence from GPS-enabled taxi data in Shanghai. *Landsc. Urban Plan.* **2012**, *106*, 73–87. [CrossRef]

© 2019 by the authors. Licensee MDPI, Basel, Switzerland. This article is an open access article distributed under the terms and conditions of the Creative Commons Attribution (CC BY) license (http://creativecommons.org/licenses/by/4.0/).

MDPI
St. Alban-Anlage 66
4052 Basel
Switzerland
Tel. +41 61 683 77 34
Fax +41 61 302 89 18
www.mdpi.com

Applied Sciences Editorial Office
E-mail: applsci@mdpi.com
www.mdpi.com/journal/applsci

www.ingramcontent.com/pod-product-compliance
Lightning Source LLC
Chambersburg PA
CBHW061845210926
51850CB00006B/970
9783039364893